Irrigation Principles and Practices

IRRIGATION PRINCIPLES

AND PRACTICES

Third Edition

ORSON W. ISRAELSEN, Ph.D.

Emeritus Professor, Civil and Irrigation Engineering

VAUGHN E. HANSEN, Ph.D.

Director, Engineering Experiment Station

UTAH STATE UNIVERSITY, LOGAN, UTAH

John Wiley and Sons, Inc.

New York · London · Sydney

ISBN 0 471 42999 6
Library of Congress Catalog Card Number: 62-15179
Printed in the United States of America

The Third Edition is dedicated to THE INTERNATIONAL COM-MISSION ON IRRIGATION AND DRAINAGE in appreciation of its great contribution toward world progress. The work of this commission in its congress activities in India, 1951 and 1954; in the United States, 1957; and in Spain, 1960, has contributed to real advancement in irrigation and drainage education, research, and practices in the more than fifty member countries. The commission's activities are greatly advancing irrigation and drainage and thereby providing an increased and more dependable supply of food for the world's increasing population.

Orson W. Israelsen
Vaughn E. Hansen

PREFACE

In the third edition of *Irrigation Principles and Practices*, the basic underlying principles governing irrigation practices are stressed. Generalized concepts are outlined, and newly formulated practices are included. Specific examples and local experience have been replaced in the third edition by general concepts and practices.

The scope of irrigation is shown to be worldwide, and to extend from the sources of the water supplies to the drainage outlets. Sources and the storage and measurement of water are outlined. Wells, pumping, and conveyance of water are included. Basic soil and water relations including storage, movement, and drainage of waterlogged soils are presented. Saline and alkali problems are stressed, and principles of consumptive use of water are outlined. Questions of when to irrigate and how much water to apply are answered in a manner that applies to any crop in any locality. New concepts of irrigation efficiency are outlined, and principles of surface and sprinkler irrigation are discussed. The importance of social, administrative, and legal aspects of irrigation is stressed, and guiding concepts are included. The interrelationship of soil, water, climate, and irrigation management is considered to be of fundamental importance.

In an appendix of metric units, not only the usual conversion tables are included, but also the key equations, tables, and figures are shown in their metric counterpart for the aid of readers who prefer the metric system.

The problems have been revised to present broader coverage of the subject matter, and new problems have been added to illustrate the new subject matter and expanded coverage of the third edition.

I have accepted gladly the major responsibility for the third edition of *Irrigation Principles and Practices*. After the manuscript was prepared, Dr. Orson W. Israelsen, who had just returned from an extensive consulting trip abroad, carefully reviewed the manuscript and offered many helpful suggestions.

I wish to express my deep appreciation for the privilege of being closely associated professionally during the last quarter of a century with Dr. Israelsen. His inspiration and leadership have been deeply appreciated by those of us who have been fortunate enough to be his students. This gratitude should be expressed now while he is still dynamically active in his endeavors to improve irrigated agriculture throughout the world.

I am grateful to Norma Y. Schiffman and Lillian A. Nielsen for their excellent secretarial assistance and to Nella W. Lauritzen for her very capable editorial review. Searching reviews of an early draft of the third edition were made by Guy O. Woodward and Lyman S. Willardson. Their comments were very helpful. The support and encouragement of Dr. Dean F. Peterson, Dean of Engineering at Utah State University, have been extremely helpful.

Vaughn E. Hansen

Logan, Utah
May 1962

CONTENTS

CHAPTER 1 IRRIGATION—WORLDWIDE

Irrigation is an age-old art. Historically, civilization has followed the development of irrigation. Civilizations have risen on irrigated lands; they have also decayed and disintegrated in irrigated regions. Most men who are well informed on irrigation are certain of its perpetuity, as long as it is intelligently practiced. Others think that a civilization based on agriculture under irrigation is destined sooner or later to decline, because some ancient civilizations based on irrigation have declined. The duration of civilized peoples is probably dependent on many factors, of which a permanently profitable agriculture is vitally important. Some of the principles and practices essential to permanent and profitable agriculture under irrigation are considered in this volume.

1.1 CENTURIES OF IRRIGATION

The antiquity of irrigation is well documented throughout the written history of mankind. Genesis mentions Amraphel, King of Shinar, a contemporary of Abraham, who is probably identical with Hammurabi, sixth king of the first dynasty of Babylon. He developed laws, bearing the name of Hammurabi, indicating that the people had to depend upon irrigation for existence. One of the laws of Hammurabi states that if a man neglects to strengthen his bank of the canal and waters carry away the meadow, the man in whose bank the breach is opened shall render back the corn which he has caused to be lost.

The letters of Hammurabi about 2000 B.C. reveal a busy, governmental administrator who wastes no words when instructing his officials:

To Sid-Indiannam, Hammurabi speaks as follows: Gather the men that have fields along the Damanum Canal to clear out the Damanum Canal. Within this month, let them complete the digging of the Damanum Canal.

Further mention of irrigation is found in Second Kings 3:16–17:

And he said, Thus saith the Lord, Make this valley full of ditches. For thus saith the Lord, Ye shall not see wind, neither shall ye see rain; yet that valley shall be filled with water, that ye may drink, both ye, and your cattle, and your beasts.

An ancient Assyrian Queen, supposed to have lived before 2000 B.C., is credited with directing her government to divert the water of the Nile to irrigate the desert lands of Egypt. The inscription on her tomb is:

I constrained the mighty water to flow according to my will and led its waters to fertilize lands that had before been barren and without inhabitants.

Irrigation canals supposed to have been built under this Queen of Assyria are still delivering water. Thus, there are records and evidence of continuous irrigation for thousands of years in the valleys of the Nile and for comparatively long periods likewise in Syria, Persia, India, Java, and Italy.

Egypt claims to have had the world's oldest dam, 355 feet long and 40 feet high, built 5000 years ago to store water for drinking and irrigation. Basin irrigation introduced on the Nile about 3300 B.C. still plays an important part in Egyptian agriculture.

In China, where reclamation was begun more than 4000 years ago, the success of early kings was measured by their wisdom and progress in water-control activities. King Yu, of Hsia-Dynasty (2200 B.C.), was elected king by the people as a reward for his outstanding work in water control. The famous Tu-Kiang Dam, still a successful dam today, was built by a man named Mr. Li and his son in the Chin-Dynasty (200 B.C.) and provides irrigation water for about one-half million acres of rice fields. The water ladder, a widely used pumping device in China and neighboring countries, is believed to have been invented about the same time. Its inventor was worshipped as a god by country carpenters. The Grand Canal, 700 miles long, was built by the Sui Empire, 589–618 A.D.

The practice of irrigation in India antedates the historical epic by an indeterminate period. There are reservoirs in Ceylon to the south of India more than 2000 years old. Writings at 300 B.C. indicate that the whole country was under irrigation and very prosperous because of the double harvests which the people were able to reap each year.

The Spaniards on their first entrance into Mexico and Peru found elaborate provisions for storing and conveying water supplies which had been used for many generations. Their origin was almost lost even to tradition. Extensive irrigation works also existed at that time in the southwestern United States. The early Spanish missionaries brought knowledge of irrigation from their Mediterranean homes. Irrigation was practiced also by trappers, miners, and frontiersmen in many places in the West, although no effort was made to develop an agricultural economy based on irrigation until the Mormon Pioneers entered Salt Lake Valley in July, 1847. Under the Mormons, irrigation was a cooperative undertaking, with communities being located on the streams issuing from the mountains. Community ditches were constructed to serve both outlying agricultural areas and garden plots in the towns.

1.2 IMPORTANCE OF IRRIGATION

The importance of irrigation in the world today is well stated by N. D. Gulhati of India:

Irrigation in many countries is an old art—as old as civilization—but for the whole world it is a modern science—the science of survival.

The pressure of survival and the need for additional food supplies are necessitating a rapid expansion of irrigation throughout the world. Even though irrigation is of first importance in the more arid regions of the earth, it is becoming increasingly important in humid regions.

1.3 ARID REGIONS OF THE WORLD

Areas requiring irrigation are very extensive and encompass portions of every continent of the world. The arid belt is roughly divided into two parts: The northern belt extends across the western part of the United States and Mexico, across Spain, southern France, Italy, and Greece into the Asia Minor Area, thence over most of India and into China. The southern arid belt encompasses the portion of South America on the west side of the Andes Mountains, the southern part of that continent and most of South Africa, joins with the upper belt through the Arabian peninsula and India, and then extends south, encompassing essentially all of Australia.

Civilization has existed and now exists in these areas solely because of the art and science of irrigation. The irrigated acreage of the world, however, is not restricted to these areas. Some of the most profitably irrigated agricultures in the world are located in areas normally thought to have sufficient rainfall. These are areas such as Central Brazil, Central America, the West Indies, and the western part of Africa, including Portuguese West Africa and parts of South Africa, which have an ample annual rainfall, but during 6 months of the year have practically no rainfall. Other areas have periods of

2 weeks to 2 months of drought which necessitate irrigation if a profitable and diversified agriculture is to be practiced. Therefore, irrigation is no longer a regional practice of arid countries, but is becoming a basic part of well-developed agriculture throughout the world. Having ample water available for crop production in humid regions as well as arid regions, contributes to profitable and productive agriculture. The basic principles of irrigation are the same, whether practiced in arid or humid climates.

1.4 IRRIGATION DEFINED

Irrigation generally is defined as the application of water to soil for the purpose of supplying the moisture essential for plant growth. However, a broader and more inclusive definition is that irrigation is the application of water to the soil for any number of the following six purposes:

1. To add water to soil to supply the moisture essential for plant growth.
2. To provide crop insurance against short duration droughts.
3. To cool the soil and atmosphere, thereby making more favorable environment for plant growth.
4. To wash out or dilute salts in the soil.
5. To reduce the hazard of soil piping.
6. To soften tillage pans.

Irrigation may be accomplished in four different ways: (1) by flooding; (2) by means of furrows, large or small; (3) by applying water underneath the land surface through sub-irrigation, thus causing the water table to rise; (4) or by sprinkling.

Irrigation water is supplied to supplement water available from the following four sources, none of which should be ignored when irrigation water requirements are estimated:

1. Precipitation.
2. Atmospheric water other than precipitation.
3. Flood water.
4. Ground water.

Failure to consider all four sources and the proportion of water that each supplies to total plant needs may result in faulty design of an irrigation system. In some areas one of the four sources may supply the major portion of plant needs; in other areas two, or even all four, will contribute appreciable amounts of water to the plants.

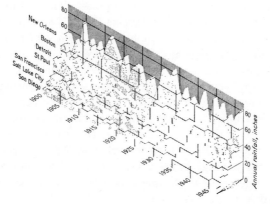

Fig. I.I Precipitation varies from year to year. Long-time averages are of little help in predicting the amount of rain to expect in a given month or year. Every area has its "wet" and "dry" years. (Reprint from *Power*, September 1952.)

Precipitation

To be of greatest benefit, precipitation should have the following characteristics:

1. Amounts should be sufficient to replace moisture depleted from the root zone.
2. Frequency should be often enough to replenish the soil moisture before plants suffer from lack of moisture.
3. Intensity should be low enough so that water can be absorbed by the soil.

In only a few locations will precipitation fulfill all of the above requirements at all times to produce maximum yields. The failure of precipitation to fill these requirements is resulting in increasing amounts of irrigation in arid and in humid areas.

When planning irrigation projects, special precautions should be taken not to misinterpret "average" precipitations during the month and year. Long-time averages can be very misleading, since every area has its wet and dry months, and wet and dry years, which are either above or below normal. Figure 1.1 shows the variations in precipitation that can occur from year to year. Variations from month to month are usually more extreme. Hence, long-time averages have little value in predicting how much moisture to expect in any month or year. Irrigation systems must be designed to provide for expected frequencies of drought periods.

One of the most interesting things in nature is its variation, its changes from time to time and place to place. There is no uniformity in nature or in the rainfall; they are continuously changing from day to day, week to week, month to month, and from year to year. These changes are of vital concern to agriculture in both arid and humid regions. Rainless periods of 2 or more weeks during the crop-growing season frequently occur in humid-climate states. During a 10-year period in Michigan, there were on the average seven periods each year, from 1 to 2 weeks, in which there was no rainfall. In Iowa there were eight such periods; in Wisconsin, Minnesota, Illinois, and Indiana, six; and in Ohio, five. A rainless period of 2 to 3 weeks' duration occurred on the average twice each year in Minnesota and once in each of the other states. Rainless periods of 3 weeks or more are comparatively rare in this "humid" region.

The need for irrigation has been brought forcibly to the attention of farmers throughout the United States because of severe droughts that have affected much of the area. Although sufficient rainfall may be available for the growing of crops in normal years, it has been found through costly experience that short periods without rainfall have ruined crops which would otherwise have brought ample returns to the farmers.

In bulletins concerning irrigation for Missouri and regions of similar rainfall, Rubey reports that during the 77-year period from 1870 to 1947, one-fourth of the years have been very dry and have caused excessive losses of crops. From 1870 to 1930 there were 15 dry years in which the Missouri corn crop was much below the average. In 3 of these years it was less than half of the average yield, and during 10 of the 15 years it was less than three-fourths of the average.

Since precipitation records are always somewhat limited, care should be exercised not to extrapolate the records too far without adequate consideration of variations which may occur. Wide variations, of course, are readily observable through vegetative changes and from the knowledge of local residents. However significant this extreme, variations may well occur without being noticeably evident.

Atmospheric Water Other than Precipitation

In some parts of the world the contribution of atmospheric water in forms other than precipitation is significant. In portions of western Australia, sufficient dew forms to produce good pasture. In the Negeb desert southwest of the Dead Sea, dew is largely responsible for summer growth of grapes. Here the scant rainfall, 4 to 11 inches annually, is augmented with 100 to 250 nights of dew a year.

Archeologists have found rock lattices in the Negeb which were built around vines and trees to "wring" water out of the air in sufficient quantities to irrigate the plants.

The atmospheric conditions which generally prevail to make this source of water significant are: (1) considerable dew formation, (2) fog and clouds, and (3) high humidity. These conditions reduce the plant's water needs by reducing the forces causing water to transpire from the plant. Dew is especially effective in reducing the amount of water moving through the plant, and in some cases, dew is absorbed by the plant. Water which evaporates from the ground and foliage normally reduces by a like amount the water which would have been withdrawn from the soil by the plant. Fog, drip, and dew amounted to 11 inches depth of water at Cascade Head, Oregon, during 142 foggy days. Over a 6-year period 9 inches were deposited annually at Coshocton, Ohio. Fifteen inches are estimated to be a reasonable upper limit for the amount of atmospheric moisture that may condense during the year. The contribution of atmospheric moisture, in forms other than precipitation, should not be overlooked when considering the need for additional water for agricultural production.

Flood Water

Flood water is similar in some respects to irrigation water, but it is not supplied by man. As floods pass over the surface of the land, water is absorbed by the soil and stored for subsequent use by plants. In some regions agricultural production is wholly dependent upon flood water.

Ground Water

Ground water is water beneath the soil surface where voids in the soil are substantially filled with water. Upward movement of ground water by capillarity from the water table into the root zone can be a major source of water for plant growth.

To be most effective without seriously restricting growth, ground water should be near but below the depth from which the major portion of the plant's water needs are extracted. If ground water is within the normal root zone, it definitely restricts plant growth. If ground water is too near the surface, the land's ability to economically produce most crops becomes almost nil. However, a water table within the lower portion of the root zone may supply a considerable amount of water and thereby reduce the cost of irrigation more than it offsets the loss of production. The optimum depth of the water table is that depth which gives the maximum economic return.

1.5 IRRIGATION IN HUMID-CLIMATE AREAS

Irrigation in humid-climate areas, as stated by Powers: (a) controls soil moisture and overcomes drought; (b) provides a green pasture and green feed late in summer; (c) saves the clover stand and makes a cutting the first season; (d) makes double cropping possible (the areas that have long growing seasons produce late crops after early crops); (e) aids the beneficial bacterial and chemical activities in the soil; (f) improves plant quality and aids control of crop pests and diseases, especially of vegetables and berries; (g) increases soil moisture during the best growing weather; (h) aids in deep or early fall plowing and extensive cropping; (i) softens clods and dissolves plant food; (j) pays in increased yields, net profits, and productive values.

It is important that humid-climate farmers understand these and other advantages of irrigation. They should also be informed concerning the economic advisability of irrigation. They often wonder about the question, "Does irrigation pay?"

Farmers in arid regions seldom, if ever, ask this question. They study the probable profits in the growing of different irrigated crops, and since in these places nearly all crops must be irrigated in order to produce, or even to live, the question is of no significance to them.

A well-established fact is that farmers in humid climates can assure themselves of larger and more dependable crop yields insofar as influenced by available soil moisture, by providing irrigation systems and dependable, adequate water supplies. But this fact does not prove that the farmers' profits will thus be increased—they may or they may not be. The humid-climate farmer should consider irrigation as a possible means of improving his economic status.

Cost is the greatest deterrent to expansion of irrigation in humid regions. In humid-climate states, the major responsibility for determining the extent to which irrigation may be economically attractive rests largely with each farm owner, or with small groups of neighbor farmers who may develop and use water from a common source. Analysis of cost factors and trial will give the answer. Public agencies concerned with the general welfare should assist humid-climate farmers in the analysis and solution of problems of the economic advisability of irrigation.

Investments of funds for irrigation facilities in humid and arid regions are basically long-term investments. Some irrigation systems have increased profits enough in the first year of operation to pay the entire cost of installation, but most systems require a number of years to recover the original investment. Climatic conditions, marketing conditions, and management are among the factors that determine the rapidity with which the initial cost is repaid. The crops under irrigation, the depth of their rooting system, the readiness with which they are damaged by drought or respond to irrigation, and their values are also important. Some soils show the effect of dry weather much more quickly than others. In wet years, however, a large investment is tied up in irrigation equipment without increased financial return.

When deciding whether or not it is desirable to install an irrigation system, the average probable use should be carefully considered. Irrigation, even in humid areas, should be thought of as a normal farm operation. Seldom is it possible to secure and install equipment in short enough time to save a crop already suffering from lack of water.

1.6 EXTENT OF IRRIGATION

The area of land irrigated in the world is more than 400 million acres (162 million hectares). The results of comprehensive studies of irrigated land areas by the International Commission for Irrigation and Drainage are given in Table 1.1. The total cultivated and irrigated areas in 26 countries and also the percentages of the cultivated areas that are irrigated are shown. The irrigated area in each of the 26 countries exceeds 1 million acres. The total area cultivated in these 26 countries is estimated to be 1500 million acres; and the area irrigated is about 363 million acres, which is about one-fourth of the cultivated area. Five countries—China, India, the United States, Pakistan, and the Soviet Union—have the largest irrigated areas, the total being about 272 million acres, which is almost 68 percent of the world's irrigated area.

1.7 FUTURE GROWTH OF IRRIGATION

As the population of the world continues to increase, the demand for food and fiber for the people will also increase. Men and women with knowledge of irrigation will be called upon to find the solution of world food and fiber problems. Water must be provided for more of the desert and arid lands of the earth. With irrigation water many of these lands will become highly productive. Also, the productivity of the lands now producing food under natural rainfall can be increased considerably by the application of irrigation water. However, the capital investment necessary to reclaim the remaining lands of the earth will far exceed the investment that has been made to

TABLE 1.1

IRRIGATED AREAS OF EACH OF THE COUNTRIES IN WHICH MORE THAN
ONE MILLION ACRES ARE IRRIGATED
SHOWN IN THOUSANDS OF ACRES

Country	Total Area	Area Cultivated Annually	Irri-gated	Percentage of Cultivated Land That Is Irrigated
1 Afghanistan	160,000	22,267	8,645	39
2 Argentina	686,528	75,000	2,500	3
3 Australia	1,903,732	21,000	1,600	8
4 Burma	167,545	20,000	1,300	6
5 Chile	183,294	13,620	3,367	25
6 China	2,458,646	276,640	131,820	48
7 Egypt	247,166	6,604	6,604	100
8 France	136,102	78,219	6,178	8
9 India	782,003	318,000	63,630	20
10 Indonesia	470,954	35,000*	11,115	32
11 Iran	401,958	....	5,000	..
12 Iraq	110,080	....	8,150	..
13 Italy	74,478	54,856	5,190	9
14 Japan	91,320	15,055	8,307	55
15 Korea (South)	23,953	4,790	1,610	34
16 Mexico	486,639	57,700	5,330	9
17 Netherlands	8,224	....	2,528	..
18 Pakistan	233,432	52,376	27,000	52
19 Peru	328,998	39,500	3,212	8
20 Philippines	74,085	16,245	1,242	8
21 Sudan	619,200	17,537	3,500	20
22 Taiwan	8,887	2,160	1,337	62
23 Thailand	128,095	13,400	3,264	24
24 Union So. Africa	302,310	19,027	1,350	7
25 U.S.A.	2,322,016	340,998	33,022	10
26 U.S.S.R.	5,503,857	....	16,062	..
Total	17,473,502	1,499,994	362,863	Avg. 26.64

* Arable area.

supply water for areas now being irrigated. The largest reservoirs, the longest canals, and the most expensive tunnels and inverted siphons are yet to be built. As long as the population is materially increasing each decade, the demand for further utilization of water for irrigation will also increase.

1.8 SCOPE OF IRRIGATION SCIENCE

Irrigation science is not restricted to application of water to soil. The scope of irrigation science extends from the watershed to the farm and on to the drainage channel. The watershed yielding the irrigation water, the stream conveying the water, the management and distribution of the water, and the drainage problems arising from irrigation practices are all of concern to the irrigationist. Observing one portion of an irrigation system without considering its other components will lead to faulty design and inadequate preparation.

Watershed characteristics, including the nature of vegetation and retentive ability of soils and subsoils, are all important in their influence upon the yield of irrigation water. Likewise, characteristics of the stream conveying water are of importance. Oftentimes management of this stream, the diversion structures built upon it, and remedial measures to reduce seepage losses and consumptive use along the stream and canals are of importance to the irrigation engineer. On the farm, diversion structures, measuring devices, and conveyance channels are extremely important. Layout of the irrigation system on the farm, method of control, and disposition of excess and waste waters also have vital significance.

Proper disposition of excess waters is of equal importance to acquisition of irrigation water. Too often irrigation projects are designed without adequate preparation to utilize waste waters. Water is often needlessly lost in conveyance and by excess application. Both surface and sub-surface waste water should be utilized on lower lands of the project or pumped to higher lands for re-use as long as these waters are suitable for agricultural production. The design of surface and sub-surface drainage systems is of vital importance in maintaining high productivity of irrigated lands.

The natural hydrologic balance of a valley always is upset by the introduction of irrigation water to the land of the valley. Consequently, drainage problems will arise, usually both surface and sub-surface problems, which have not heretofore been present. Wise judgment is necessary to anticipate these problems and to include their consequences in original economic considerations.

1.9 ECONOMICS OF IRRIGATION

Economics is important in evaluating irrigation practices, for irrigation is largely for the purpose of increasing profit. Higher profits resulting from more efficient production will ultimately result in lower prices for consumers, and lower prices result in more consumption of food and fiber; greater availability of food and fiber results in higher standards of living for the peoples of the earth. These factors should be kept constantly in mind. Irrigation projects, as well as other engineering and agricultural works, are for the purpose of making the world a better place in which to live. The most direct way to accomplish this is to make economical developments.

REFERENCES

Christiansen, J. E., "Irrigation in Relation to Food Production," Vol. 34, pp. 400–407, *Agr. Eng.*, June 1953.

Finch, J. Kip, "Master Builders of Mesopotamia," *Civil Eng.*, Vol. 50, pp. 256–260, April 1957.

Gulhati, N. D., "Worldwide View of Irrigation Developments," *Proc. Am. Soc. Civil Eng.*, No. 1951, September 1958.

Harris, Karl and Vaughn E. Hansen, "Relative Productive Value of Land," *Eng. Exp. Sta. Bul.* 6, Utah State University, 1959.

Israelsen, O. W., "Irrigation Science—The Foundation of Permanent Agriculture in Arid Regions," Second Annual Faculty Research Lecture, Utah State Agr. College, 1943.

Israelsen, O. W., "The Engineer and Worldwide Conservation of Soil and Water," *Proc. Am. Soc. Civil Eng.*, No. 1775, September 1958.

Israelsen, O. W., "The Historical Background of Reclamation," reprinted from *Agr. Eng.*, Vol. 32, 6, pp. 321–324, June 1951.

Kimball, Frank, "Soil and Water Conservation in Irrigated Areas," *Agr. Eng.*, Vol. 25, p. 285, 1944.

Matson, Howard, "More Production from Improved Irrigation Practices," *Agr Eng.*, Vol. 24, p. 119, 1943.

Powers, W. L., "The New Reclamation Era in Venezuela," *Agr. Eng.*, Vol. 24, p. 345, 1943.

Robertson, G. Scott, "FAO and the World Food Problem," *J. Am. Soc. Agron.*, Vol. 40, pp. 2–120, 1948.

Rubey, Harry, "Supplemental Irrigation for Missouri and Regions of Similar Rainfall," Univ. of Missouri Bul. 9, Vol. 49, 1947. *Eng. Exp. Sta. Bul.* 33, 1947 (revised).

Thomas, George, "Early Irrigation in the Western States," University of Utah, 1948.

Wolfe, Thomas F., Water, Spec. Edit. Report, *Power*, pp. 73–117, September 1952.

CHAPTER 2 SOURCES AND STORAGE OF IRRIGATION WATER

Basically, rain and snow are the sources of all water. Melting snow and rainfall are not entirely used consumptively. That portion which is not used at the point where it falls, flows over the surface of the land or seeps into the ground and augments the ground-water supply. Therefore, the rain or snow which is not used becomes a potential source of either surface or underground water for irrigation. Irrigation water may also come from waste water, which is water that is not used consumptively by agriculture, industry, and municipalities.

The success of every irrigation project rests largely on the adequacy and dependability of its water supply. In irrigated regions public agencies should make continuous long-term records of precipitation, stream flow, and ground-water storage as a basis for intelligent and complete utilization of all water resources.

2.1 VALLEY AND MOUNTAIN PRECIPITATION

Rain and snow which fall in irrigated valleys are valuable as sources of water to be stored directly in the soil. In some valleys winter precipitation provides enough water to germinate seeds and maintain growth of young plants for several weeks. In these valleys, perennial crops also make substantial growth early in the season by using winter precipitation which has been stored in the soil. Other arid region valleys receive so little winter precipitation that farmers must irrigate the soil before seeding their crops in order to insure a sufficient amount of moisture to germinate seeds and start satisfactory growth. Such valleys must depend almost wholly on rain and snow that falls in adjacent mountain areas as a source of their water supply for irrigation. As a rule, in nearly all irrigated regions, valley precipitation is relatively unimportant as a direct source of irrigation

13

Fig. 2.1 Snow stored on the watershed represents a potential irrigation water supply. (Photograph by Bert V. Allen.)

water.   Precipitation in mountain areas, as illustrated in Fig. 2.1, constitutes a major source of water supply.   Conveyance of water from mountainous sources to valley lands presents to the people of arid regions very interesting, and yet perplexing, problems.   Painstaking study of seasonal and annual water yields of each mountainous area on which rain and snow fall is urgently needed.   Complete and economical utilization of nearly all western natural resources are inseparably connected with the water yield of the watershed, its conveyance to places of use, and its economical use, whether for power, irrigation, or domestic and industrial purposes.   There is urgent need for a wider recognition of the fact that intelligent solution of watershed-yield problems and of economical conveyance of water will appreciably advance the general welfare.

2.2 WATER-SUPPLY STUDIES

Accumulation of dependable information concerning water supply demands intelligent and painstaking endeavor and continuous effort. Inadequate water supply has probably contributed most to financial stress and failure of irrigation projects. Over-optimism and conclusions based on insufficient knowledge of watershed yield have been common and expensive follies among leading citizens in both private and public places. Overestimates of water supply for various projects are frequently reflected in the small areas of land actually irrigated when compared to the area of land intended to be irrigated in the original project. Occurrence of climatic cycles, which cannot as yet be predicted with precision, together with wide variations of precipitation and stream flow from one time of year to another, complicates the problem of economically using all the available water every year. Yet arid-region communities may intelligently adjust their irrigation practices, to some extent, on the basis of reliable information concerning water supplies. Investigations on a watershed should be conducted to determine the yield and supply of water available for further irrigation expansion.

2.3 SNOW SURVEYS AND THEIR BENEFITS

Snow remaining on the surface of the land provides a storage of water far greater than any man-made reservoir. For example, 1 foot of snow normally holds approximately 1 inch of water and may hold as much as 4 inches. Hence, a snow fall of 1 foot, considering 1 inch of water per foot depth on a 10-square-mile area, will provide for the storage of 5000 acre-feet of water. This amount of water would fill a surface reservoir covering 1000 acres to a depth of 5 feet. Not infrequently does the snowfall lie 10 feet deep on a watershed. Also in the spring of the year the water held per foot of snow usually is in excess of 1 inch. When and how fast this vast quantity of water is released by the melting of the snow is of utmost importance to water users. As the stream becomes more fully developed, it becomes increasingly important to know how much water can be expected.

It is important to know whether or not the water supply will be adequate or inadequate and in what manner runoff will occur. In years of limited water supply, cropping and irrigation plans may be modified, less land may be irrigated, crops that use less water may be planted, and early-maturing crops may be substituted for those requiring a longer season.

When water supplies are above normal, additional lands may be brought under irrigation or more intensive farming may be practiced. These and other benefits from snow surveys have resulted in a great expansion of snow survey network to obtain more reliable data more quickly and safely.

Since many remote locations must be sampled once a month, vehicles (see Fig. 2.2) have been developed that will travel over muddy roads, rocks, snow, and ice, carrying the surveyors into and out of the area safely. Also considerable effort is now being devoted to developing sensing elements that will modulate an electrical circuit and thereby permit transmission of snow depth and water content by radio to central locations.

Since forecasts must be made quickly after surveys are made, techniques of analysis have been developed utilizing high-speed computers.

Representative snow courses are usually established in small mountain meadows where the snow cover is least influenced by shifting winds. The course consists of ten to fifteen observation points spaced about 50 to 100 feet apart. Depth and water content of the snow are determined at each observation point.

Snow is sampled with a lightweight seamless aluminum tube. The bottom of the tube is tipped with a circular saw-edge cutter for pene-

Fig. 2.2 The trackmaster developed by the Utah Scientific Research Foundation for off-highway travel to snow courses.

Fig. 2.3 Driving a sampling tube into the snow, measuring the depth of snow, and weighing the tube and snow core to determine water content. (Photos by Bronstead, courtesy Soil Conservation Service, U.S. Dept. of Agriculture.)

trating hard snow or icy crusts. The tube is forced into the snow and then removed, holding within the tube a core of snow, and weighed as shown in Fig. 2.3. Data collected are correlated with one or more of the following variables to predict the peak and total runoff: present and antecedent stream flow, temperature, water content of the soil mantle, and rainfall.

2.4 SURFACE RESERVOIRS

Surface reservoirs are built to store irrigation water for use when the natural flow of a stream is not sufficient to meet irrigation demands.

Fig. 2.4 Government reservoirs like this one contribute to irrigation advancement. (Courtesy Bureau of Reclamation.)

Winter and spring runoff can be impounded until needed for crop growth. In large reservoirs excess runoff in wet years can be stored until needed in dry years.

All storage dams must be built with spillways large enough to convey the maximum anticipated flood flows. The larger the structure and the reservoir, the greater the danger to life and property, and the safer must be the spillway.

The capacity of each reservoir is fixed largely by the natural conditions of the canyon or valley in which water is to be stored, together with the height a dam must be to store the quantity of water needed and economically available. Capacities vary from a few thousand acre-feet for reservoirs on small streams (as shown in Fig. 2.4), to millions of acre-feet. Likewise, dams constructed for storage of irrigation water vary from a few feet in height, built at a low cost, to massive masonry structures over 700 feet high and built at a cost of several millions of dollars. Increasing demands for water will necessitate construction of higher and more expensive dams than have thus far been built.

2.5 SMALL EARTH DAMS

Small dams are valuable for impounding water from springs and small streams so that it can be more efficiently utilized. Small reservoirs built for irrigation frequently have multiple advantages. Often they supply water for fish, provide stock water, and supply nesting cover and food for ducks and geese. Ponds can also add general beauty and interest, as well as utility, to the farming units.

When selecting a reservoir site, several factors should be considered. The dam should be as short as possible and located on dry firm soil void of roots and brush. The height of the dam should be consistent with water supply, economy of the project, and diversion and spillway requirements. A good supply of medium-textured soil is advisable for the dam. When heavy clay is used the dam may crack when dry, and sand cannot be used because it will not hold water. The site should be readily accessible and should permit construction of a good spillway and reservoir. Consideration should be given to the hazard to life and property which would occur in case an excess flood washes out the dam and releases the water in the reservoir.

The reservoir should be located on material that will not allow excessive seepage. Shale and rock outcroppings frequently have cracks that will develop solution channels and result in serious loss of water. Trees and shrubs should be removed from the area to be inundated. The size of the reservoir should be consistent with the water supply and the amount of water needed.

The spillway is a very critical segment of the design and is the feature which frequently causes failure of dams. Design considerations for spillways are complex and must fit site conditions. Therefore, competent professional engineering help should be secured to design and construct the spillway and dam to insure that the most economical design is used consistent with materials at hand, site, and flood-water flows.

2.6 SEDIMENTATION OF RESERVOIRS

The useful life of irrigation reservoirs may be shortened by accumulation of sediment. Once sediment has accumulated, the site is essentially of no further value for storage of water. Since good damsites are very limited, it is not only a serious financial loss to the owner when a site is no longer of value, but it is also a grave, often irreplaceable, loss of a natural resource. In irrigation-project planning, allowance should be made for the effect of sediment accumulation upon the usefulness of the reservoir, and all practical steps should be undertaken to minimize the rate of sedimentation.

Devices and methods used to control silting of reservoirs are: settling basins, by-pass canals, off-channel locations, vegetated streams, venting density currents, flood sluicing, dredging, draining, and flushing. Most of these methods are useful only under special site conditions. Watershed protection and special reservoir design which will permit utilizing one or more of the foregoing means of prevention are the two most useful approaches to controlling sediment.

2.7 REDUCING EVAPORATION LOSSES

Loss of water by evaporation is a very serious problem in arid regions. Critical shortage of water in desert areas has stimulated interest in ways to reduce evaporation. Extensive research is being conducted to discover practical methods. In the seventeen western states of the U. S. A. estimates are that 45 percent as much water evaporates as is used for irrigation. Hence, the incentive is great to reduce evaporation in arid regions, particularly with the increasing competition for water supplies.

Two methods that appear to hold the most promise are: using a monomolecular film on the reservoir water surface and making more use of ground-water reservoirs for storage.

Several technical problems must be solved before using monomolecular film becomes a practical method for general use to conserve irrigation water. Materials are costly, micro-organisms consume the material, and wind drags the film across the water surface often depositing the chemical on the shore of the reservoir. Suitable chemicals will undoubtedly be found and better techniques of application be developed so that this method can be used more widely to conserve water supplies. It is now being used in a number of areas to conserve water supplies for livestock.

The use of underground reservoirs for storage of water is increasing and will continue to increase as the water supplies become more valuable and surface reservoir storage sites become more limited and more costly. As in the case of the monomolecular films, technical advances are needed to facilitate full use of underground reservoirs. Nevertheless, the absence of an exposed water surface from which evaporation can occur is a major asset. Development of ground-water basins, and techniques used for their exploration, are discussed in a later section.

2.8 PHREATOPHYTE PROBLEMS

Phreatophytes are water-loving plants growing along stream courses and on wet soils having high water tables where an abundant supply of water is available. They have little or no economic value, and they consume water which normally could be used beneficially for agricultural and industrial purposes. The problem is particularly acute in arid regions where phreatophytes abound and where water is critically needed for economic development. Estimates have been made that for every 10 acre-feet of water used for agricultural crops in the southwestern United States, 8 acre-feet are used by native vegetation such as cattails, tules, willows, salt cedar, cottonwoods, salt grass, greasewood, baccharis, and mesquite.

Water use by phreatophytes is high and increases as the depth to the water table decreases. The occurrence and growth of most species are controlled by the depth to the water table. The thicker the capillary fringe above the water table, the more favorable is the growing condition. Therefore, even though the water table may be 10 feet below the surface, a normally shallow-rooted plant like salt grass may be able to obtain ample water from a thick capillary fringe which brings water near enough to the ground surface. Cattails and tules prefer water on or near the surface. Salt grass does best when the water table does not exceed 6 to 8 feet; willows, salt cedar, and cottonwood prefer a capillary fringe not more than 10 feet below the surface. Greasewood does best where the water table is less than 15 feet, whereas mesquite has been known to send its roots to a depth of 40, and even 100 feet to water.

Salt cedar, mesquite, and baccharis grow best in warm climates. Willows, cottonwood, and salt grass grow rather generally, and greasewood is usually found in cold or dry climates.

Two general approaches are used to control phreatophytes. One is to remove the water supply by lowering the water table, channeling the water, or piping the water around or through critical areas. The other approach is to mechanically or chemically prevent plant growth. Both methods have been useful and successful in varying degrees, but better techniques are needed and more knowledge is required for effective control. Methods of eradication must be adapted to the local physical and economic conditions.

An excellent summary of the problem and approaches to control are given by Fletcher and Elmendorf in Water, *The U.S.D.A. Yearbook of Agriculture, 1955.*

2.9 RAINMAKING OR CLOUD-SEEDING

One of the most interesting developments in man's search for additional water supplies is cloud-seeding, or rainmaking. The question is, "Can man increase natural rainfall, and if so, by how much?"

To understand the efforts of rainmaking, it is necessary first to understand how nature precipitates water from the skies. By a process known as nucleation, ice crystals are formed upon foreign matter, such as salt particles, dust, or other substances in the air. These ice crystals then combine with each other and grow in size until they fall toward the earth as snow or melt and fall as rain.

Rainmaking or cloud-seeding operations consist primarily of introducing particles into the air to initiate formation of drops. Silver iodide and dry ice (solid carbon dioxide) have been found to be much more effective than particles normally carried in the atmosphere, and both have been used extensively. Silver iodide has the advantage that it can be vaporized on the ground surface and carried aloft into clouds by naturally rising airstreams existing in storm situations suitable for cloud-seeding. Dry ice has the disadvantage that it must be distributed directly into the proper clouds, usually by airplanes, and is rapidly dissipated as it falls through air. Therefore, silver iodide can be applied to clouds as a nucleating agent more economically than can dry ice.

But does cloud-seeding really work? There seems to be little doubt that under proper conditions additional precipitation can be secured by seeding clouds with dry ice, silver iodide, and other nucleating products. This has been confirmed by laboratory and field experiments. However, the question of economic feasibility still remains. Considerable effort is being devoted toward improving techniques of control and measurement. A better understanding of storm activity is needed. Surely rainmaking efforts and success will increase with the need for weather control and with increased technical knowledge of the process. Mankind will ever strive to control the weather, and by so doing will make available the greatest source of water yet undeveloped—an essentially untouched resource, the clouds.

2.10   SALINE-WATER CONVERSION

Another source of usable water is found in the saline waters of the world. Until recently these waters were reclaimed only by the natural processes of evaporation and subsequent condensation as precipitation. Here, as in rainmaking and reduction of evaporation, the pressure for conservation and development of our water resources has led to significant technological advances. Great emphasis now is being placed on conversion of saline water to water fit for human, animal, plant, and industrial use.

Inorganic salts dissolved in water have definite physical and chemical properties which are well understood. These known properties are utilized by a variety of techniques to develop practical and useful ways of reclaiming saline waters. The challenge is to develop practical and efficient techniques of removing the salt.

Theoretical calculations showing the cost of sea-water conversion under optimum conditions with 100 percent efficiency are very informative. To remove the salt from 1000 gallons of sea water requires at least 2.8 kw-hrs (3.8 hp-hrs) of electrical energy. At one cent per kilowatt hour the cost is 2.8 cents for 1000 gallons of water. Likewise, 900 kw-hrs (more than 1200 hp-hrs) are required to obtain one acre-foot of fresh water at an equivalent cost of $9.00.

When efficiency, capital investment, and operating expenses are also considered, economical conversion of saline water presents a major challenge to man's ingenuity. New and cheaper sources of energy and the increasing value of water will certainly promote practical development. Major support has been given to this development by many governments of the world, and results should yield techniques of increasing value.

Water need not be pure for all uses. With proper methods of management irrigation water containing considerable salt can be used for permanently irrigated agriculture. Water with a satisfactory degree of purity may be reclaimed from saline water at a fraction of the cost that is needed to obtain pure water. Table 2.1 shows the general relationship of energy consumption to purity of feedwater reclaimed with a commercial membrane demineralizer. It may not be long before practical, usable methods are available to reclaim the brackish waters of inland areas at a cost that is compatible with the economy of agriculture in areas of short water supply.

2.11   IMPORTANCE OF GROUND WATER

Silting of reservoirs, availability of good ground water storage sites and growing lack of surface storage sites, prevention of evaporation,

TABLE 2.1

INCREASE OF ENERGY CONSUMPTION WITH INCREASE OF SALINITY
FOR A COMMERCIAL MEMBRANE DEMINERALIZER, BASED ON A
PRODUCT PURITY OF 500 PPM.   (COURTESY OF IONICS, INC.)

Feedwater Salinity (ppm)	Approximate Production Rate (gal/day)	Approximate Energy Consumption (kwhr/1000 gal)
900	28,000	3
1,500	14,000	7
2,000	10,000	10
3,000	7,000	15
4,000	5,500	20
5,000	4,500	25
6,000	4,000	30
8,000	3,000	40
10,000	2,500	50
15,000	1,800	80
20,000	1,400	100
35,000	900	180

and development of additional water will all contribute to more extensive and intensive development of underground water. Numerous natural advantages are inherent in utilizing ground water and underground storage reservoirs.

Many countries are blessed with copious water supplies. However, almost every country and section of the earth is plagued with the problem that water supplies are not evenly distributed, neither geographically nor seasonally. Full utilization of the total water supply will require complete water control. Surface water control is well underway. Numerous dams, diversions, and conveyances have been built. Nevertheless, the control of sub-surface water is still in its infancy. For economic reasons sub-surface water supplies generally are not developed until surface supplies have been reasonably well utilized.

Pumping water from underground sources is widely practiced and is a well established method of obtaining irrigation water. Lowering the ground-water table following extensive pumping for irrigation in some places has proved valuable as a means of drainage. In other localities the ground-water surface has been lowered so much by irrigation pumping that deepening of wells has become necessary.

Lowering the ground-water levels increases the pumping lift and makes the water so obtained increasingly expensive. In some areas shortages have become sufficiently acute to justify curtailed use.

2.12 RECHARGING GROUND-WATER RESERVOIRS

Critical shortages of underground water due to limited natural recharge, small storage capacity, and overuse have stimulated efforts to recharge ground-water reservoirs with surface waters. Flood flows which would otherwise have been lost are diverted and applied to the land, thus providing water to seep into underground reservoirs. Winter flow of streams, sewage, and industrial water may also be utilized to excellent advantage for recharging these reservoirs. Full conservation and use of available water supplies requires an integrated use of surface and sub-surface waters and storage facilities.

Systematic flooding of land surfaces overlying or draining into underground reservoirs is a recognized method of water storage. Usually the water is spread over the land to expedite infiltration into the soil and percolation downward to the ground water. Figure 2.5 shows a typical canyon debris fan with rock dams to retard flood flow in terraced basins causing infiltration to recharge a ground-water basin. Rock dams, or dikes, with soil on the upstream face form ponds, shown in Fig. 2.6, from which the water percolates into the ground-water reservoir, to be stored until needed for irrigation.

Infiltration has been increased by using organic residues and grasses to condition ground surfaces to prolonged periods of submergence. Grasses appear to maintain a more pliable and open condition on soil surfaces than do presently used soil conditioners, particularly where the sediment content of the water is considerable. High infiltration rates during the recharging operation are essential.

A less common method of recharging utilizes shafts, pits, trenches, or wells to convey water to the gravel and rock materials underground. The Peoria recharge pit (Suter, 1956) has been very successful in obtaining an infiltration sufficient to absorb 35 to 41 cfs per acre (70 to 80 feet depth per day) by using pea gravel as a filter. When sand is used the rate is only 9 cfs per acre. Suter emphasizes from the results of the Peoria pit that, since many unexpected facts have been observed which are not as predicted, recharging operations still offer a wide field for research.

2.13 SAFE YIELDS FROM GROUND-WATER RESERVOIRS

Early withdrawals of ground water were based upon water needs and costs. Serious lowering of the water level (or pressure) brought alarm and resulted in attempts to define "safe yield." Demands from

Fig. 2.5 Aerial view of San Antonio Creek debris fan with snow-covered Mount San Antonio in background. In the foreground, to the right of the diagonal light streak that marks the course of the stream, the white strips extending from left to right represent rock dams which form terraced basins and hold the storm water. (*U.S.D.A. Tech. Bul. 578.*)

Fig. 2.6 Soil placed against upper face of dikes retards flow of water through them and encourages ponding of water from crest of lower dike to toe of dike next above. (*U.S.D.A. Tech. Bul.* 578.)

adjacent water users resulted in many courts restricting water use to the extent that the level (or pressure) of the water in the aquifer is maintained. In essence, these court decisions guarantee not only the right of the user to the water, but also the mechanical energy represented by the elevation of water in the well. Ground-water resources cannot be effectively utilized if the natural water levels and pressures have to be maintained. Usually only a small fraction of the available water can be withdrawn under these restrictions. For full utilization of surface and sub-surface water resources, ground-water levels and pressures must be lowered during periods of drought or periods of limited surface supplies and elevated by recharging the reservoirs during periods of water surplus.

In some areas where recharging possibilities are limited, the concept of "mining" the water is being accepted as a necessary and desirable procedure. Water is being considered, therefore, in the same category as other natural resources—resources which may be expended in the interest of current development.

An outstanding example of this philosophy exists in the Central Valley development in California. Depletion of the ground water was not only permitted, but encouraged, by some of the more progressive irrigation leaders. The plan was to utilize the "mined" water to build a strong taxable economy which could support the later development and transportation of distant waters to areas of the valley which would develop a water shortage. This plan of ultimate development and integration of water resources of the state is now being implemented, and supplemental waters are being diverted and stored in the ground-water reserves which were seriously depleted.

With the development of new techniques of exploration for ground water, with increasing control and understanding of weather, and with decreasing cost of mechanical energy incident to the atomic age, water supplies can be made available to areas of shortage which at the present time appear to be impractical. Therefore, the definition of "safe-yield" is changing and will continue to change as economic and technological conditions change. Certainly, surplus and reclaimed surface waters cannot be stored in sub-surface reservoirs until capacities are created in these reservoirs by withdrawal of existing waters.

Safe-yield also implies a wise balance where excessive and uneconomical pumping lifts are not imposed, nor serious deterioration in water quality permitted. It is important to consider "safe-yield" in the light of total water resources: surface and sub-surface, developed and potential.

2.14　FEASIBILITY OF GROUND-WATER DEVELOPMENT

Undue lowering of ground water results in higher pumping lifts and sometimes prohibitive pumping costs. Wells may need to be deepened and pumps lowered in order to obtain sufficient quantities of water. The extent of irrigation pumping from ground-water supplies should therefore be determined on the basis of thorough, long-time investigations of the quantity of annual inflow or recharge to ground-water streams, basins, or reservoirs.

Essential decisions concerning development of ground-water supplies for irrigation can be made after checking "yes" or "no" on twelve important points, listed in the following questionnaire. The answers shown are generally considered as indicating that ground-water supplies can be developed satisfactorily.

	Yes	No
A. Availability, quality, and depth of water:		
1. Is a plentiful supply of water available?	yes	
2. Is water of the right quality available to permit production of the desired crops?	yes	
3. Is water available at a depth that will permit economical pumping?	yes	

	Yes	No
B. Trend of the water table:		
4. Is the water table stable?	yes	
5. Is it rising?		no
6. Is it declining?		no
7. Is development in the area likely to bring about withdrawals of water seriously in excess of the natural recharge?		no
C. Legal or natural protection:		
8. Do the statutes or court decisions provide legal or administrative protection in your area against excessive depletion of water supplies, or	yes	
9. Is the area protected against overexpansion and depletion by "natural" controls?	yes	
D. Cost of operation:		
10. Will the prospective production under irrigation bring enough returns to pay the increased costs of irrigation farming?	yes	
E. Land requirements:		
11. Is the land physically suitable for irrigation from the standpoint of contour, productivity, water-holding ability?	yes	
12. Is the land suitable for the types of crops to be produced?	yes	

2.15 CHANGES IN GROUND-WATER STORAGE

The change that can be expected to occur in ground-water storage can be shown in terms of the components to that change. The general ground-water equation is helpful in relating and focusing attention upon the various sources of supply and disposition of water to and from the ground-water reservoir:

$$\Delta S = (Q_i + S_i + P) - (Q_o + S_o + W + C) \qquad (2.1)$$

in which ΔS = change in storage
              Q_i = surface inflow
              Q_o = surface outflow
              S_i = sub-surface inflow
              S_o = sub-surface outflow
              P = precipitation
              W = withdrawal
              C = consumptive use (evapo-transpiration)

Usually all of these elements can be changed and moderated by man. Therefore, all must be considered when a plan of operation of a ground-water reservoir is being developed. This is true not only because these factors can be varied by management, but because each one either contributes water to or withdraws it from the reservoir. Certainly all are involved when safe yields are being considered.

2.16   GROUND-WATER INVESTIGATIONS

The greater the detailed knowledge of the ground-water reservoir, the more effectively and efficiently ground waters can be developed, removed, and utilized. Physical features of ground-water basins are hard to ascertain under ideal conditions. Therefore, all available methods for gleaning factual data should be used. Four approaches can normally be made: (1) geological observations and investigations; (2) physical aspects of the existing ground water, such as pressures, elevations, movements, temperatures; (3) geophysical methods of exploration; and (4) logs of existing wells. All of these are complementary, and all are needed to obtain the complete picture of an underground reservoir.

Geological investigations and observations can outline physical features of the surface and are a good index to sub-surface conditions. Physical surface features and geological patterns are often indicative of what can or cannot be reasonably expected beneath the surface, particularly with regard to the occurrence of a suitable aquifer from which water can be withdrawn. Coarse-textured, highly permeable material or fractured sedimentary deposits are essential in the aquifer if available water is to be economically withdrawn for irrigation.

Aerial maps are very helpful to outline physical features. When the aerial map is viewed through a stereoscope, surface relief is portrayed clearly, and physical characteristics of the terrain become very evident.

Knowledge of local well-drillers should not be overlooked. Well-drillers accumulate a wealth of information through their well-drilling activities.

Frequently, existing wells are available wherein water levels, water temperatures, and ground-water gradients can be ascertained. Well-logs are also extremely informative. These data, when plotted in plan and in profile, often permit extrapolation with reasonable confidence to new locations.

Geophysical methods are based upon detecting differences in the physical properties of surface and sub-surface materials. Electrical resistivity and seismic refraction are the two methods most commonly used. Both require special equipment. Experience is essential to interpret the usually complex results properly. More compact and reliable equipment has extended the use of both methods to studies of water supplies, whereas formerly these concepts were employed only in oil-well exploration.

Electrical resistivity methods involve measurement of changes in resistivity. The resistance of rock formations to electrical current varies widely. In the more porous formations water content is the major variable affecting resistivity. The apparent resistivity is compared to the electrode spacing as an indication of the change in properties with depth. Wells drilled on the basis of the results of resistivity surveys, conducted by competent and trained personnel, consistently yield from 75 to 90 percent success. This method has also been used for detecting the advance of salt-water intrusion and locating perched water tables. Resistivity has been used also as an index to the seepage from canals.

Seismic refraction involves creating waves in the earth's surface by impact or explosion, and measuring the time required for the wave to travel a known distance. Waves are reflected and refracted by changes in the properties of conducting media in a manner similar to light waves. The time interval for the first shock to reach varying distances from the point of origin is analyzed to determine sub-surface properties. Because the seismic method requires special equipment and a greater degree of training of operators and is more complex to interpret, it has been applied to ground-water explorations only on a relatively limited basis.

REFERENCES

Banks, Harvey O., "Utilization of Underground Storage Reservoirs," *Proc. Am. Soc. Civil Eng.*, Vol. 78, Sep. No. 114, January 1952.

Baumann, Paul, "Ground Water Movement Controlled through Spreading," *Trans. Am. Soc. Civil Eng.*, Paper No. 2525, Proc. Separate No. 86, August 1951.

Beasley, R. P., "Determining the Effect of Topography and Design on the Characteristics of Farm Ponds," *Agr. Eng.*, pp. 702–704, November 1952.

Brown, Carl B., "Factors Affecting the Useful Life of Reservoirs," *Proc. Am. Soc. Civil Eng.*, No. 1503, January 1958.

Butler, Charles C., "Public Recognition of the Nation's Water Problems," Second Annual Water Conference, New Mexico A. & M. College, November 1957.

Clyde, George D., "Utilization of Natural Underground Water Storage Reservoirs," Paper delivered at Soil Conservation Society meeting in Detroit, October 26–28, 1950. Reprinted by permission of editor of *J. Soil and Water Conservation*, Vol. 6, No. 1, January 1951.

Colby, C. B., "Snow Surveyors," Coward-McCann, New York, 1959.

Conkling, Harold, "Utilization of Ground Water Storage System Development," *Trans. Am. Soc. Civil Eng.,* Vol. 3, p. 275, 1946.

Corfitzen, William E., "Economic Effects of Reservoir Sedimentation," *Trans. Am. Soc. Civil Eng.,* Paper No. 2458, published as *Proc.* Separate No. 30, August 1950.

Harbeck, G. Earl, Jr., "Can Evaporation Losses Be Reduced?" *Proc. Am. Soc. Civil Eng.,* Paper 1499, January 1958.

Hauk, Ivan E., "Projects, Conduits, and Structures," John Wiley and Sons, New York, 1956.

Kazmann, Raphael G., "Safe-Yield in Ground Water Development, Reality or Illusion," *Proc. Am. Soc. Civil Eng.,* Paper 1103, Vol. 82, November 1956.

Magin, George B., Jr. and Lois E. Randall, "Review of Literature on Evaporation Suppression," *Geol. Survey Prof. Paper,* 272-C, U.S. Govt. Printing Office, Washington, 1960.

Muckel, Dean C., "Research in Water Spreading," *Trans. Am. Soc. Civil Eng.,* Paper No. 2543, published as Proc. Separate No. 111, December 1951.

Peterson, Dean F., Jr., "Hydraulics of Wells," *Trans. Am. Soc. Civil Eng.,* Paper No. 2869, Vol. 122, 1957.

Sandbach, E. K., "Engineering Seismology," reprinted from the *Explosives Engineer,* November-December 1954.

Schaefer, Vincent J., "The Production of Ice Crystals in a Cloud of Super-Cooled Water Droplets," *Science* 104:457, 1946.

Schiff, Leonard, "The Status of Water Spreading for Ground Water Replenishment," *Trans. Am. Geophys. Union,* Vol. 36, No. 6, December 1955.

Schiff, Leonard, "Water Spreading for Storage Underground," *Agr. Eng.,* November 1954.

Suter, Max., "The Peoria Recharge Pit: Its Development and Results," *Proc. Am. Soc. Civil Eng.,* No. 1102, November 1956.

Todd, David K., "Investigating Ground Water by Applied Geophysics," *Proc. Am. Soc. Civil Eng.,* Vol. 81, Separate No. 625, February 1955.

Todd, David K., "Ground Water Hydrology," John Wiley and Sons, New York, 1956.

Todd, David K., "Sea-Water Intrusion in Coastal Aquifers," *Trans. Am. Geophys. Union,* Vol. 34, No. 5, October 1953.

U.S. Department of Agriculture, "Water—the Yearbook of Agriculture," Superintendent of Documents, Washington 25, D.C., 1955.

Water Resources Development Corporation, "Why and How Cloud Seeding Works," 1956.

Wisler, C. O. and E. F. Brater, "Hydrology," John Wiley and Sons, New York, 1949.

Work, R. A., "Stream-flow Forecasting from Snow Surveys," *U.S.D.A. Circ.* 914, March 1953.

Wolfe, John W., "A Pumping Manual for Irrigation and Drainage," *Ore. Agr. Exp. Sta. Bul.* 481, August 1950.

CHAPTER 3 WELLS FOR IRRIGATION WATER

Throughout man's history wells have been an integral part of his life and activity, supplying clean pure water where surface supplies were not adequate. The Bible frequently refers to wells, wherein they are a source of life-giving water and a symbol of security and well-being. Joseph's well in Cairo, Egypt, constructed some 3700 years ago, is still in operation. The ancient Chinese drilled wells as deep as 3600 feet with primitive churn-drill methods. Some were cased with bamboo. The ancient ghanats of Persia (Iran) have been and still are the basis of existence in many parts of that land. Milligan (1955) clearly and interestingly states the historical significance of wells as follows:

The well was and still is, in many areas, the center of religious social activity. These areas include Southern Europe, the Middle East, the Far East, Latin America, and many islands of the sea. The pagans worshipped their wells; the Greeks consecrated them to their Gods; the Moslems constructed mosques over them; and the Hindus constructed temples near important wells.

The ancient ghanats of Persia were shafts started from the surface and dug slightly upgrade toward the ground-water supply, Fig. 3.1. The essentially horizontal shaft served as a conveyance channel and continued sometimes for 20 or 30 miles to the source of the ground-water supply. Approximately every 100 to 1000 feet, depending upon the depth, a vertical shaft was dug to the tunnel through which earth was removed and air, food, and general access provided to the diggers. The skill developed by diggers of ghanats is still passed from father to son and the accuracy attained by these artisans is truly phenomenal.

The Maui tunnels were developed on the island of Maui in the Hawaiian Islands, Fig. 3.2. These wells are vertical shafts with hori-

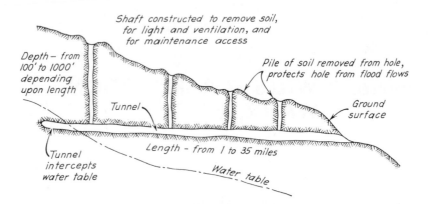

Shaft constructed to remove soil,
for light and ventilation, and
for maintenance access

Depth – from
100' to 1000'
depending
upon length

Pile of soil removed from hole,
protects hole from flood flows

Tunnel

Ground
surface

Tunnel
intercepts
water table

Length – from 1 to 35 miles

Water table

Fig. 3.1 Ghanats constructed in Persia (Iran) to intercept water source and convey water to agricultural lands.

Fig. 3.2 Maui tunnels constructed in lava formation in Hawaii to collect water for irrigation. (Courtesy of Hawaiian Commercial and Sugar Company.)

zontal tunnels or galleries extending in one or two directions from the central shaft. Frequently these tunnels intercept porous lava formations and tubes which result in extremely copious flows of water.

Dug wells have been the principal source of water until recent times, when modern well-drilling methods have been developed and used extensively. The dug wells were usually shallow wells, generally less than 50 feet in depth and several feet in diameter. In the past they were usually excavated by hand, and even today in many areas this is the principal method of construction.

The Ranney well is a modification of the dug well, wherein a large central shaft is constructed to a depth below the water table. Wells are then drilled horizontally in a radial direction from the bottom of the shaft. Large quantities of flow can usually be obtained from a well of this type.

Drilling methods developed by the French in the early 1800's are still used extensively. The most common well is the vertical cased well. It varies in size from the small-diameter well of shallow depth with sufficient capacity to supply domestic needs for a family and water for livestock, to the large-diameter well of great capacity for agricultural, municipal, and industrial purposes. Some vertical wells extend 1000 or 2000 feet into the earth. Because of the great variety of types, sizes, capacities, and drilling conditions, many different methods are used to drill vertical wells.

The importance of wells and ground-water development is indicated by the size of the well-drilling industry. It is estimated that there are more than 9000 well-drilling contractors in the United States, operating approximately 22,000 drilling machines. The number of wells drilled per year by these contractors is estimated to be about 500,000. Approximately 200,000 are new wells and 300,000 are replacements. The number of wells in operation in 1957 has been estimated to be 13,500,000. Approximately 15 percent of the total water supplies of the United States are obtained from ground water.

3.1 DRILLING METHODS AND EQUIPMENT

Excavation by hand is the oldest method used for construction of wells. When mechanical equipment is not available, the material is removed from the hole by hand and bucket. When power is available, the material is removed by clamshell or orange peel buckets attached to excavating equipment. Lining or cribbing is lowered as the well deepens, to prevent cave-ins, particularly in unconsolidated sedimentary materials. Occasionally large shallow open cuts through

Fig. 3.3 Basic well-drilling tools for the cable tool method. (After Todd.)

water-bearing materials serve satisfactorily as collecting pits for ground water.

Cable Tool Method

A cutting tool is suspended from a cable, and vertical oscillation of the tool cuts and loosens material at the bottom of the hole. A special tool, sometimes a modification of a cutting tool, is used to remove the material from the hole. Generally, water is required for drilling, and the holes are usually less than 14 to 16 inches in diameter. A typical cable tool outfit is shown in Fig. 3.3.

The material at various depths in the hole can be determined quite accurately by this method. Noting the material and the depth from which it comes is referred to as logging the well. Such information is required often by the governmental agency responsible for groundwater development and is very helpful for interpreting the performance of the well and anticipating what can be expected at other locations in the vicinity.

Difficulty is experienced when drilling through unconsolidated sand, especially quicksand, with the cable tool because of material caving in around the bit. The variety of tools used by the cable tool method can be grouped into five general classes, Fig. 3.4. The cutting tool or bit cuts, softens, and grinds the materials so that the second type of tool, the bailer, can pick up this material and remove it from the hole. Three auxiliary tools are used to facilitate the operation of the cutting tool and the bailer: the rope socket for attaching the cable to the other tools, the jars which allow vertical movement in order to loosen tools should then be stuck in the hole, and the drill stem to add weight and length to the drill so that it will cut rapidly and vertically.

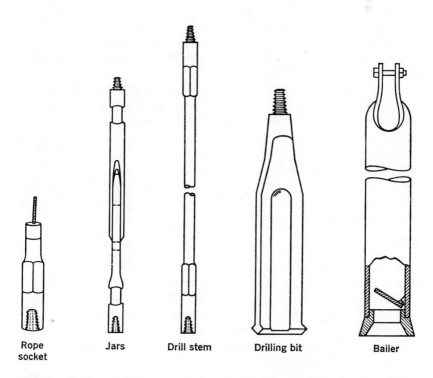

Rope socket   Jars   Drill stem   Drilling bit   Bailer

Fig. 3.4 Basic well-drilling tools for the cable tool method. (After Todd.)

Hydraulic Rotary Method

A bit is rotated rapidly on the bottom of a string of drill pipe. A mud of clay and water, the consistency of which is varied with drilling conditions, is forced through the drill pipe and carries the cuttings around the outside of the pipe to the ground surface. With the proper consistency of rotary mud, and correct pressure, the sides of the hole do not slip in during drilling and no casing is used during the operation. The fluid circulating through and around the drill cools the bit.

The rotary method permits the drilling of a test hole without having to invest in casing. Hence, exploratory work is generally cheaper when done by this method. If favorable conditions are found in the hole, casing is lowered and perforated sections inserted, or perforations are made at the depths of the more favorable water-bearing aquifers. Clay which is deposited on the walls of the hole from the rotary mud is removed in those sections opposite the perforations by washing and development procedures. If sub-surface conditions are found to be very favorable, the well can be reamed to a large size very rapidly and a large casing inserted. Because of ease and rapidity of drilling, rotary holes are usually considerably larger than the casing with the space between being backfilled with graded gravel.

One of the principal disadvantages of the rotary method is that it is sometimes difficult to remove all of the mud from the water-bearing aquifer. If the consistency is not ideal, the mud may penetrate the aquifer for some distance and then be difficult to remove by normal development methods. Also, accurate well-logs are more difficult to obtain by the rotary method than by other methods of drilling. Since the mud mixes with the drillings, the amount of fine material is difficult to ascertain in some circumstances. An experienced operator can minimize both problems.

Reverse Rotary Method

When drilling mud is forced down the outside of the drill pipe and comes up inside the pipe carrying the drillings with it, the method is called the reverse rotary method. One of its particular advantages is that the velocity of the upward flow is much greater for a given discharge and thereby can carry larger size sediments and rocks to the surface. This becomes a distinct advantage when drilling in gravel.

Care must be taken not to seal with mud the low pressure zones which may nevertheless be excellent aquifers. Procedures for casing and developing wells drilled by the reverse rotary method are the same as those used in the hydraulic rotary method.

3.2 WELL CASING PERFORATIONS AND SCREENS

Only in consolidated formations can a well be left uncased. When the formation is unconsolidated, a casing is necessary, and that casing must have suitable openings to permit the water to enter with a minimum of restriction and yet not permit more than a small portion of the surrounding material to enter the well. The casing can be perforated while in the well by a perforating tool, or perforating can be done before placement by a cutting torch or punch. Factory perforated casings are frequently used when the depths to water-bearing formations are known prior to the placement of the casing.

When considerable fine sand is in the aquifer, a screen may be advisable in place of perforations. A screen ordered for the existing gradation of materials will prevent entry of excessive amounts of sand into the well. Excessive entry of sand is extremely dangerous in that a cave-in might result which would collapse the well and possibly cause the ground surface to subside. The slot opening should pass about 70 percent of the surrounding material. The reduced velocity through the screen should be about 0.2 of a foot per second or lower if feasible. A low value of velocity of this order insures that the energy loss through the screen will be negligible. High velocities result in lower pressures as water passes the screen, producing increased incrustation because of release of calcium carbonate and mineral salts. Excessive incrustation and corrosion will also result if the well screen extends above the pumping level of water in the well.

3.3 GRAVEL-PACKED WELLS

A gravel envelope around the outside of the casing is frequently used where excessive quantities of fine sand exist in the aquifer. At times the openings in a screen would have to be so small that high velocities would occur or insufficient strength would result, were sufficient open area to be secured. Gravel-packed wells are well adapted to rotary-drilled wells. Seldom is it necessary to place more than a 6-inch thickness of gravel around the well.

For deep wells the gravel can be placed best if dropped to the desired depth through a 2- or 3-inch pipe outside the casing. This practice will prevent the gravel bridging at some intermediate depth.

When local materials permit, the gravel should generally be graded from sand up to one-quarter inch gravel. Additional gravel should be placed as the pack settles during development and subsequent operation. The criterion mentioned above for the relationship of extreme size to surrounding material will apply to perforated pipe and the surrounding gravel pack. Any gravel pack extending to the surface will permit contamination of the sub-surface waters by the generally less pure surface water. Hence, if the gravel-packed well is to be used for domestic supply, care must be taken to prevent contamination.

3.4 DEVELOPMENT OF WELLS

Developing the irrigation well is almost as important as drilling it. The primary purpose of developing a well is to increase the water yield. This is accomplished by increasing the permeability of the formation through which water moves toward the well. The secondary, but nevertheless very important, reason for developing a well is to determine the water supply available and the needed characteristics of pump and power unit to be installed.

The method used to increase the yield of a well depends upon the formation in which the well is drilled and upon the equipment available. If the water-bearing material is sand and gravel, marked improvement is obtained by removing the smaller sand fractions of silt and clay from the water-bearing formation in the area immediately around the perforated portion of the casing.

When the well penetrates a consolidated sedimentary formation, it may be necessary to dissolve part of the cementing agent to permit greater cavities and larger openings adjacent to the well. When a well penetrates a rock formation, it is often desirable to fracture the rock and thereby create additional crevices through which water can flow toward the well. In all cases it is essential to remove by means of a bailer or sand pump the material which moves into the well and settles to the bottom during the developmental operation.

Care should be taken in the development of a well not to remove excessive quantities of fine materials. If excessive amounts of fine sand are removed, there is considerable danger that overbearing layers might cave in and crush the well, making it completely inoperative. If excessive quantities of sand cannot be prevented from entering the well after it has been constructed, it may be necessary to install a special well screen.

Development by Pumping

Developing a well by pumping requires a variable speed pump of large capacity. A new pump should not be used because of the damage and lowering of efficiency caused by the wearing action of sand as it moves through the pump. Pumping should be started slowly. Since a constant speed will cause bridging of sand particles and a marked lowering in the capacity of the well, the speed should be increased in steps and held constant between steps until no further sand is removed. Pumping should be continued until the maximum discharge is reached and no further sand is withdrawn. The pump is then shut off and the water level allowed to return to normal position. This cycle is repeated until no further sand is removed by this method.

A process known as "raw hiding" is often used, wherein the pump is started and stopped very quickly. This allows for a decrease in pressure followed by a back-surge on the well which gives a flushing action. No foot valve is installed during this program so that the flow can move both ways in the well.

Development by Surging

Surging is one of the commonest and most effective methods of developing a well in sand and gravel formations. The plunger is moved up and down opposite the perforated casing causing water to move alternately into and out of the well. Surging is started slowly at first and increased in speed as development proceeds. Solid surging blocks are most commonly used. However, on wells of low yields, a valve-type block should be used. On the downward stroke, water passes through the valve and also is forced into the aquifer. On the upward stroke, water is drawn rapidly into the well.

Development by Compressed Air

Developing with air is best suited to small-type wells. The depth of water in the well should be at least two-thirds the depth of the well. Pressures of 100 to 150 psi are necessary. The power available on the compressor should be no less than the power required to pump the maximum capacity from the well. The most effective development with air is a combination of surging and pumping. For irrigation wells a 2-inch diameter air pipe is normally lowered inside a 6- to 8-inch discharge pipe, the end of which is near the bottom of the well. With the air pipe extending below the end of the discharge pipe, a large quantity of air is suddenly released causing a surge of water outward into the aquifer. By raising the air pipe within the discharge pipe an air jet pumping action is created which causes water to flow into the well. Hence, raising and lowering the air line with respect to the end of the discharge pipe, causes a reversal in the direction of flow which dislodges sand, silt, and clay and carries it into the well. When the flow of sand stops, the discharge pipe is raised and the process repeated until the entire perforated portion of the well has been subjected to the surging and pumping action.

"Backwashing" is also used to develop a well. In this process air is forced into the capped well until the water level is lowered to the top of the perforations. Then air is suddenly released causing a reversal of flow within the well. Surging action removes fine material and improves yield.

Yield in wells is also improved by using dry ice, chemicals, and explosives with varying degrees of success. Greater skill and experience is usually required when these methods are used than is required in the procedures described previously. Dry ice (solid carbon dioxide) is also used to create pressure surging in wells and operates much the same as backwashing. Muriatic acid (hydrochloric—usually a 15 percent solution) is sometimes added to loosen the sand, when cemented with lime, and to widen interstices in rock. Other chemicals are also used as dispersing and cleansing agents to remove clay and silt and for the purpose of removing lime, iron, and other incrustations. Explosives are used to shatter rock formations and remove incrustations. Special low-power vibratory explosives have been developed which cause sufficient vibrations to shake loose incrustations in well casings.

3.5 CONTRACTS FOR DRILLING

A definite agreement between the owner and driller is essential. A written contract minimizes misunderstanding, but a rigid contract will increase cost. The best assurance of obtaining a good job is to employ a reliable driller. Cut-rate operations usually result in the owner's paying excessively for a job poorly done. The following items should be clearly understood before drilling is commenced.

1. Starting and completion dates.
2. Diameter and thickness of casing.
3. Vertical alignment, essential for installation of a deep-well pump.
4. Well record, showing an accurate log of material encountered in drilling, water-bearing formations, perforations, static water level, draw-down, discharge, and development work.
5. Perforating or screening procedure and cost.
6. Procedure for developing the well and cost.
7. Test discharge and draw-down procedure and cost.
8. Price for drilling, including time for moving and setting up equipment.

9. Price if a change of diameter is made.

10. Method and time of payment.

3.6 HYDRAULICS OF WELLS

The general behavior of a well and the hydraulic formulas which apply to it depend upon the formation in which the well is drilled. Figure 3.5 illustrates four wells and a spring in a typical valley. The stream entering the valley from the canyon on the left passes over a gravel fan where considerable seepage occurs. Well A drilled in this zone is not flowing, with the water level in the well at the level of the water table adjacent to the well. Well B flows because the top of the well is below the piezometric surface of the confined aquifer.[1]

Well C is similar to well A, and both are referred to as unconfined wells, often called gravity wells or water-table wells. Well D is similar to well B, except that well D does not flow without a pump, since the top of the well is above the piezometric surface in the aquifer. Wells B and D are called confined wells, and are also referred to as artesian wells and as pressure wells. Well B may also be called a flowing well.

[1] The piezometric surface is the height water will stand in a piezometer or pipe open at the end which extends into the aquifer. The height water will rise in the pipe above the bottom of the pipe is equal to the pressure divided by the unit weight of the water, $h = p/w$.

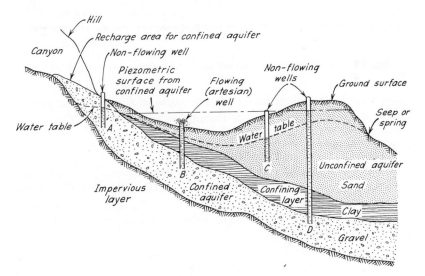

Fig. 3.5 Wells in typical valley.

Water Yield of Wells

The stream of water obtained for irrigation from a well with a pumping plant is determined by one or both of two major factors: (1) the capacity of the pump and the horsepower of the motor or engine, (2) the capacity of the well, which depends on the draw-down of the water surface or pressures, the depth and effective diameter of the well, and the permeability of the water-bearing material. Pump capacities and energy requirements to lift given quantities of water through specified heights are well understood and may be predicted with fair precision. It is far more difficult to predict the energy requirement to move specified quantities of water through water-

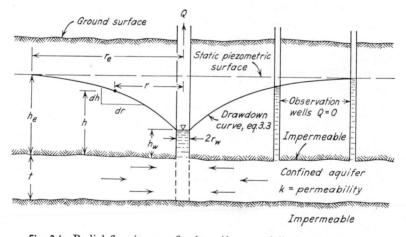

Fig. 3.6 Radial flow in a confined aquifer to a fully penetrating well.

bearing sands or gravels into a well, because of the uncertain permeability of the sands and gravels through which the water flows.

Confined Wells

Radial horizontal flow in a confined layer of thickness t to a fully penetrating well is shown in profile in Fig. 3.6. The flow Q into the well can be represented as $Q = AV$. The area A of a cylinder of radius r and height t is $2\pi rt$. The velocity V through the cylinder is given by the Darcy Law:

$$V = k\frac{dh}{dr} \qquad (3.1)$$

where k is permeability and dh divided by dr is slope of the draw-down curve at radius r. Therefore, the quantity of radial flow toward a well is:

$$Q = AV = 2\pi rtk \frac{dh}{dr} \tag{3.2}$$

Separating the variables and integrating this differential equation between limits of h_e and h_w in the vertical plane and r_e and r_w in the horizontal plane, the result is:

$$Q = \frac{2\pi kt(h_e - h_w)}{2.3 \log_{10} (r_e/r_w)} \tag{3.3}$$

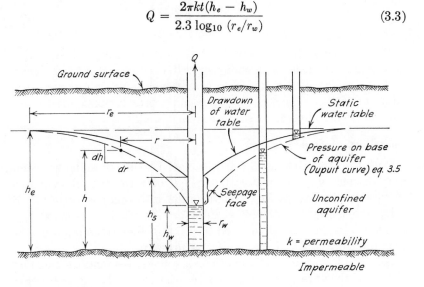

Fig. 3.7 Radial flow in an unconfined aquifer to a fully penetrating well.

The draw-down $(h_e - h_w)$ substantially is directly proportional to the discharge, whereas the discharge varies as the logarithm of the well radius. Hence, when the well radius is doubled, the discharge increases only 10 percent, and increasing the radius four times will increase the discharge only about 20 percent.

Unconfined Wells

A confined well becomes an unconfined well when the flow is not restricted by the impervious layer above the flow, Fig. 3.7. In unconfined wells the surface of the flow is the water table, which is the plane of atmospheric pressure below which the pores of the aquifer are essentially saturated.

The hydraulics of the unconfined well can be developed in the same manner as that of the confined well, except that the constant thickness of the aquifer t is replaced by a variable thickness h. Hence, Equation 3.2 becomes:

$$Q = AV = 2\pi rhk \frac{dh}{dr} \qquad (3.4)$$

and integrating between the same limits, the classical Dupuit equation is obtained:

$$Q = \frac{\pi k (h_e^2 - h_w^2)}{2.3 \log_{10} (r_e/r_w)} \qquad (3.5)$$

The discharge Q now varies as the difference of the squares of h_e and h_w, whereas in a confined well discharge it varied linearly with h, provided that in each case there is no interference from adjacent wells. The same relationship exists between discharge Q and the well radius as exists in confined wells.

Limitations of Equations

Both Equations 3.3 and 3.5 are developed assuming radial horizontal flow through uniform material perpendicular to vertical cylindrical surfaces. Also, steady flow (no variation with time) is assumed. By using the Darcy equation, $V = k(dh/dr)$, the flow is assumed to be laminar. In the case of confined flow the further assumption is made that the thickness t of the aquifer is constant.

Seldom, if ever, in field conditions are these assumptions met fully. However, they are reasonably close in most cases. Being aware of the assumptions and knowing the effect of the error involved, allowance can generally be made so that results from theoretical equations are nevertheless very useful.

Unconfined well flow near the well is not horizontal as assumed. Hence, the nearer the well the greater the error that can be expected. Equation 3.5 shows that the depth of water outside the well approaches the depth within the well h_w as the radius approaches the well radius r_w. However, in actuality a seepage face h_s always occurs above the water surface in an unconfined well. Hence, the correct boundary condition is given by:

$$Q = \frac{\pi k (h_e^2 - h_s^2)}{2.3 \log_{10} r_e/r_w} \qquad (3.6)$$

In Equations 3.3, 3.5, and 3.6, h_e and r_e can be replaced by h and r at any intermediate point. Since h_s generally is not known and there

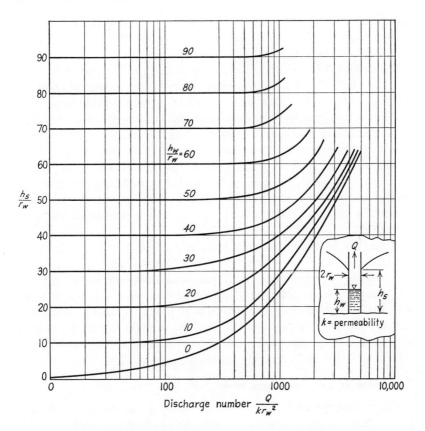

Fig. 3.8 Seepage face and depth of water in the well related to discharge number for an unconfined radial well.

is no mathematical expression relating the commonly measured h_w to h_s, a functional relationship has been developed as shown in Fig. 3.8. A dimensionless discharge number Q/kr_w^2 characterizes the draw-down of the water table around the well—the greater the number, the greater the draw-down. For an unconfined pumped well, the water table adjacent to the well may be considerably higher than the water surface inside the well. When wells are being used to lower the water table, the height of the seepage face h_s becomes very important.

Unsteady Flow from Wells

Flow conditions in a well are seldom steady. Either the discharge or the depth of water is usually changing with time. Nevertheless,

Equations 3.3, 3.5, and 3.6 can be applied to unsteady flow because the velocity of flow through the aquifer toward the well is so small that kinetic energy is negligible.

Difficulty is experienced in selecting a value of the radius of influence r_e, since the draw-down curve approaches asymptotically the original water-table condition. However, a considerable error in r_e will not seriously affect the calculated draw-down or discharge, since the logarithm of the radius is used in the formulas. A difference of less than 10 percent results when a radius of 500 feet is used in lieu of 1000 feet and 2000 feet causes another 10 percent change in Q or h. A reasonable value of r_e for a short period of pumping or for a tight aquifer is 500 feet. For long periods of pumping or very permeable aquifers a value of 2000 feet may be used as a reasonable radius of influence.

When the formulas are used, the correct mathematical definition for r_e is the intercept of the draw-down curve with the unaffected water table. For a confined well, plot h against log r and extrapolate the straight line resulting over the major portion of the radius to the water-table line. For the unconfined well, plot h^2 against the logarithm of r and similarly extrapolate the resulting straight line to the original water-table line. The intercept of the draw-down line and water-table line is the radius of the influence r_e.

Conditions at an Unconfined Well

Three dimensionless parameters $(Q/kr_w{}^2, h_s/r_w, h_w/r_w)$ have been found useful in describing flow conditions to an unconfined well. The discharge number, $Q/kr_w{}^2$, is indicative of the cone of influence, with large numbers indicating steep cones and small numbers indicating shallow cones of depression of the water table. The parameters h_s/r_w and h_w/r_w index the geometry adjacent to the well. Figure 3.8 relates the seepage face and the depth of the water in the well to the discharge number. These parameters avoid the necessity to refer to the radius of influence and permit calculating the height of the seepage face, $h_s - h_w$.

3.7 DRAW-DOWN DISCHARGE RELATIONS

In pumping from thick unconfined water-bearing formations, or from confined ground-water or artesian formations, the discharge is directly proportional to the draw-down as shown in Fig. 3.9, provided that there is no interference from the pumping of other wells in close proximity. Shallow unconfined water-bearing formations sustain a gradual decrease in well discharge per foot of draw-down as the draw-down increases. This relation is shown in the upper curve of Fig. 3.9.

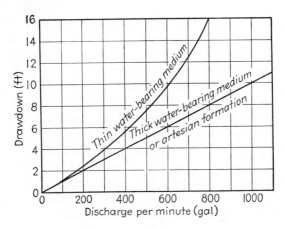

Fig. 3.9 Typical discharge draw-down relation of a well in a thin water-bearing formation and one in a thick water-bearing formation or an artesian stratum. (*U.S.D.A. Circ.* 678.)

Fig. 3.10 A pump installation where the water is pumped 2 feet higher than necessary. The stilling basin with three baffles is very effective. (Courtesy Soil Conservation Service.)

In general, water flows radially through rather homogeneous water-bearing sands and gravels toward the well through a series of concentric cylindrical surfaces having the well as a vertical axis. Therefore, for constant yield as the water approaches the well, its velocity must increase because the cross-sectional area through which it flows is continuously decreasing. Consequently the hydraulic slope (driving force per unit weight) must increase as the water approaches the well, and the draw-down curve becomes steeper as the well is approached. Where the capacity of the pump exceeds the capacity of the well, the draw-down is excessive in the immediate vicinity of the well. It is therefore desirable, if practical, to provide a large "effective" diameter well to avoid excessive draw-down (and excessive power requirements) in order to drive the water into the well.

3.8   BATTERIES OF WELLS

In places where water-bearing materials have low permeability, it is sometimes advantageous to draw water with one pump from two or more wells. The most economical spacing of wells, where two or more are constructed to supply one pump, is a problem to be given thorough scientific study. Wells should be far enough apart to reduce interference to the minimum.

REFERENCES

American Water Works Association, Inc., "Southwest Water Wells," Las Cruces, New Mexico.

Bennison, E. W., "Ground Water, Its Development, Uses and Conservation," Edward E. Johnson, Inc., St. Paul, Minn., 1947.

Code, W. E., "Construction of Irrigation Wells in Colorado," *Colo. Exp. Sta. Bul.* 350, 1929.

Fischback, P. E., P. E. Schleusener, and V. H. Dreeszen, "Nebraska Minimum Standards for Artificially Gravel Packed Irrigation Wells," Approved by Nebraska Well Drillers Association, June 5, 1957.

Hansen, Vaughn E., "Unconfined Ground Water Flow to Multiple Wells," *Am. Soc. Civil Eng.,* Vol. 118, p. 1098, 1953.

Hickok, R., W. V. Morris, D. B. Simons, "Ground Water," Colo. Agr. and Mech. College, Ft. Collins, Colo., July 1952.

Jacob, C. D., "Draw-down Test to Determine Effective Radius of Artesian Well," *Trans. Am. Soc. Civil Eng.,* Vol. 112, p. 1047, 1947.

Johnson, Edward E., Inc., "The Yield of Water Wells," *Bul.* 1238, revised, June 1947.

Johnson National Drillers Journal, "Judging Proper Gravel-pack Thickness," Vol. 27, No. 2, Mar.-Apr. 1955.

Johnson National Drillers Journal, "Improve Yield and Life of Small Wells," Vol. 27, No. 3, May-June 1955.

Johnston, C. N., "Irrigation Wells and Well Drilling Methods in California," *Calif. Agr. Exp. Sta. Circ.* 361, 1945.

Johnston, C. N., "Irrigation Wells and Well Drilling," *Calif. Agr. Exp. Sta. Circ.* 404, May 1951.

Karplus, Walter J. and Otto J. M. Smith, "The Application of Electrical Transients to Well-Logging," *Tech. Note* 326, revised 1956.

McLaughlin, Thad G., "Hydrologic Aspects of Ground Water Law," *Johnson National Drillers Journal*, Vol. 27, No. 3, May-June 1955.

Meeks, Tom O., "Developing and Testing Irrigation Wells," *U.S.D.A. Bul.* 114, March 1952.

Milligan, Cleve H., "Pumping Ground Water for Irrigation and Drainage," *Trans. Am. Soc. Civil Eng.* Paper No. 2857, published as Proc., February 1955.

Mylander, Harvey A., "Well Improvement by the Use of Vibratory Explosives," October 25, 1951.

Peterson, Dean F., "Hydraulics of Wells," *Proc. Am. Soc. Civil Eng.*, Vol. 81, June 1955.

Peterson, Dean F., O. W. Israelsen, and Vaughn E. Hansen, "Hydraulics of Wells," *Agr. Exp. Sta. Utah State Agr. College Bul.* 351, March 1952.

Peterson, Jack S., Carl Rohwer, and Maurice L. Albertson, "Effect of Well Screens on Flow into Wells," *Trans. Am. Soc. Civil Eng.*, published as Proc., December 1953.

Rohwer, Carl, "The Hydraulics of Water Wells," U.S.D.A., March 1947.

Rohwer, Carl, "Putting Down and Developing Wells for Irrigation," *U.S.D.A Circ.* 546, revision, March 1941.

Rohwer, Carl, "The Hydraulics of Water Wells," U.S.D.A., 1950.

Rorabaugh, M. I., "Graphical and Theoretical Analysis of Step-Draw-Down Test of Artesian Well," *Proc. Am. Soc. Civil Eng.*, Vol. 79, December 1953.

Todd, David Keith, "Ground Water Hydrology," John Wiley and Sons, Inc., New York, 1959.

Vaadia, Yoash and Verne H. Scott, "Hydraulic Properties of Perforated Well Casings," *Proc. Am. Soc. Civil Eng.*, Paper 1505, January 1958.

CHAPTER 4 PUMPING WATER FOR IRRIGATION AND DRAINAGE

There are large areas of land in arid regions so situated that available water cannot be brought to them by gravity. Other areas may be reached by gravity, but their location and topography with respect to the water supply are such that the cost of building the necessary gravity canals, flumes, inverted siphons, tunnels, and other conveyance structures is so great that water cannot be provided economically. For many of these areas, water is raised by some mechanical device from its natural sources, whether surface or underground, to the elevation of the higher parts of the land, or to still higher elevations if at distant points, so that it will flow over the land by gravity for irrigation purposes. This practice of raising water, known as irrigation pumping, is widely followed in arid regions of the world. Pumping is also important for irrigation by sprinkling.

Pumping for drainage may be as important as pumping for irrigation. Drainage waters must frequently be lifted to a higher elevation where they can be used for irrigation. Also, pumps are frequently required to lift drainage water into a higher channel where it can flow out of the area by gravity.

4.1 POWER REQUIREMENTS AND PUMPING-PLANT EFFICIENCIES

Work is defined as the product of force and distance, and mechanical power is defined as the time rate of doing work. The power units commonly used in irrigation are foot-pounds per second and horse-power. To lift 2 cu ft of water (125 lb) a vertical distance of 1 ft per sec requires 125 ft lb per sec, provided the lifting device (pumping plant) is 100 percent efficient. If the pumping-plant efficiency is only 50 percent it requires 125 ft lb per sec of power for overcoming friction and generating heat. The unit of power most commonly used in the United States is the horsepower, which is 550 ft lb per sec, or 33,000 ft lb per min. One horsepower would lift 1 cfs a vertical distance of 8.8 ft if it were possible to get 100 percent efficiency as shown below:

$$\text{Horsepower} = \frac{1 \times 62.5 \times 8.8}{550} = 1 \qquad (4.1)$$

Because it is impossible to obtain an efficiency of 100 percent, the computed horsepower to lift 1 cfs any height, as illustrated above, is designated "theoretical horsepower."

TABLE 4.1

HORSEPOWER REQUIRED TO LIFT DIFFERENT QUANTITIES OF WATER
TO ELEVATIONS OF 10 TO 80 FT

(Efficiency of pumping plant 50 percent of theoretical.
Use for estimating only.)
(*N. Mex. Agr. Exp. Sta. Bul.* 237)

Gallons per Minute	Cubic Feet per Second	Horsepower Required for Elevations of							
		10 ft	20 ft	30 ft	40 ft	50 ft	60 ft	70 ft	80 ft
100	0.22	0.5	1.0	1.5	2.0	2.5	3.0	3.5	4.0
150	0.33	0.8	1.5	2.3	3.0	3.8	4.6	5.3	6.1
200	0.45	1.0	2.0	3.0	4.0	5.0	6.1	7.1	8.1
250	0.56	1.3	2.5	3.8	5.0	6.3	7.6	8.8	10.1
300	0.67	1.5	3.0	4.6	6.1	7.6	9.1	10.6	12.1
350	0.78	1.8	3.5	5.3	7.1	8.8	10.6	12.4	14.1
400	0.89	2.0	4.0	6.1	8.1	10.1	12.1	14.1	16.2
450	1.00	2.3	4.6	6.8	9.1	11.4	13.6	15.9	18.2
500	1.11	2.5	5.0	6.7	10.1	12.6	15.2	17.7	20.2
600	1.34	3.0	6.1	9.1	12.1	15.2	18.2	21.2	24.2
700	1.56	3.5	7.1	10.6	14.1	17.7	21.2	24.8	28.3
800	1.78	4.0	8.1	12.1	16.2	20.2	24.2	28.3	32.3
900	2.01	4.6	9.1	13.6	18.2	22.7	27.3	31.8	36.4
1000	2.23	5.0	10.1	15.2	20.2	25.2	30.3	35.4	40.4
1250	2.78	6.3	12.6	18.9	25.2	31.6	37.9	44.2	50.5
1500	3.34	7.6	15.2	22.7	30.3	37.9	45.4	53.0	60.6

Pumping-plant efficiency is defined as the ratio of power output to power input. The electricity, gas, oil, or coal consumed by the motor or engine is the input. Table 4.1, taken from New Mexico Agricultural Experiment Station Bulletin 237, shows actual horsepower requirements for streams from 0.22 to 3.34 cfs, and pumping lifts from 10 to 80 ft with a 50 percent pumping-plant efficiency. To determine the horsepower required for lifting a stream of water of any size to a given elevation the reader need only multiply the observed value for 1 cfs in the table by the size of the stream selected. For example, Table 4.1 shows that to lift 1 cfs 40 ft would require 9.1 hp. Therefore, 5 cfs would require 45.5 hp for a 40 ft lift. To obtain the theoretical horsepower, multiply the actual requirement of 45.5 hp by the assumed efficiency of 50 percent expressed as a decimal.

$$\text{Theoretical hp} = 0.50 \times 45.5 = 22.7$$

The horsepower delivered by an electric motor or by an engine to the shaft it turns is known as the brake horsepower. The ratio of useful water horsepower delivered by a pump (the output) to brake horsepower (the input to the pump) is defined as the pump efficiency.

It is helpful to note that, by definition:

$$\text{Power} = \frac{\text{Work}}{\text{Time}} \tag{4.2}$$

and hence that Work = Power × Time

The expression horsepower-hour is used to designate the continuous consumption or delivery of 1 hp for a period of 1 hr, and is therefore equal to $550 \times 60 \times 60$ ft lb of work.

The "water horsepower" is defined as the power theoretically required to lift a given quantity of water each second to a specified height. In irrigation pumping it may be termed the "output." Then:

$$\text{WHP} = \frac{62.5Qh}{550} = \frac{Qh}{8.8} \tag{4.3}$$

where WHP = water horsepower
    Q = discharge in cfs
    h = vertical lift in feet

If Q is measured in gallons per minute rather than cubic feet per second, then:

$$\text{WHP} = \frac{8.33Qh}{33,000} = \frac{Qh}{3960} \tag{4.4}$$

Equations 4.3 and 4.4 are useful in determining water horsepower when Q and h are known.

The efficiency of energy converters, pumps included, is the ratio of energy output to energy input. The power input to a pump is referred to as brake horsepower, BHP, which is the horsepower delivered to the motor or other power unit. Hence, for pumps the pumping plant efficiency is:

$$E_p = \frac{\text{output}}{\text{input}} = \frac{\text{WHP}}{\text{BHP}} \tag{4.5}$$

Occasional field tests of pumping-plant efficiencies aid the irrigator to guard against low efficiencies and expensive operation. Field tests have been made by Johnston of the efficiencies of ninety-one irrigation pumping plants in California. The results of these tests show averages of 49.8 percent for centrifugal pumps, 40.5 percent for deep-well turbine pumps, and 44.5 percent for deep-well screw pumps. The maximum plant efficiency found was 70 percent and the minimum 15.2 percent. It is important to the farmer to keep pumping equipment in good condition. Low efficiencies are largely the result of failure on the part of pump owners to keep equipment in good running order.

4.2 PUMPING LIFTS

The vertical distance water is lifted for irrigation purposes varies widely. In some localities water is lifted only a few feet; in others, it is raised several hundred feet. In irrigation practice the maximum economic height of lift is determined by cost limitations, not by mechanical or power limitations. The discussion in the preceding section, and also Table 4.1, indicates that for any given size of irrigation stream the power requirement is proportional to the lift. The difference in elevation of the water surface in a pond, lake, or river from which pumped water is taken, and the water surface of the discharge canal into which water flows from a submerged discharge pipe is known as the "static head." In pumping from ground-water sources the static head is the difference in elevation between the water surface in the well and the water surface of the discharge canal. "Draw-down" around the well is the difference in elevation between the ground-water table and the water surface at the well when pumping. This is the head required to drive the same volume of water per second from the soil into the well as is received by the pump and delivered to the surface of the land. It is nearly always desirable to avoid excessive draw-down in order to reduce excessive power requirements. In addition to the static head that pumps must work against, consumption of a certain amount of power is essential to drive a given stream against pipe frictional resistance, sharp curves in pipes, and other factors that retard water motion. Poorly designed pumps, wear, and plugging of well perforations are the main reasons why pumping-plant efficiencies range from about 75 percent under very favorable conditions down to 20 percent or less under unfavorable conditions.

Because of the large number of variable factors influencing profits from pumping water for irrigation, it is impracticable to set specific limits of profitable pumping lifts that will apply in different localities for any length of time. Those who contemplate irrigation pumping should keep in mind the fact that the cost of pumped water is roughly proportional to the height of lift. Proposals to pump water through high lifts should be carefully considered before investments are made; on the other hand, good dependable water supplies that may be made available for irrigation by pumping only a few feet should not be overlooked.

4.3 PRIMITIVE IRRIGATION PUMPING

Pumping water for irrigation has been practiced for centuries in Egypt, India, and other older countries where irrigation is essential to agriculture. One of the devices used in early Egypt and India, known as the shaduf, is illustrated in Fig. 4.1. This device makes use of the principle of the lever with a suspended fulcrum and a counterweight. The bucket, suspended from the long end of the pole, is sometimes made of leather, stiffened near the top with a wooden hoop. The operator throws his weight on the sweep, the bucket fills, and the counterweight raises it to the next higher channel into which the water is poured. A single shaduf is operated by one man, and with it he can lift water only 5 or 6 feet, but the devices are sometimes installed in series of three or four, thus raising the water 20 feet or more. With the shaduf one man can raise approximately 20 gpm from 5 to 6 feet.

4.4 MODERN IRRIGATION PUMPING

In contrast to the primitive methods of pumping water for irrigation, pumping machinery of high efficiency is used on many irrigated farms. Substantial advancement has been made in the design and operation of pumps. Costs are greatly reduced by obtaining the necessary energy for pumping from fuels, rather than using the energy of men or animals.

For example, assume that for irrigation pumping, 1 kw-hr of electric energy may be purchased at a cost of 1 cent. As 1 kw-hr equals approximately $\frac{4}{3}$ hp-hr, 1 hp-hr on the basis of 1 cent per kw-hr would cost $\frac{3}{4}$ of a cent. A strong healthy man, in an hour, can generate about $\frac{1}{8}$ hp-hr of work. At the rate of $1.00 per hour

Fig. 4.1 The shaduf, used in Egypt and India to lift water. (*U.S.D.A. O.E.S. Bul.* 130.)

for man-labor, the cost of a man working 8 hours to generate 1 hp-hr would be $8.00 as compared to ¾ cent for electricity. Hence, manpower would cost about 1100 times as much as electrical energy to do the same pumping.

This is a major reason why irrigation and industry are being mechanized in countries of high labor cost. The degree of mechanization economically justified for irrigation depends markedly upon the comparative ratio between the cost of mechanical energy and of man-

power energy. In those countries where labor will cost less than $1.00
per day, power may cost 4 cents per kw-hr or 3 cents per hp-hr. The
cost of manpower over machinery becomes only 33 times assuming
an 8-hour day compared to 1100 times in the previous example. In
these areas machinery is expensive, repairs are costly and time-
consuming, and trained dependable operators are not readily available. These are formidable economic reasons which retard mechanization in countries of high fuel and machinery costs and low labor
costs.

Modern irrigation pumping methods are based on years of painstaking laboratory research, together with careful study of field pumping conditions by competent engineers. Out of these investigations
there have come into use pumps of many different classes and types,
each suited to different conditions of operation.

4.5 PUMP CHARACTERISTICS

In order to use modern pumps most profitably to obtain irrigation
water, it is essential to select pumps well adapted to the particular
conditions of operation and to obtain a relatively high efficiency. If
the quantity of water pumped is appreciably less than the quantity
for which the pump is designed, and the head is excessive, a low
efficiency results. Likewise, a pump may deliver more water than it
is designed to deliver at a head lower than normal and cause the
efficiency to be low. The interrelations between speed, head, discharge,
and horsepower of a pump are usually represented by curves which
are designated "characteristic" curves, Fig. 4.2. Knowledge of pump
characteristics enables one to select a pump which fits operating
conditions and thus attain a relatively high efficiency with low operating cost.

The head-capacity curve shows how much water a given pump will
deliver with a given head. As the discharge increases, the head
decreases. The resulting efficiency is observed to increase from 0
when the discharge is 0 to a maximum of 82 percent when the discharge is 1125 gpm and the head is 92 feet (and then decreases to 0
at 0 head). The brake horsepower curve for a centrifugal pump
usually increases over most of the range as the discharge increases,
reaching a peak at a somewhat higher rate of discharge than that
which produces maximum efficiency. The curves shown in Fig. 4.2
vary with the speed of the pump. Therefore speed must be considered when selecting a pump to secure maximum efficiency. Each of
the curves also varies with the type of pump. The variation of
characteristics is shown in Fig. 4.2.

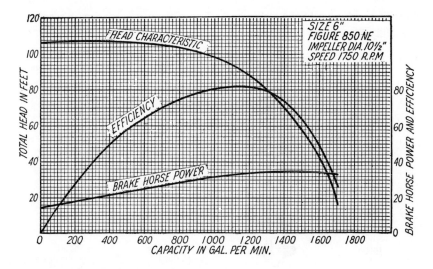

Fig. 4.2 Characteristic performance curves from test records. (Courtesy Fairbanks-Morse Company.)

4.6 TYPES OF PUMPS

Pumps for irrigation purposes are of many different makes. In general they range from pumps with small discharge and high heads to large discharges with low heads. Small-discharge and high-head pumps are centrifugal pumps used for sprinkler irrigation and where water is lifted considerable distances. Pumps with large discharge and low heads are used in drainage and where large quantities of water are lifted only a few feet.

A good index to operating characteristics of pumps is specific speed n_s, expressing the relationship between speed in revolutions per minute, discharge in gallons per minute, and head H in feet:

$$n_s = \frac{\text{RPM} \times \sqrt{\text{gpm}}}{H^{3/4}} \qquad (4.6)$$

Low specific speeds characterize centrifugal pumps (radial flow). High specific speeds are characteristic of propeller-type pumps (axial flow). These relationships are shown in Fig. 4.3.

Pump Characteristics

The characteristics of impeller-type pumps can be summarized by illustrating the relationship between discharge Q, speed N, diameter D,

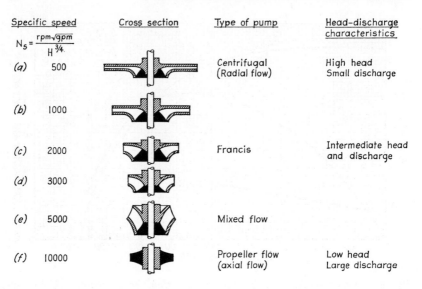

Specific speed $N_s = \dfrac{rpm\sqrt{gpm}}{H^{3/4}}$	Cross section	Type of pump	Head-discharge characteristics
(a) 500		Centrifugal (Radial flow)	High head Small discharge
(b) 1000			
(c) 2000		Francis	Intermediate head and discharge
(d) 3000			
(e) 5000		Mixed flow	
(f) 10000		Propeller flow (axial flow)	Low head Large discharge

Fig. 4.3 Variation in impeller design, specific speed, and head-discharge characteristics of pumps.

width of impeller W, head H, and power P with subscript zero referring to original conditions.

$$\frac{Q}{Q_0} = \frac{N}{N_0} = \frac{D}{D_0} = \frac{W}{W_0} \tag{4.7}$$

$$\frac{H}{H_0} = \left(\frac{N}{N_0}\right)^2 = \left(\frac{D}{D_0}\right)^2 = \left(\frac{W}{W_0}\right)^2 \tag{4.8}$$

$$\frac{P}{P_0} = \left(\frac{N}{N_0}\right)^3 = \left(\frac{D}{D_0}\right)^3 = \left(\frac{W}{W_0}\right)^3 \tag{4.9}$$

Hence, Equation 4.7 indicates a 50% increase in impeller speed, diameter, or width will increase discharge 50%. Whereas, Equation 4.8 shows that a 50% increase in impeller speed, diameter, or width will increase the head $(1.5)^2$ or 2.25 times. Note, also, from Equation 4.9 that when speed, diameter, or width increases 50%, the power required increases $(1.5)^3$ or 3.37 times.

Centrifugal Pumps

Water is drawn into a centrifugal pump axially and leaves regularly as shown in Fig. 4.3(a). Centrifugal pumps are built on both horizontal and vertical shafts. Horizontal-shaft centrifugal pumps have the advantages of being efficient, simply constructed, relatively free of trouble, low cost, easy to install, and capable of high speeds; therefore, they are usually connected directly to electric motors. One of their principal disadvantages is their limited suction lift and susceptibility to losing prime.

The practical suction lift will decrease with elevation, being not over 20 feet at sea level and generally less than 15 feet at a 5000-foot elevation. If pumping seals are in excellent condition the lift can be greater, but it is not wise to design for larger lifts unless the pump can be maintained in very good condition. Because of the priming problem and limited lift, it is usually advantageous to set pumps as near the water surface as is convenient, yet protect them from submergence during high water. It is especially important to avoid submerging electric motors.

Before a horizontal-shaft centrifugal pump can be started, it is necessary to fill the suction pipe and pump case with water and thus expel all air. This operation of filling suction and pump case is designated "priming the pump." Priming is essential on a horizontal-shaft centrifugal pump because of the non-positive action of the impeller. These pumps must be primed before they will lift water from the source of supply. Priming can be accomplished by adding water to the intake line in the pump from an outside source, such as a storage tank. In this case a foot valve on the end of the intake is essential to hold the water in the intake line while it is being filled. A good foot valve will also hold water in the line so that priming will not be necessary each time the motor is started. Priming can also be accomplished by sucking out the air with various devices, thereby lowering the pressure in the line causing the water to rise from the source to the pump by atmospheric pressure. Hand primers, exhaust primers, manifold primers, and dry vacuum pumps are used to remove air when priming a pump.

Deep-Well Turbine Pumps

When the impeller is suspended vertically on the drive shaft within a long discharge pipe, the pump is termed a deep-well turbine unit. The impeller may be a centrifugal unit, an axial-flow unit, or any design between these extremes, depending upon the desired head-discharge relationship. The bowl of the pump houses the impeller and guide vanes. When several bowls are connected in series to obtain the desired total head, the pump is referred to as a multiple-stage pump. These bowl assemblies are nearly all located beneath the water surface.

High efficiency is dependent upon a close fit of the impeller against the pump housing. Adjustments in position are made by raising or lowering the impeller shaft. Many deep-well pump heads are equipped with a clutch or ratchet to prevent the pump from turning in the wrong direction because of the backflow of water when the power is shut off, which may unscrew the fittings.

Deep-well turbine pumps are used for irrigation when the water surface is below the practical lift of centrifugal pumps. Successful installations have been made when water is as much as 1000 feet below the ground surface.

The pump is driven by an electric motor or other source of power located at the ground surface and connected by a long vertical shaft held in position by bearings built in the discharge pipe or column. Being submerged, the deep-well pumps have the advantage of requiring no priming and of meeting rather wide fluctuations of water surface without having to reset the pumps. They have the disadvantage of the operating parts being inaccessible and consequently difficult to inspect.

Low efficiency is more common in vertical-shaft pumps than in horizontal-shaft pumps. This is because deep-well pumps are frequently permitted to continue running after bearings are worn badly, and sometimes until the pumps fail to operate, before repairs are provided. Figure 4.4 shows a typical section of a deep-well turbine pump.

Submersible Turbine Pumps

A deep-well turbine pump close-coupled to a small diameter submersible electrical motor is termed a submersible pump. The characteristics of the pump unit are similar to the conventional deep-well submersible pump having the same flexibilities and design possibilities. Efficiency is increased by direct coupling and effective cooling resulting from complete immersion which permits a reduction in amount of iron and copper in the core. Submersible pumps have been used in wells over 12,000 feet deep. Units with more than 250 stages have been used. Submersible motors as large as 250 hp are in use in 8-inch casings, and larger motors are available for larger casing sizes. Pumps are available for 4-inch and larger well casings.

The principal advantage of submersible pumps is that they can be used in very deep wells where long shafts would not be practical. These pumps are also excellent in crooked wells where the ordinary deep-well turbine might not operate. Bending in the pump column

Construction
Berkeley vertical turbine pumps

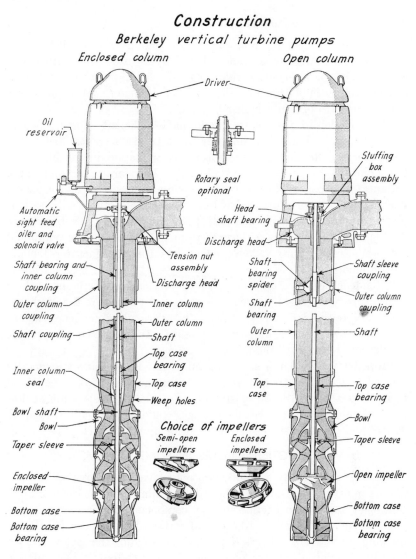

Enclosed column   Open column

Driver

Oil reservoir

Rotary seal optional

Stuffing box assembly

Automatic sight feed oiler and solenoid valve

Head shaft bearing

Discharge head

Shaft bearing and inner column coupling

Tension nut assembly

Discharge head

Shaft bearing spider

Shaft sleeve coupling

Outer column coupling

Inner column

Shaft bearing

Outer column coupling

Shaft coupling

Outer column

Shaft

Inner column seal

Shaft

Top case bearing

Outer column

Shaft

Top case

Bowl shaft

Top case

Weep holes

Top case bearing

Bowl

Choice of impellers
Semi-open impellers   Enclosed impellers

Taper sleeve

Taper sleeve

Enclosed impeller

Open impeller

Bottom case

Bottom case

Bottom case bearing

Bottom case bearing

Fig. 4.4 Vertical turbine pump showing main features. (Courtesy Berkeley Pump Company.)

of a conventional turbine will cause the shaft to crystallize and rupture. Submersible pumps are also useful where the installation may be flooded or where an above-ground pump house would be inconvenient, unsightly, or hazardous.

Propeller and Mixed-Flow Pumps

Propeller (axial-flow) and mixed-flow pumps are used for low-head high-discharge operation. By varying the vane pitch and curvature of the impeller, a wide range of head-capacity requirements can be met without changing the impeller diameter. Multiple stages may be used to obtain a higher head.

The impeller of the pump should have sufficient submergence to minimize cavitation. The impeller must be deep enough below the water surface to minimize the formation of vapor on the blades at the point of maximum local velocity. The formation and rapid collapse of vapor bubbles occurring several times a second causes a fluctuating local pressure which results in metal fatigue with serious pitting of the blade surface and reduced efficiency. Submergence depths, as well as clearance depths from the bottom of the pump and between adjacent units, as recommended by the manufacturers, should not be ignored.

A tendency of an axial-flow pump to overload makes it inadvisable to use this type of pump when it is necessary to throttle the discharge to secure reduced delivery. Also, the overloading tendency at increased heads makes it important to select a motor which will provide ample power under all pumping conditions.

Propeller pumps have a low initial cost. Each stage is generally limited to less than 10 feet of lift. Additional stages may be added to increase the head to 30 or 40 feet. For higher heads a mixed-flow pump should be used, since single stages could economically lift up to 25 feet and multiple stages can be used for up to 125 feet.

4.7 FRICTION LOSSES IN PUMP SYSTEMS

As water is lifted by a pump from one elevation to another, several factors must be considered when determining horsepower required. The head loss due to friction in each component of the pumping and pipe system must be added to the total vertical height water is lifted to the point of delivery. These sources of loss include the foot valve, intake pipe, pump, discharge pipe, and all intermediate valves, elbows, and other fittings. Figure 4.5 lists the loss of head due to friction in elbows, valves, and other fittings in terms of the equivalent length of pipe. The combined equivalent pipe losses due to fittings can then be added to the losses in the straight section of the pipe to obtain the total equivalent length of pipe. With this total loss of head, Table 4.2 then gives the total head required in the entire system due to change of elevation and fittings. Table 4.2 is for pipe 15 years old. Since

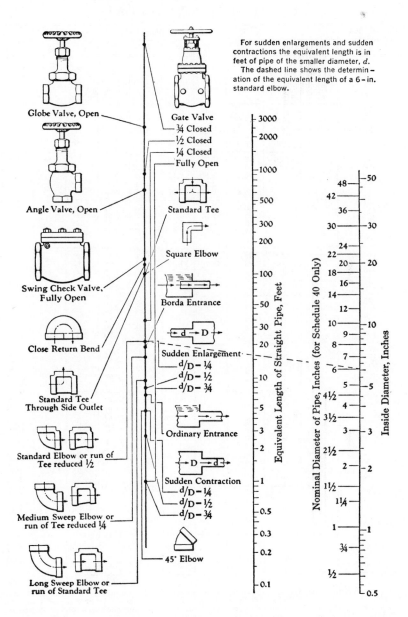

For sudden enlargements and sudden contractions the equivalent length is in feet of pipe of the smaller diameter, d. The dashed line shows the determination of the equivalent length of a 6-in. standard elbow.

Fig. 4.5 Minor friction losses caused by elbows, valves, and other fittings. (From *Oregon Agr. Exp. Sta. Bul.* 181, by Wolfe, 1950.)

TABLE 4.2

Loss of Head in Feet Due to Friction, per 100 Feet of 15-Year-Old Ordinary Iron Pipe[1]

Rate of Flow in Gallons per Minute	Nominal Diameter of Pipe									
	$\frac{3}{4}''$	$1''$	$2''$	$2\frac{1}{2}''$	$3''$	$4''$	$5''$	$6''$	$8''$	$10''$
1										
2	1.9									
3	4.1	1.26								
4	7.0	2.14								
5	10.5	3.25								
10	38.0	11.7	0.50	0.17	0.07					
15	80.0	25.0	1.08	0.36	0.15					
20	136.0	42.0	1.82	0.61	0.25					
25		64.0	2.73	0.92	0.38					
30		89.0	3.84	1.29	0.54					
35		119.0	5.1	1.72	0.75					
40		152.0	6.6	2.20	0.91	0.22				
45			8.2	2.80	1.15	0.28				
50			9.9	3.32	1.38	0.34				
70			18.4	6.20	2.57	0.63	0.21			
75			20.9	7.1	3.00	0.73	0.24			
100			35.8	12.0	4.96	1.22	0.41	0.17		
120			50.0	16.8	7.0	1.71	0.58	0.24		
125			54.0	18.2	7.6	1.86	0.64	0.27		
150			76.0	25.5	10.5	2.62	0.88	0.36		
175			102	33.8	14.0	3.44	1.18	0.48		
200			129	43.1	17.8	4.40	1.48	0.62		
225				54.3	22.3	5.45	1.86	0.77		
250				66.0	27.2	6.72	2.24	0.95	0.23	
270					31.3	7.70	2.60	1.08	0.26	
275					32.5	8.05	2.70	1.12	0.27	
300					38.0	9.30	3.14	1.32	0.32	
350						12.40	4.19	1.75	0.42	
400						16.00	5.40	2.21	0.55	
450						19.80	6.70	2.37	0.69	0.23
470						21.60	7.22	2.99	0.74	0.25
475						22.00	7.42	3.02	0.76	0.25
500						24.00	8.10	3.34	0.83	0.28
550							9.60	4.02	0.98	0.33
600							11.30	4.91	1.16	0.40
650							13.20	5.40	1.34	0.45
700							15.10	6.20	1.54	0.52
750							17.20	7.12	1.77	0.60
800								8.00	1.90	0.67
850								8.95	2.20	0.75
900								10.11	2.46	0.83
950								11.20	2.71	0.92
1000								12.04	2.98	1.01
1050								13.25	3.22	1.09
1100								14.55	3.55	1.21
1150								15.60	3.83	1.31
1200								17.10	4.17	1.42
1250								18.50	4.43	1.51
1500									6.29	2.14
2000									10.80	3.60
2500										5.40
3000										7.72
3500										10.00
4000										

Courtesy Goulds Pumps, Inc.

[1] Pipe coefficients: The values of friction given in this table are for commercial wrought iron or cast iron pipe of 15 years' service when handling soft clear water. In order to be able to use that table for other classes of pipe, the values taken from the table should be multiplied by a coefficient, selected from the list below, to correspond to the required condition:

New smooth brass and steel pipe (Durand) coefficient	=0.6
New smooth iron pipe (Williams & Hazen C = 120)	=0.7
15-year-old ordinary pipe (Williams & Hazen C = 100)	=1.0
25-year-old ordinary pipe (Williams & Hazen C = 90)	=1.2
Portable aluminum with couplers (Williams & Hazen C = 120)	=0.7

(Bulletin 481, Oregon Agricultural Experiment Station.)

the system will be operating for many years, it is wise to figure the power requirement on the average, or longer, life to insure reasonably adequate power during the entire period of operation of the irrigation system.

The following example will illustrate the use of Fig. 4.5 and Table 4.2 for estimating total horsepower requirements. Assume, for simplicity, that the entire pumping system is composed of 6-inch diameter pipe and fittings; assume further that the discharge is 500 gpm with a static lift of 75 feet and a pump efficiency of 70 percent.

1. Foot valve, assuming equivalent to an open globe valve   170 ft
2. Gate valve, fully open   4 ft
3. Three 90° medium sweep elbows, 14 ft each   42 ft
4. Length of 6-inch pipe in intake and discharge lines   130 ft
    Total equivalent length of pipe $\overline{346 \text{ ft}}$

Loss of head in feet (500 gpm through 6 in. pipe, results in 3.34 ft loss per 100 ft given in Table 4.2), $3.46 \times 3.34 = 12$ ft.

Total dynamic head = static lift of 75 ft + friction loss of 12 ft = 87 ft.

Use 90 ft to allow for losses in entrance and exit.

Required power output at a pump which is also referred to as the water horsepower is found by Equation 4.4 to be:

$$\text{WHP} = \frac{QH}{3960} = \frac{500 \times 90}{3960} = 11.4$$

The required power input to the pump, referred to as brake horsepower, is the water horsepower divided by the efficiency of the pump, Equation 4.5, assumed to be 70 percent:

$$\text{BHP} = \frac{\text{WHP}}{E_p} = \frac{11.4}{0.7} = 16.3$$

4.8 SELECTION OF POWER PLANT

The power plant must be capable of delivering the required power under varying conditions. The variations of operating conditions which must be accommodated by the power unit will influence the choice of unit. Features to be considered in selecting a power unit include:

1. Brake horsepower required.
2. Initial cost.
3. Availability and cost of energy or fuel.
4. Depreciation.

5. Dependability of unit.
6. Portability required.
7. Maintenance and convenience of operation.
8. Labor availability and quality.

In all cases it is good insurance to obtain a power plant from a reputable dealer. Occasionally old power units are used to reduce the initial cost. Frequently, the very low efficiencies which result in very high fuel costs, more than offset low initial cost. Make allowance in the design for local conditions that will reduce the horsepower rating given by the manufacturer. The operator should carefully follow recommendations of the manufacturer.

4.9   ELECTRIC MOTORS

Electric motors which are properly selected, installed, and mounted may give many years of trouble-free power. Advantages of electrical motors are relatively long life, dependability, ease of operation, and low maintenance cost. In some areas frequent power interruptions are a major disadvantage associated with electrical motors. The necessity to construct power lines to each power site and the cost of transformers at pumping locations are disadvantages. Lightning striking an electrical installation can cause serious loss of time and money. Variation in line voltage will cause motors to burn out.

Single-phase motors are not usually feasible for loads of more than $7\frac{1}{2}$ hp. In some areas three-phase power is not readily available. The best electric motor for irrigation pumping is the squirrel-cage induction motor operating on 60 cycle, 220 to 440 volt, three-phase energy. The most common speed is 1760 rpm. Direct connection betweeen motor and pump should be used whenever possible for minimum cost and maximum efficiency.

4.10   INTERNAL COMBUSTION ENGINES

Two general types of internal combustion engines are used: (1) spark ignition and (2) compression ignition, better known as diesel. Ignition systems generally use gasoline, LP gas, and natural gas fuels while the diesel units use the well-known diesel fuel. Since the initial cost of a diesel engine is much higher than for spark ignition, it is necessary that diesel units be used many more hours per season than are normally required for spark ignition units to be economically feasible. Generally, spark ignition power units or single-phase electrical units are used for horsepower up to $7\frac{1}{2}$. Spark ignition units and three-phase electricity are practical between $7\frac{1}{2}$ and 40 hp; and above

40 hp, diesel units generally become more practical. Natural gas is competitive with diesel fuel in many areas for irrigation pumping.

4.11 ELECTRIC SERVICE SCHEDULES AND COSTS

Power companies usually base charges for electric service in part on the maximum demand of the consumer, regardless of the energy consumed, and in part on the energy actually used. Power companies encourage consumers to avoid high demands for electricity over short periods of time and to strive to use power as many hours per day and days per month as conditions may justify. Consideration of a typical power-rate schedule will clarify the low energy costs of low demand and continuous use, as compared to high energy costs resulting from high demand and short-time use.

A sample irrigation service schedule provides a monthly power charge of $1.50 per kw for the first 100 kw of demand and $1.00 per kw for all additional kilowatts of demand. The monthly charge for kilowatt-hours used by the motor is then added to the monthly demand charge, on the following step rate:

1.5¢ per kw-hr first 100 kw-hr per kw of demand
0.9¢ per kw-hr next 5000 kw-hr
0.6¢ per kw-hr next 20,000 kw-hr
0.4¢ per kw-hr all additional kw-hr

Most irrigation installations are of a permanent nature. Usually, the irrigator signs a 10-year contract for service which entitles him to a 5 percent term discount on his monthly power bill. Also, when the irrigator makes his original permanent installation, he includes the transformers with his pump and motor installation, which further entitles him to a discount on his monthly power bill, as follows:

20¢ per kw for first 100 kw of demand, and
10¢ per kw for all additional kw of demand

Example

For a 200-hp motor, if it runs 40 hr (using a measured demand of 160 kw), the monthly charge would be:

Demand:		
100 kw at $1.50 per kw	=	$150.00
60 kw at $1.00 per kw	=	60.00
Energy:		
6400 kw-hr at 1.5¢ per kw-hr	=	96.00
Gross charge		306.00

Less term discount, 5%	15.30
Less voltage discount	
20¢ for first 100 kw	20.00
10¢ for next 60 kw	6.00
Net monthly charge	264.70

Average cost per kw-hr = $264.70/6400 = 4.14 cents

If the irrigator runs the same 200-hp motor with a measured demand of 160 kw continuously for 30 days, 24 hr each day, then 720 hr use equals 115,200 kw-hr, and his monthly charge would be:

Demand:		
100 kw at $1.50 per kw	=	$150.00
60 kw at $1.00 per kw		60.00
Energy:		
16,000 kw-hr at 1½¢ per kw-hr	=	240.00
5,000 kw-hr at 0.9¢ per kw-hr	=	45.00
20,000 kw-hr at 0.6¢ per kw-hr	=	120.00
74,200 kw-hr at 0.4¢ per kw-hr	=	296.80
Gross charge		911.80
Less term discount, 5%		45.59
Less voltage discount		
20¢ for first 100 kw		20.00
10¢ for next 60 kw		6.00
Net monthly charge		840.21

Average cost per kw-hr = $840.21/115,200 = 0.73 cent

Thus by operating his 200-hp motor 720 hr per month, instead of only 40 hr, the irrigator reduces the unit cost of electricity from 4.14 cents to 0.73 cent per kw-hr, or a reduction of 82.5 percent.

A farmer cannot as a rule use irrigation water continuously to advantage. The advantages in the use of large streams obtained by large motors and pumps partly compensate the irrigator for higher costs for electricity. On the other hand, it is frequently advantageous, where electricity is used for pumping, to provide small reservoirs in which to store the water during the night, thus making it possible to irrigate with a stream approximately twice the size of the pump discharge.

4.12 IRRIGATION PUMPING COSTS

In order to estimate costs of irrigation water obtained by pumping, and to compare these costs with costs of water from gravity systems, it is customary to compute all pumping costs in terms of the volume of water delivered annually to the irrigated farm. The factors that determine the annual costs of pumped water are:

a. Interest on capital investment, i.e., on first cost.
b. Taxes.
c. Depreciation on pumping machinery and on housing.
d. Fuel or power and lubricating oils.
e. Attendance.

The application of these factors to arrive at the cost of water obtained by pumping is most clearly presented by working out a typical example.

A well is 14 inches in diameter and 140 feet deep, the depth of the water table being 56 feet. While pumping 2.5 cfs, the lift is 77 feet. A turbine pump is driven by a 45-hp, high-speed diesel engine. The costs during the 1960 season for pumping 503 acre-feet during the season are presented in Table 4.3.

It will be noted from Table 4.3 that operating costs make up approximately ⅗ of the total cost of $4.99 per acre-foot of water.

TABLE 4.3

Cost of Operating a Modern Pumping Plant

Cost of Plant:			
Well	$1960.00		
Pump	2455.00		
Engine (installed)	2200.00		
Shelter	225.00		
	$6840.00		
Annual Fixed Costs:			
Interest on $6840.00 @ 6%	$ 410.40		
Taxes (estimated)	161.22		
Depreciation on engine @ 7%	140.00		
Depreciation on pump @ 7%	171.85		
Depreciation on well and shelter @ 4%	87.40		
		$ 970.87	
Annual Operating Costs: (2415 hrs oper/year)			
Fuel—7150 gal @ 16¢/gal	$1144.00		
Lubrication @ $3.32/100 hrs	80.18		
Sales tax on above @ 2%	24.48		
Repairs @ $7.85/100 hrs	189.58		
Attendance @ $4.17/100 hrs	100.71		
		$1538.95	
		$2509.82	
Total Cost per Acre-Foot	$4.99		
Operating Cost per Acre-Foot	$3.06		

4.13   EFFICIENCY OF PUMPING PLANTS

Irrigation pumping plants invariably operate at considerably less than peak efficiency. Several reasons for low efficiency exist, such as the pipe being too small or having many bends. The most common error is discharging water considerably above the necessary level. The pump may also be worn, and frequently the bowls are not set properly in turbine-type pumps. The drive or coupling between pump and motor may not be an efficient unit. Too frequently a pump or motor is not the best pump or motor for the job to be done. Correct matching of pump, motor, and drive is important.

Efficiency will also be reduced by elevation, heat, accessories, and continuous operation. Hence, industrial-type engines used for irrigation are derated from the maximum rating given by the manufacturer in order to determine safe continuous operating horsepower available for pumping. Following are some of the factors which must be considered:

	Decrease in efficiency, %
1. For each 1000 feet above sea level	3
2. For each 10° operating air temperature above 60°F	1
3. For accessories, using heat exchangers	5
4. Radiator, fan	5
5. For continuous load operation	20
6. Drive losses	0–15

Hence, if these losses total 40 percent and the water horsepower energy imparted to the water including pump and conveyance losses are 30 horsepower, a unit rated at $50 = 30/(1.0 - 0.4)$ brake horsepower would have to be selected.

A good measure of the efficiency of a pumping-plant is the fuel consumed by the plant. Standards shown in Tables 4.4 and 4.5 have been developed by Schleusener, Sulek, and Schrunk. Standards of common fuels and electricity are shown in Table 4.4. Fuels listed are mixtures of hydrocarbons which may vary in composition. However, this tabulation is a good relative guide to what can be expected from an efficient pumping plant. Table 4.5 lists fuel requirements for various pumping conditions which can readily be compared to fuel consumed by a given pumping plant. Table 4.6 converts any discharge pressure to an equivalent lift which must be added to the distance the water must be lifted while pumping in order to obtain the correct fuel requirement from Table 4.5.

By installing high-compression pistons or cylinder heads, a portion

TABLE 4.4

PERFORMANCE STANDARDS FOR NEW DEEP-WELL PUMPING PLANTS
(AFTER SCHLEUSENER ET AL.)

Energy Source	Rated Load hp-hr/gal for Representative Power Units	Performance Standard in whp-hr/gal[1]
Diesel	14.75[2]	11.06
Gasoline	11.30[2]	8.48
Tractor fuel	11.08[2]	8.31
Propane	8.92[2]	6.69
Natural gas	81.9[3] per 1000 cu ft	61.4 per 1000 cu ft
Electric	88 percent efficient	0.885[4] per kw-hr

[1] Based on pump efficiency of 75 percent.
[2] Taken from Test D of Nebraska Tractor Test Reports. Drive loss for flat belt is included.
[3] Manufacturer's data corrected for 5 percent drive loss.
[4] Not corrected for drive loss. Assume a direct connection.

TABLE 4.5

MAXIMUM FUEL REQUIREMENTS FOR A GOOD PUMPING PLANT[1]
(AFTER SCHLEUSENER ET AL.)

Flow of Water in Gallons per Minute	Lift in Feet[2]	Water Horse-power	Fuel Required in Gallons per Hour				
			Pro-pane	Die-sel	Gasoline or Tractor Fuel	Natural Gas[3]	Elec-tricity[4]
	100	13	2	$1\frac{1}{4}$	$1\frac{1}{2}$	210	14
500	150	19	$2\frac{3}{4}$	$1\frac{3}{4}$	$2\frac{1}{4}$	310	21
	200	25	$3\frac{3}{4}$	$2\frac{1}{4}$	3	410	29
	100	18	$2\frac{3}{4}$	$1\frac{3}{4}$	2	290	20
700	150	27	4	$2\frac{1}{2}$	$3\frac{1}{4}$	430	30
	200	35	$5\frac{1}{4}$	$3\frac{1}{4}$	$4\frac{1}{4}$	580	40
	100	20	3	$1\frac{3}{4}$	$2\frac{1}{2}$	330	23
800	150	30	$4\frac{1}{2}$	$2\frac{3}{4}$	$3\frac{1}{2}$	490	34
	200	40	6	$3\frac{3}{4}$	$4\frac{3}{4}$	660	46
	100	25	$3\frac{3}{4}$	$2\frac{1}{4}$	3	410	29
1000	150	38	$5\frac{3}{4}$	$3\frac{1}{2}$	$4\frac{1}{2}$	620	43
	200	50	$7\frac{1}{2}$	$4\frac{1}{2}$	6	820	57

[1] Based on standards in Table 4.4.
[2] If you have a pressure at the pump discharge, add the number of feet shown in Table 4.6 to the distance the water must be lifted while pumping.
[3] Cubic feet per hour.
[4] Kilowatt-hours per hour.

TABLE 4.6

RELATIONSHIP BETWEEN DISCHARGE PRESSURE AND LIFT
(AFTER SCHLEUSENER ET AL.)

Discharge Pressure in Pounds per Square Inch	Equivalent "Lift" in Feet
30	69
50	116
70	162
110	254

of the loss due to altitude can be regained. If air temperatures over 110°F are encountered, provision should be made for cooling the intake manifold. Fuel consumption per brake horsepower-hour increases if the engine is not fully loaded. If the engine is loaded to a small portion of the rated horsepower for the speed, the fuel consumption will be considerably increased. Under these conditions it will be more economical to reduce the engine speed until the proper horsepower is developed. By changing the pulley drive the proper pump speed can be obtained.

REFERENCES

Code, W. E., "Equipping a Small Irrigation Pumping Plant," *Colo. Agr. Exp. Sta. Bul.* 433, 1946.

Molenaar, Albert, "Irrigation Pumping with Electric Power," *Agr. Eng.*, Vol. 22, p. 257, 1941.

Rohwer, Carl and M. R. Lewis, "Small Irrigation Pumping Plants," *U.S.D.A. Farmers' Bul.* 1857, (supersedes Bul. 1404) 1940.

Rohwer, Carl, "Design and Operation of Small Irrigation Pumping Plants," *U.S.D.A. Circ.* 678, 1943.

Schleusener, Paul E., J. J. Sulek, and J. F. Schrunk, "How Efficient is Your Pumping Plant?" *Exp. Sta. Qtrly,* Lincoln, Nebraska, Spring 1955.

CHAPTER 5 CONVEYANCE OF IRRIGATION AND DRAINAGE WATER

Irrigated lands are often located great distances from the sources of their water supplies. Water obtained from natural streams and from surface reservoirs, as a rule, must be conveyed farther than water obtained from underground reservoirs. Main conveyance canals of irrigation projects vary from a few miles to 100 or more miles in length. Some projects convey water several hundred miles from storage reservoirs in the mountains by commingling stored water with water of natural rivers and then again diverting it into large canal systems in the valleys. Days are required on some projects to convey the water from points of diversion to points of use.

The same principles governing flow in irrigation canals also govern flow in drainage ditches. Therefore, the material of this chapter applies to all conveyance channels associated with irrigation projects.

5.1 HYDRAULICS OF FLOW

Irrigation water is conveyed in either open or closed conduits. Hydraulically, the two methods are similar; however, slightly different forms of the equations are used because with pipe flow, pressure head and elevation head, differences are usually measured to determine the flow rate while in an open-channel flow the pressure head does not change and the slope of the water surface is the criterion of flow. Derivation of hydraulic equations and details of application are thoroughly covered in hydraulics texts and are not duplicated here. Only those results applicable to irrigation and drainage are summarized. Older empirical equations of flow are by-passed in favor of expressions which have a sounder physical basis and which are being used extensively in irrigation practices.

The most basic equation of water flow is the continuity equation which relates the flow Q past a cross section to the velocity and area

A at two different sections:

$$Q = A_1 V_1 = A_2 V_2 \qquad (5.1)$$

When area is measured in square feet and velocity in feet per second, discharge will be in cubic feet per second (cfs).

The second fundamental equation in the hydraulics of irrigation and drainage water conveyance is the Bernoulli equation:

$$\frac{V_1^2}{2g} + \frac{p_1}{w} + y_1 = \frac{V_2^2}{2g} + \frac{p_2}{w} + y_2 + h_L \qquad (5.2)$$

where p = pressure intensity at any point

y = elevation of the point above a common datum

w = weight of unit volume of water

h_L = head loss, energy loss per unit weight of fluid between points 1 and 2

g = acceleration due to gravity

The third basic equation is for head loss h_L as water flows between points 1 and 2 in the system. This is given by the Darcy-Weisbach equation:

$$h_L = f \frac{L}{d} \frac{V^2}{2g} \qquad (5.3)$$

where f = coefficient of friction loss

L = the length between the two points

d = diameter of the conduit

V = average velocity

The fourth relationship that is also very useful relates the head h to the pressure intensity p and the unit weight of water w:

$$h = \frac{p}{w} \qquad (5.4)$$

Flow through a pipe system in which these four equations apply is shown in Fig. 5.1. Hydraulic terminology commonly used is also shown. At all points within the pipe line, Equations 5.1 and 5.2 can be applied. Equation 5.3 will give the loss of head between any two points of the line. Loss of energy and flow through valves, bends, and transitions of a pipe system can be calculated by these procedures and by utilizing the tables outlined in Chapter 4.

The fifth relationship which has extensive value in open channel

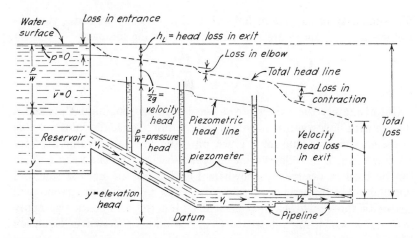

Fig. 5.1 Illustrating the flow of water from a large reservoir through a pipe line showing how the energy changes from point to point.

flow is the hydraulic radius R. This is defined as:

$$R = \frac{A}{P} \tag{5.5}$$

where A = cross-sectional area of the flow

P = wetted perimeter of the flow in contact with the channel

When the hydraulic radius is calculated for a full round pipe, it is found to have the value of $R = d/4$; this value changes somewhat with shape, but for practical purposes in most irrigation problems the value of hydraulics radius can be assumed as $d/4$. Substituting this quantity in Equation 5.3, replacing g with its value of 32.2 ft per second squared, and solving for V, the following equation is obtained:

$$V = \frac{16}{\sqrt{f}} \sqrt{R \frac{h_L}{L}} = \frac{16}{\sqrt{f}} \sqrt{RS} \tag{5.6}$$

where S is the slope of the water surface or piezometric head line.

An equation of similar form was obtained by Chezy, wherein C is the coefficient of roughness.

$$V = C\sqrt{RS} \tag{5.7}$$

Another equation was derived by Manning utilizing the Chezy equation but defining differently the coefficient of roughness n.

$$V = \frac{1.49}{n} R^{\frac{2}{3}} S^{\frac{1}{2}} \tag{5.8}$$

The similarity of Equations 5.6, 5.7, and 5.8 readily indicates that the coefficients f, C, and n are related in the following manner:

$$C = \frac{16}{\sqrt{f}} = \frac{1.49}{n} R^{1/6} \qquad (5.9)$$

Unfortunately the values of the coefficients of roughness C, f, and n change with flow conditions within the conduit; hence, it is not possible to assign a fixed value for a given pipe and have that value hold for all flow conditions. The variation which occurs and the values which can be used in design are shown in Fig. 5.2 for the normal range of turbulent flow in irrigation channels. The following discussion will deal with the variations in coefficient C, but the comments also apply to f and n which are related to C in the manner shown in Equation 5.9.

The abscissa in Fig. 5.2 is the Reynolds number N_R, a dimensionless quantity where V is average velocity, d is diameter, R is hydraulic radius, and ν is kinematic viscosity. When Reynolds number is computed, care should be taken to insure that proper dimensions of the variables are used so that all dimensions will cancel. The ratio of e/d is relative roughness of channel and e is absolute roughness. Thus, only in the wholly-rough region is the value of the roughness coefficient a constant with changes in velocity. Furthermore, absolute roughness e of one kind of pipe, for example cast iron, is reasonably constant; however, relative roughness e/d changes as size of conduit changes. Hence, coefficient of friction varies as flow conditions defined by the Reynolds number vary, and relative roughness varies more than absolute roughness. Values of absolute roughness for different kinds of pipe are tabulated in Table 5.1.

In solving practical flow problems it becomes obvious that the velocity required in the Reynolds number is not known; therefore, it is necessary to estimate the velocity, obtain a solution, and check the assumed value. If the assumed velocity is considerably in error, the computed velocity should be used and a new velocity calculated. The process should be repeated until satisfactory accuracy is obtained.

The roughness of open channels has not been defined as well as the roughness of pipes, nor can it be, because of the large variation which occurs in nature. This condition presents a challenge to irrigation engineers to develop criteria that will better define the absolute roughness of channels. It is important to remember that under the best conditions, values vary with flow conditions; hence, allowance must be made when designing for extreme conditions.

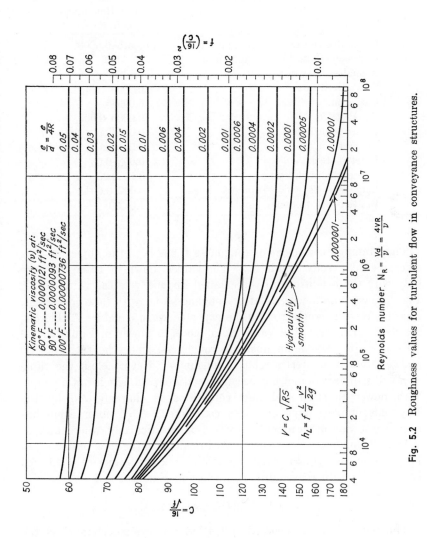

Fig. 5.2 Roughness values for turbulent flow in conveyance structures.

TABLE 5.1

VALUES OF ROUGHNESS e FOR VARIOUS CONDUIT MATERIALS
(AFTER ALBERTSON, BARTON, SIMONS)

	Relative Roughness (e in ft)
Glass, drawn brass, copper, and lead	Smooth–0.000005
Commercial steel or wrought iron	0.00015
Asphalted cast iron	0.0004
Galvanized iron	0.0005
Cast iron	0.00085
Wood stave	0.0006–0.003
Cement-lined steel	0.0013
Concrete	0.001–0.01
Riveted steel	0.003–0.03
Corrugated metal pipe	0.10–0.20
Large tunnels, concrete or steel lined	0.002–0.004
Blasted-rock tunnels	1.0–2.0

The Manning coefficient of roughness n has been the object of considerable field study, Table 5.2. The values shown in Table 5.2 vary only with the conditions of the channel surface, whereas values of C, Fig. 5.2, vary both with relative roughness and with flow conditions. For this reason some error can be expected when n values, Table 5.2, are used. However, the value R to the $\frac{1}{6}$th power in Equation 5.9 does in part make n an absolute roughness, since the ratio of $R^{\frac{1}{6}}$ (power) to n is in effect a relative roughness. Nevertheless, the value of n will still vary somewhat because relative roughness is not a simple $\frac{1}{6}$th power function and flow conditions as characterized by the Reynolds number have not been considered. This observation has been borne out in practical construction, particularly where normal values of roughness have been used to calculate the flow occurring in large structures. For normal irrigation work the values of Table 5.2 can be used with reasonable confidence in Equation 5.8.

5.2 EARTH CANALS

The most common type of irrigation conveyance channel is the one excavated in natural material along the line that water must be conveyed. When used without artificial lining of bed or sides, such a channel is called an earth canal. Excessive velocities of water in earth canals cause erosion. Very few natural materials will stand

TABLE 5.2
HORTON'S VALUES OF n. TO BE USED WITH KUTTER'S
AND MANNING'S FORMULAS

Surface	Best	Good	Fair	Bad
Uncoated cast-iron pipe..............	0.012	0.013	0.014	0.015
Coated cast-iron pipe................	.011	.012*	.013*	
Commercial wrought-iron pipe, black....	.012	.013	.014*	.015
Commercial wrought-iron pipe, galvanized	.013	.014	.015	.017
Smooth brass and glass pipe...........	.009	.010	.011	.013
Smooth lockbar and welded "OD" pipe..	.010	.011*	.013*	
Riveted and spiral steel pipe...........	.013	.015*	.017*	
Vitrified sewer pipe..................	$\left\{\begin{array}{l}.010\\.011\end{array}\right.$	.013*	.015	.017
Common clay drainage tile............	.011	.012*	.014*	.017
Glazed brickwork....................	.011	.012	.013*	.015
Brick in cement mortar; brick sewers....	.012	.013	.015*	.017
Neat cement surfaces.................	.010	.011	.012	.013
Cement mortar surfaces...............	.011	.012	.013*	.015
Concrete pipe.......................	.012	.013	.015*	.016
Wood stave pipe.....................	.010	.011	.012	.013
Plank Flumes:				
Planed...........................	.010	.012*	.013	.014
Unplaned.........................	.011	.013*	.014	.015
With battens......................	.012	.015*	.016	
Concrete-lined channels...............	.012	.014*	.016*	.018
Cement-rubble surface................	.017	.020	.025	.030
Dry-rubble surface...................	.025	.030	.033	.035
Dressed-ashlar surface................	.013	.014	.015	.017
Semicircular metal flumes, smooth......	.011	.012	.013	.015
Semicircular metal flumes, corrugated...	.0225	.025	.0275	.030
Canals and Ditches:				
Earth, straight and uniform.........	.017	.020	.0225*	.025
Rock cuts, smooth and uniform.......	.025	.030	.033*	.035
Rock cuts, jagged and irregular.......	.035	.040	.045	
Winding sluggish canals.............	.0225	.025*	.0275	.030
Dredged earth channels.............	.025	.0275*	.030	.033
Canals with rough stony beds, weeds				
on earth banks..................	.025	.030	.035*	.040
Earth bottom, rubble sides...........	.028	.030*	.033*	.035
Natural Stream Channels:				
(1) Clean, straight bank, full stage, no				
rifts or deep pools................	.025	.0275	.030	.033
(2) Same as (1), but some weeds and				
stones..........................	.030	.033	.035	.040
(3) Winding, some pools and shoals,				
clean...........................	.033	.035	.040	.045
(4) Same as (3), lower stages, more				
ineffective slope and sections........	.040	.045	.050	.055
(5) Same as (3), some weeds and				
stones..........................	.035	.040	.045	.050
(6) Same as (4), stony sections........	.045	.050	.055	.060
(7) Sluggish river reaches, rather				
weedy or with very deep pools.......	.050	.060	.070	.080
(8) Very weedy reaches..............	.075	.100	.125	.150

From *Handbook of Hydraulics*, by King, McGraw-Hill Book Company.

    * Values commonly used in designing.

Fig. 5.3 Diesel tractors owned by North Side Canal Company, Jerome, Ida., pulling a large chain weighing 2300 pounds to loosen moss and weeds in bottom of a canal. (Courtesy Caterpillar Tractor Company.)

velocities in excess of 5 feet per second. The low initial cost constitutes the major advantage of earth canals. The disadvantages are:

(a) excessive seepage losses.
(b) low velocities and therefore, relatively large cross-sectional areas.
(c) danger of breaks due to erosion and burrowing of animals.
(d) favorable conditions for growth of moss and aquatic weeds which retard the velocity and cause high annual maintenance costs, Fig. 5.3.

The sides of earth canals are usually built as steep as the earth will stand when wet. The slopes of the sides vary from three horizontal and one vertical to one horizontal and one vertical, for very stable materials. The relation of bed width b, to depth of earth canals d, is determined according to the topographic conditions. The bed width may be less than the depth, or it may be ten or more times the depth. The best hydraulic cross section under favorable structural conditions is:

$$b = 2d \tan \frac{\theta}{2} \qquad (5.10)$$

where θ is the angle between the side slope and the horizontal. This relation applies also to lined canals. For rectangular channels $\tan \theta/2 = 1$, and hence, the bed width equals twice the depth for best hydraulic characteristics.

5.3 STABLE CHANNELS

Irrigation channels should be stable—having negligible scour, and negligible deposition of sediments. Scour can be eliminated by low velocities, but Equation 5.1 ($Q = AV$) shows that as velocity decreases the area must increase if a given quantity of water is to be conveyed. Large canals are costly to construct, costly to maintain, and have greater loss of water from seepage and evaporation than small canals.

Another factor to be considered is that as the velocity doubles, the head loss increases by four, Equation 5.3. Since most irrigation canals are constructed at a minimum grade in order to reach as much arid land as possible, it is usually desirable to maintain small velocities. The requirement that the velocity cannot be less than a certain minimum to keep silt in suspension works against low energy loss and promotes scour. Therefore, a compromise is always necessary with the compromise being tailored to fit the conditions that exist at the site.

The recommendations of Fortier and Scobey on limiting velocity, together with corresponding values of roughness, n, are shown in Table 5.3. The values shown apply to water depths of 3 feet or less in canals that have been brought to capacity gradually. For depths of water over 3 feet, the velocity should be increased by 0.5 fps. For canals with sinuous alignment, a reduction of 25 percent is recommended.

The concept of traction force is more rational and accurate for determining maximum allowable canal velocities. An excellent summary of data and procedure of application of the tractive force concept is presented by Lane (1955). The details of application are beyond the scope of this book.

5.4 LINING CANALS

Irrigation canals are lined for purposes of (1) decreasing conveyance-seepage losses, (2) providing safety against breaks, (3) preventing weed growth, (4) retarding moss growth, (5) decreasing erosion from high velocities, (6) cutting down maintenance costs, (7) reducing drainage problems, and (8) increasing the capacity to convey water.

TABLE 5.3

LIMITING VELOCITIES FOR ESSENTIALLY STRAIGHT CANALS AFTER AGING
(AFTER FORTIER AND SCOBEY)

Material	Value of Manning's n	Velocity in Feet per Second	
		Clear Water	Water Transporting Colloidal Silts
Fine sand, colloidal	0.020	1.50	2.50
Sandy loam, non-colloidal	0.020	1.75	2.50
Silt loam, non-colloidal	0.020	2.00	3.00
Alluvial silts, non-colloidal	0.020	2.00	3.50
Ordinary firm loam	0.020	2.50	3.50
Volcanic ash	0.020	2.50	3.50
Stiff clay, very colloidal	0.025	3.75	5.00
Alluvial silts, colloidal	0.025	3.75	5.00
Shales and hardpans	0.025	6.00	6.00
Fine gravel	0.020	2.50	5.00
Graded loam to cobbles when non-colloidal	0.030	3.75	5.00
Graded silts to cobbles when colloidal	0.030	4.00	5.50
Coarse gravel, non-colloidal	0.025	4.00	6.00
Cobbles and shingles	0.035	5.00	5.50

Detailed cost studies are essential to the determination of economic advisability of lining canals. The annual cost of lining is the true cost, not the initial cost. Annual cost includes annual depreciation, maintenance, and interest. All of these factors should be considered when determining economic feasibility.

From the viewpoint of an irrigation project, the most important single factor in a study of advisability of lining is the annual value of water saved by decreasing conveyance losses. In localities where water is very limited, public interests are advanced by lining canals, thus contributing to a more economical use of the available water supply.

Excess canal seepage contributes to waterlogging of farm lands, salt and alkali concentration in the soils, costly road maintenance and drainage activities, ground-water seepage into basements of buildings, and other conditions that concern the public. Although it is difficult, costly, and sometimes quite impractical to measure accurately the degree of contribution to these adverse conditions by any one canal, the public should encourage reduction of canal seepage to protect public interests.

5.5 MATERIALS FOR CANAL LINING

The most used materials for canal lining include: concrete, rock masonry, brick, bentonite-earth mixtures, natural clays of low permeability, and different rubber, plastic, and asphalt compounds. A few canals have been lined under low-cost methods with clay, Fig. 5.4, and with asphalt membrane. For the main canal of the Yakima Project in the state of Washington, Fig. 5.5, concrete prevents both seepage and erosion.

For smaller irrigation canals and laterals, precast concrete slabs made at regular concrete mixing plants, hauled to, and placed in canals, are helpful, Fig. 5.6. Both frequent inspection of the concrete slab joints and careful maintenance are essential to prevent erosion of soil materials under the slabs with resulting settlement and damage to the lining. Masonry rock and brick work make good canal linings, Fig. 5.7.

Concrete lining has a high initial cost but several advantages. It is durable, maintenance costs are at a minimum, and capacity is increased because of the relatively smooth surface. Placement costs, which are normally fairly high, are reduced when mechanized methods are used such as shown in Figs. 5.8 and 5.9.

Shotcrete (Gunite) is used to line ditches and canals. This term

Fig. 5.4 All-American Canal clay blanket lining looking downstream. The white sand in which the canal is built has a permeability 1800 times higher than the clay used for lining.

Fig. 5.5 The concrete-lined Kittitas Main Canal, looking downstream, Yakima Project in the State of Washington, carrying about 600 cfs of water. (*Reclamation Era*, August 1946.)

Fig. 5.6 Placing 8-inch precast concrete slabs with tongue and groove joints. Yakima Project, Washington, 1947. (Courtesy Bureau of Reclamation.)

is used to designate pneumatically applied portland cement mortar. Compressed air drives a mixture of cement and sand through a hose to a nozzle where water is added. The mixture is blown onto the surface to be covered by air pressures of from 30 to 90 psi. Hoses of 50 to 150 feet in length operate best although they can be used for horizontal distances up to 500 feet and vertically up to 150 feet. For ordinary purposes one part of portland cement is mixed with four parts of sand by volume. When applying Shotcrete the nozzle is held perpendicular to the surface with the nozzle from 3 to 5 feet from the surface. An experienced operator, or nozzle man, is essential.

Earth materials are used for canal linings, especially when lower initial cost is desirable or necessary, and when suitable materials are available. Three methods are employed: (1) placing a blanket of relatively impervious material over or within the permeable canal bed; (2) dispersing clay in the water and having it filter out as the water seeps through the bed; (3) chemically stabilizing the earth material to make it less pervious.

Dirmeyer (1959) outlines specifications for the dispersion methods using bentonite, a montmorillonite clay that swells from eight to twelve times its original volume upon wetting, to seal earth and rock canals.

Fig. 5.7 A serviceable and attractive small Utah canal lined to solve the seepage, erosion, and weed problems. (Courtesy Work Projects Administration.)

Fig. 5.8 Transit-mix trucks are frequently used for mixing concrete and conveying it to the slipform for placement. This method is particularly favored for reasons of economy when the concrete can be discharged from the mixer directly into the slipform. The tractor and cable keep the form moving forward. (Courtesy Bureau of Reclamation.)

1. The Dispersion Method shall be used in sandy channels having sufficient flat grade for ponding and having bottom and sides responsive to harrowing so that the bentonite can be made to penetrate the soil rather than to form a temporary surface seal. The ditch section shall be effectively dammed by either waterproofing and existing check structure or constructing a temporary earth dam. The bentonite used shall have a grit or sand content of seven percent or less, and a colloidal mixture shall be ponded above the high water line at the upper end of the pond and be kept ponded for at least 48 hours. After the maximum ponding depth is attained, the ditch sides and then the bottom shall be stirred by a harrow, disk or similar implement. This shall be done at least twice each day for two days and in a manner producing a minimum of bank sloughing.

Known extreme high loss areas shall be blanketed with a tamped soil-bentonite mixture prior to ponding. Additional bentonite shall be added whenever an unexpected insufficiency occurs because of bentonite flocculation, or excessive loss of mixture during filling.

2. The Multiple Dam Method shall be used in gravelly or rocky channels having steep grade and ditch bottom and sides unsuitable for harrow opera-

tion. The ditch section shall be divided by an adequate number and spacing of dams built of a mixture of granular high-swell bentonite (same bentonite quality as above) and some supplemental bridging agent such as a local low-grade bentonite or wet sawdust. Composition of this mixture material shall be acceptable to the responsible technician. Water shall be released from the upstream end. As the first dam is over-topped, equipment such as a dragline or back-hoe shall be used to help break up the dam and obtain a lumpy mixture at maximum speed. The same process shall be repeated for the remaining dams.

In case of extremely rocky and open materials, additional quantities of the sedimenting mixture shall be spread or water jet sluiced on the ditch bank areas and whenever deemed necessary by the technicians.

Fig. 5.9 Fullerform continuous concrete pipe laid in place in a preformed ditch. An inflated rubber tube forms and holds the inside shape until the concrete sets. (Courtesy Fullerform Continuous Pipe Corporation.)

The most commonly used method of stabilizing soils is to mix about 10 percent of portland cement with a moist soil and compact the mixture. Thorough mixing, often difficult to secure in practice, is essential for a satisfactory, relatively durable lining.

Asphalt, plastic, and rubber compounds are becoming more widely used for linings in canals and ponds. Thin films are used where only an impermeable layer is desired. Thin layers are often covered with earth for protection. Thicker layers and built-up sections are used when resistance to abrasion and mechanical damage, as well as durability, is needed. These products are being increasingly used when lower initial cost is required. Their feasibility will continue to increase as better products and procedures are developed.

5.6   SEEPAGE FROM CANALS

Seepage of water from irrigation canals is a serious problem. Not only is water lost, but also drainage problems are often aggravated on adjacent or lower lands. Occasionally water that seeps out of a canal re-enters the river in the valley where it can be rediverted, or enters an aquifer where it can be reused. It is a more serious economic loss when the water seepage losses are not recoverable. However, serious legal problems may occur as a result of seepage water which is subsequently used. Users of a return flow to the river may have established a water right so that water saved by lining a higher canal will still have to be delivered to lower users. Also, economic and legal liabilities may result from water seepage from upper canals causing drainage problems on lower lying lands. The economic and legal problems associated with seepage from canals are sometimes very complex.

Several methods used to measure seepage from canals are inflow-outflow, ponding, seepage meters, wells, laboratory tests of permeability of soils, and special methods, including electrical resistance and tracing of natural and radioactive salts. The method best suited to a canal will depend on the depth and velocity of flow, ability to drain the canal, the material in the bed, and rate of seepage. The three most commonly used methods for field measurement are discussed briefly.

The inflow-outflow method consists of measuring the flow into and the flow from a selected section of canal. The accuracy of this method increases with the difference between the quantity of inflow and outflow rates. Water level should be held constant during the measurement and allowance should be made for rainfall and evaporation.

The ponding method consists of constructing dikes in the canal, filling the intervening section with water, and measuring the drop in water surface per hour. This is the most accurate method, provided that leaks through the dikes are either prevented or measured, and provided allowance is made for rainfall and evaporation. Measurements of water level should be made at both ends of the pond to eliminate the effect of wind on the water level. The principal disadvantages to this method are that tests can be made only when the canal is not in use and that the expense of constructing dikes and supplying and regulating the water is considerable.

Seepage meters are used to obtain measurements of seepage from relatively small areas of canal surface. Measurements can be taken without disturbing the flow in the canal unless velocity is excessive. Seepage meters generally consist of a cylinder with a dome or cone on the top to allow for trapped air to be removed from the system through a valve attached to the top of the cone. The cylinder is pressed gently into the soil and the unit is filled with water. A plastic bag attached to the cylinder is also filled with water and placed beneath the water surface. Seepage within the cylinder causes a corresponding reduction in water content in the plastic bag. The loss of weight of the bag indicates the rate of seepage through the cylindrical surface. The hydrostatic pressure on the bed of the canal within the cylinder is the same as around the cylinder, since the pressures are equalized through the plastic bag. Sometimes the plastic bag is replaced by a small-diameter container elevated so that the water surface in the container corresponds to the water surface in the canal. A Marriotte siphon can also be used to obtain a constant head.

Extensive studies indicate that several factors contribute to error in the seepage meter, so that these meters give only an indication of the order of magnitude of seepage rather than absolute rates of seepage. The surface is also disturbed by installing the meter, and the area under test is small and may not be representative of the canal section. To obtain results indicative of actual seepage, several readings are necessary with representative readings taken on the sides as well as on the bottom of the canal.

5.7 KEEPING CANALS CLEAN

One of the vital maintenance problems in conveyance of irrigation water is cleaning canals. Growth of weeds and willows on canal banks, and moss and other aquatic plants in canals, greatly retards water velocities decreasing canal capacities. Silt and clay sedimentation in canals also restricts water flow. Dense weed growth prevents proper inspection of the canal bank and provides winter cover that allows burrowing animals to ruin the canal banks.

Weed top growth falls into the water and catches floating debris, retards flow, and causes failure of banks. Large perennials such as willows, tamarisk, and cane make it almost impossible to clean a canal and remove the smaller weeds. In addition to the operational problem, many land weeds developing seed along a canal bank are a source of weed infestation to crop land.

Water weeds such as tules, cattails, pondweeds, coontail, and chara also reduce the flow. These weeds, like those on the banks, often clog measuring flumes, weirs, spillways, and other parts of the irrigation system, causing delays and additional costs in cleaning. Weed growths cause sand and silt bars to build up in the channel, which retard the velocity of flow, increase the seepage loss, and at times cause overflows of the bank.

Hand-labor methods of canal cleaning are being replaced by the use of bulldozers, dragline excavators, and tractor-drawn chains, Fig. 5.3. The most common methods of controlling weeds on canal banks are: pasturing, mowing, burning, and the application of chemical weed killers. Methods for controlling water-weed growths may be classed as: mechanical, drying, shading, and chemical.

Chemicals, primarily aromatic solvents, have been very effective in killing weeds in canals and ditches. When applied in recommended quantities, no injury to crops or to livestock has been observed, but these chemicals are deadly to aquatic animal life. Aromatic solvents are most economical in small channels, while mechanical methods of weed control are more practical and more economical on larger canals.

Weeds along rivers, canals, and ditches have been estimated to use, and thereby largely waste, 25 million acre-feet of water annually in the seventeen western states. This is a significant loss of water in an area which uses approximately 100 million acre-feet of water for irrigation. Water losses by evaporation and consumptive use are increased by weeds and growth along irrigation water courses.

5.8 WATER CONVEYANCE STRUCTURES

Structures used for water conveyance are varied to meet conditions of terrain and flow requirements. To convey water to the point of use for irrigation in the most efficient manner is a challenge to the irrigation engineer. A few structures commonly used to complement canals and pipe lines are discussed.

Flumes

For crossing natural depressions or narrow canyons, and for conveying irrigation water along very steep sidehills, flumes are constructed either of wood or metal, or both, as shown in Figs. 5.10 and

Fig. 5.10 Small flume crossing over arroyo in Upper Anton Chico. This structure saved 1700 feet of ditch length. Guadalupe County, New Mexico. (Courtesy Soil Conservation Service.)

5.11. Concrete is also used for flumes. To attain economy in the application of materials for flumes, it is desirable to give the flume sufficient slope to assure water velocity appreciably higher than in earth canals, thus making possible a proportional reduction in flume cross section.

Tunnels

To shorten the length of a diversion canal, to avoid difficult and expensive construction on steep, rocky hillsides, and to convey irrigation water through mountains from one watershed to another, many tunnels have been constructed. It is usually economical to line the bottoms and sides of tunnels through rock formations as a means of decreasing seepage losses and lessening frictional resistance. An irrigation tunnel constructed through loose material is lined with concrete as the boring of the tunnel progresses.

Drops and Chutes

Wood, concrete, or masonry bulkheads are used in places where the natural slopes down which canals must flow are so high as to cause excessive water velocities and erosion of the canal banks and bed. Water is dropped several feet, and the energy of a flowing stream is dissipated in the stilling basin without causing erosion, Fig. 5.12. A number of closely spaced drops may be used, as shown in Fig. 5.13. Chutes built of wood, concrete, or steel, Fig. 5.14, are serviceable where it is necessary to convey water down relatively steep hills which would require many drops closely spaced to control water velocity, and in which water would cause serious erosion if not controlled. Chutes may well be considered in three sections: (1) the section of transition and accelerating velocity; (2) the section of uniform high velocity; and (3) the stilling basin. In the first section the velocity is increased from approximately 3 feet per second up to 20 or more feet per second, and the cross-sectional area of the water is proportionately decreased. In the second section, because of very high velocity, the retarding forces are essentially equal in magnitude and opposite in direction to the driving forces; hence the water velocity remains constant. To dissipate the kinetic energy at the lower end of a chute, it is necessary to provide a deep stilling basin.

Inverted Siphons

For crossing wide deep hollows, depressions, or canyons it is customary to build pipe lines and convey water through them under pressure. The cost of flumes for crossing wide depressions is often so high

Fig. 5.11 Large metal flume over wash, near Price, Utah. Steel substructure set on concrete piers. (Courtesy Soil Conservation Service.)

Fig. 5.12 A masonry canal drop. Long apron reduces erosion below the structure. Suitable for small canals and low drops. (Courtesy Soil Conservation Service.)

Fig. 5.13 Series of check drops to prevent high canal-water velocities and erosion. Flashboards can be placed in notch to raise water level so that farmer can irrigate fields on both sides of the canal. (Structure in foreground appears to have apron set too high.) (Courtesy Soil Conservation Service.)

Fig. 5.14   Concrete chute conveying a stream of 282 sec-ft.   Weber-Provo diversion canal.   (Courtesy Bureau of Reclamation.)

as to prohibit their construction.   Pipes by which irrigation water is conveyed across canyons usually built on or near the ground surface are known as inverted siphons.   Such pipe lines are built either of steel, or wood staves held in place by iron bands, or of reinforced concrete.   A wood stave inverted siphon across the Bear River in Idaho resists a water pressure head of nearly 300 feet along the bottom of the canyon.

The stress caused by the water pressure in siphons is resisted by the unit strength and thickness of steel in steel pipes; by the unit strength, diameter, and spacing of the iron bands around wood stave pipes; and by the unit strength and amount of steel reinforcement in concrete pipe.   Large-diameter siphons required respectively thicker steel, larger and more closely spaced bands, or more steel reinforcement than small-diameter siphons under the same water pressure.   The velocity of water flowing through a siphon of given diameter is fixed by the hydraulic slope and by the roughness of the inside of the pipe.   The velocity is not influenced by the total water pressure inside the pipe.   Velocity may be determined by using Manning's formula,

Equation 5.8, together with Table 5.2. For example, consider a 6-foot diameter wood stave siphon in the best of condition, 1 mile long, having a drop in water surface of 9 feet. Then $R = 1.5$ and $S = 0.0017$; from Table 5.2, $n = 0.01$. Therefore, $V = 8.03$ ft per sec. As the cross-sectional area of a 6-foot diameter pipe is 28.3 sq ft, this pipe would discharge $8.03 \times 28.3 = 227$ cfs.

Flexible Tubing

Tubing made of rubber or plastic, often embedding synthetic glass fibers, is being used in increasing quantities to convey water. The tubing is generally sufficiently flexible so that it will lie flat when no water is in the pipe; hence, the term "lay-flat" tubing is often applied to this product. Flexible tubing is very easy to install and can be removed for cultivation or harvesting. The hydraulics of flexible tubing is more complex than rigid pipe because cross-sectional shape depends upon pressure, and pressure is a function of shape. The friction losses decrease as the shape of the tube approaches a round pipe.

REFERENCES

Albertson, Maurice L., James R. Barton, and Daryl B. Simons, "Fluid Mechanics for Engineers," Prentice-Hall, Englewood Cliffs, N.J., 1960.

Bruns, V. F., J. M. Hodgson, H. F. Arle, and F. L. Timmons, "The Use of Aromatic Solvents for Control of Submersed Aquatic Weeds in Irrigation Channels," *U.S.D.A. Circ.* 971, October 1955.

Dirmeyer, R. C., Jr., "Evaluation Report on Recent Bentonite Sealing Work in Wyoming Canals," *Wyoming Agr. Ext. Ser.*, March 1959.

Fortier, S. and F. C. Scobey, "Permissible Canal Velocities," *Trans. Am. Soc. Civil Eng.*, Vol. 89, 1926.

Lane, Emory W., "Design of Stable Channels," *Trans. Am. Soc. Civil Eng.*, Vol. 120, 1955.

Lauritzen, C. W., "Lay-flat Tubing," *Farm and Home Sci.*, No. 3, pp. 68–70, Vol. 18, September 1957.

Lauritzen, C. W., Frank W. Haws, and Allan S. Humpherys, "Plastic Film for Controlling Seepage Losses in Farm Reservoirs," *U.S.D.A. and Utah State Agr. College Exp. Sta. Bul.* 391, July 1956.

Lauritzen, C. W., O. W. Israelsen, and W. W. Rasmussen, "Lining Canals and Reservoirs to Reduce Conveyance Losses," *U.S.D.A. and Agr. Exp. Sta. Utah State Agr. College Circ.* 129, June 1952.

Lauritzen, C. W. and W. H. Peterson, "Butyl Fabrics as Canal Lining Materials," *Utah State Agr. College Exp. Sta. Bul.* 363, October 1953.

McCauley, Gulley, "The Economy of Canal Linings," *Irrigation Eng. and Maint.*, June 1952.

Moody, L. F., "Friction Factors for Pipe Flow," *Trans. Am. Soc. Mech. Eng.*, Vol. 66, pp. 671–684, 1944.

Portland Cement Association, "Lining Irrigation Canals," 1949.

Portland Cement Association, "Lining Irrigation Canals," 1957.

Rasmussen, W. W. and C. W. Lauritzen, "Measuring Seepage from Irrigation Canals," *Agr. Eng.*, Vol. 34, pp. 326–329, 331, May 1953.

Robinson, August R., Jr. and Carl Rohwer, "Measurement Seepage," *Trans. Am. Soc. Civil Eng.* Paper No. 2865 published as Proc. Separate No. 728, June 1955.

Robinson, A. R. and Carl Rohwer, "Measuring Seepage from Irrigation Channels," *U.S.D.A. and Colo. Agr. Exp. Sta. and Bureau of Reclamation Tech. Bul.* 1203, 1954.

Rohwer, Carl, "Canal Lining Manual," U.S.D.A. Soil Conservation Service, Fort Collins, Colorado, November 1946.

Rohwer, Carl and Oscar Van Pelt Stout, "Seepage Losses from Irrigation Channels," *Colo. Agr. Exp. Sta. Tech. Bul.* 38, 1948.

Rounds, E. C., E. L. Forte, and W. R. Fry, "Lining Lateral Canals with Precast Concrete Slabs," *Reclamation Era*, Vol. 32, pp. 89–90, April 1942.

Scott, V. H. "Prefabricated Linings for Irrigation Ditches," *Agr. Eng.*, Vol. 32, February 1956.

Steele, Thomas L., "Shotcrete Canal Linings," *Reclamation Era*, Vol. 32, No. 10, p. 189, October 1948.

Timmons, F. L. and Roscoe Fleming, "Water Wasters of the West," reprinted from *Crops and Soils*, Vol. 11, No. 3, December 1958.

U.S.B.R. "Canal Linings and Methods of Reducing Costs," 1952.

Vennard, John K., "Elementary Fluid Mechanics," third edition, John Wiley and Sons, New York, 1954.

Woodford, T. V., M. C. Lipp, and H. M. Sult, "Lower-cost Canal Linings, A Progress Report on the Development of the Lower-cost Linings for Irrigation Canals," U.S.B.R., Denver, June 1948.

Young, Walker R., "Low Cost Linings for Irrigation Canals," *Eng. News-Record.* Vol. 38, pp. 192–196, February 1947.

CHAPTER 6 WATER MEASUREMENT

Efficient use of water for irrigation depends largely on measurement of water. Increasing utilization and value of available water and the growing tendency among irrigation companies to base annual water charges on the volume of water used make an understanding of the principles and methods of water measurement necessary. Information concerning the relationships between water, soils, and plants cannot be utilized in irrigation practice without the measurement of water.

6.1 UNITS OF WATER MEASUREMENT

Units of water measurement are considered in two classes: first, those expressing a specific volume of water at rest; and second, those expressing a time rate of flow. The commonly used units of volume of water at rest are the gallon, cubic foot, acre-inch, and acre-foot. An acre-inch is a volume of water sufficient to cover 1 acre 1 inch deep, which is 3630 cu ft. An acre-foot of water will cover 1 acre 1 foot deep, and is equal to 43,560 cu ft.

The commonly used units of rate of flow are gallons per minute, cubic feet per second, acre-inches per hour, and acre-feet per day. The miner's inch is defined as the quantity of water that will flow through an opening 1 inch square in a vertical wall under a given pressure head. The pioneer miners of the West used pressure heads ranging from 4 to 7 inches. Each of the western states defines the miner's inch in terms of 1 cfs. One miner's inch is designated as $\frac{1}{50}$ cfs in southern California, Idaho, Kansas, New Mexico, North Dakota, South Dakota, Nebraska, and Utah. In Arizona, northern California, Montana, Nevada, and Oregon, 1 miner's inch is equal to $\frac{1}{40}$ cfs. In Colorado 38.4 miner's inches is considered equal to 1 cfs.

6.2 FLOW THROUGH AN ORIFICE

When the water-pressure intensity inside the pipes of a domestic water system is high, water flows out of an open tap at a high velocity, and when the pressure is low, it flows out slowly. If water pressure within the pipe were exactly equal to the pressure of the air outside the pipe, there would be no flow. Pressure intensity at any point within a static body of water with a free surface is proportional to the depth of the point below the water surface. The velocity of water flow through an opening in a vessel (or in a wall built across a stream) which is far below the water surface is greater than the velocity through an opening near the water surface. In irrigation practice it is important to know just what the velocity will be through an orifice at any vertical distance below the water surface. The basic physical law which determines the velocity of water through an orifice is the same as the law which determines the velocity of a falling body at any vertical distance below the point from which it begins to fall.

The velocity of a falling body, ignoring atmospheric friction, may be determined from the vertical distance through which the body falls from rest. Likewise, the velocity of water escaping from an opening (orifice) in a vessel, ignoring friction, may be determined from the height of water in the vessel above the opening. Stated as an equation, this very important law of falling bodies, as applied to the flow of water, is:

$$V = \sqrt{2gh} \qquad (6.1)$$

where V = velocity in ft per sec

g = the acceleration due to gravity (or the force of gravity per unit mass of water), which is 32.2 ft per sec/per sec

h = the depth of water in feet, or pressure head, causing discharge through an orifice

If the vertical dimension of the orifice opening is very long, the velocity of flow through the orifice will be appreciably greater near the bottom of the orifice than near the top. For the purpose of this discussion it is assumed that the orifice height is so small as compared to the pressure head causing the discharge that the difference between velocity near the top and near the bottom of the orifice is negligible. To illustrate the use of Equation 6.1 assume that h in Fig. 6.1 = 4 feet. Then:

$$V = \sqrt{2 \times 32.2 \times 4} = 16.04 \text{ ft per sec}$$

Therefore, theoretically, water should flow through an orifice which is 4 feet below the surface at a velocity of approximately 16 ft per sec. Owing to frictional resistance, the actual velocity is somewhat less than the theoretical velocity

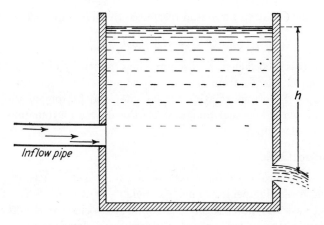

Fig. 6.1 Illustrating the discharge of water through an orifice under head, h.

The fact that the quantity of water that flows through an opening is directly proportional to the cross-sectional area of the opening and to the velocity of flow was given as Equation 5.1 and identified as the continuity equation. This basic rational equation is repeated here for convenience as:

$$Q = AV \qquad (6.2)$$

where Q = quantity of flow, cfs
  A = cross-sectional area of water, the canal or orifice, in sq ft
  V = the mean velocity, ft per sec

The theoretical discharge through an orifice may be determined by substituting the value of V from Equation 6.1 in Equation 6.2, i.e.,

$$Q = A\sqrt{2gh} \qquad (6.3)$$

If the orifice opening in Fig. 6.1 is 4 inches high by 18 inches long (perpendicular to the plane of the paper), the area would be:

$$A = \frac{4 \times 18}{144} = \frac{1}{2}\,\text{sq ft}$$

Experiment has shown that the actual discharge for a standard orifice is approximately six-tenths the theoretical discharge, so that the actual Q would be computed thus:

$$Q = \frac{6}{10} \times \frac{1}{2} \times \frac{16}{1} = 4.8\,\text{cfs}$$

Finally, the equation for actual discharge through an orifice is

$$Q = CA\sqrt{2gh} \qquad (6.4)$$

in which C is a coefficient of discharge determined by experiment. The coefficient C ranges from 0.6 up to 0.8 or more, depending on the position of the orifice relative to the sides and bottom of the vessels or of the water channel and also on the degree of roundness of the edges of the orifice.

Submerged Orifices

The water cross-sectional area of a submerged orifice, Fig. 6.2, is the length of the opening times the height, or $A = L \times H$. The loss of head as water flows through a submerged orifice is a difference in elevation of the water surface upstream and downstream, as shown by h in Fig. 6.2. Hence, from Equation 6.4 and from experiments, the discharge of a standard submerged orifice is:

$$Q = 0.61A\sqrt{2gh} \qquad (6.5)$$

The submerged orifice is used both in its standard form and as a combination head gate measuring device. Table 6.1 gives discharges for submerged orifices as computed from Equation 6.5. It shows, for example, that with a submerged orifice having an opening 6 inches high by 12 inches long (area 0.5 sq ft), if the upstream water surface is 3 inches, or 0.25, higher than the downstream surface, the discharge will be 1.22 cfs.

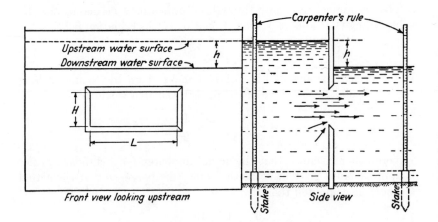

Fig. 6.2 Submerged orifice.

TABLE 6.1

DISCHARGE TABLE FOR SUBMERGED RECTANGULAR ORIFICES*

Head h		Cross-sectional Area A of Orifice, square feet							
Feet	Inches	0.25	0.50	0.75	1.00	1.25	1.50	1.75	2.00
0.09	$1\frac{1}{16}$	0.37	0.73	1.10	1.47	1.84	2.20	2.64	2.94
.10	$1\frac{3}{16}$	.39	.77	1.16	1.56	1.93	2.32	2.71	3.09
.11	$1\frac{5}{16}$	.41	.81	1.22	1.62	2.03	2.43	2.84	3.24
.12	$1\frac{7}{16}$	.42	.85	1.27	1.69	2.12	2.54	2.97	3.39
.13	$1\frac{9}{16}$	.44	.88	1.32	1.76	2.21	2.65	3.09	3.53
.14	$1\frac{11}{16}$	.46	.92	1.37	1.83	2.29	2.75	3.20	3.66
.15	$1\frac{13}{16}$	.47	.95	1.42	1.90	2.37	2.84	3.32	3.79
.16	$1\frac{15}{16}$	.49	.98	1.47	1.96	2.45	2.93	3.42	3.91
.17	$2\frac{1}{16}$	.50	1.01	1.51	2.02	2.52	3.02	3.53	4.03
.18	$2\frac{3}{16}$	.52	1.04	1.56	2.08	2.59	3.11	3.63	4.15
.19	$2\frac{1}{4}$	.53	1.07	1.60	2.13	2.67	3.20	3.73	4.26
.20	$2\frac{3}{8}$	.55	1.09	1.64	2.19	2.74	3.28	3.83	4.36
.21	$2\frac{1}{2}$	.56	1.12	1.68	2.24	2.80	3.36	3.92	4.48
.22	$2\frac{5}{8}$	.57	1.15	1.72	2.30	2.87	3.46	4.02	4.59
.23	$2\frac{3}{4}$	.59	1.17	1.76	2.35	2.93	3.52	4.10	4.69
.24	$2\frac{7}{8}$	.60	1.20	1.80	2.40	3.00	3.60	4.19	4.79
.25	3	.61	1.22	1.83	2.45	3.06	3.67	4.28	4.89
.26	$3\frac{1}{8}$	.62	1.25	1.87	2.49	3.12	3.74	4.37	4.99
.27	$3\frac{1}{4}$	.64	1.27	1.91	2.54	3.18	3.81	4.45	5.08
.28	$3\frac{3}{8}$	.65	1.29	1.94	2.59	3.24	3.88	4.53	5.18
.29	$3\frac{1}{2}$	.66	1.32	1.98	2.64	3.30	3.96	4.62	5.28
.30	$3\frac{5}{8}$	.67	1.34	2.01	2.68	3.35	4.02	4.69	5.36
.31	$3\frac{11}{16}$	.68	1.36	2.05	2.73	3.41	4.09	4.77	5.45
.32	$3\frac{13}{16}$	.69	1.38	2.07	2.76	3.46	4.15	4.84	5.53
.33	$3\frac{15}{16}$	.70	1.41	2.11	2.81	3.51	4.22	4.92	5.62
.34	$4\frac{1}{16}$	.71	1.43	2.14	2.85	3.57	4.28	4.99	5.70
.35	$4\frac{3}{16}$	.72	1.45	2.17	2.89	3.62	4.34	5.06	5.78
.36	$4\frac{5}{16}$	.73	1.47	2.20	2.93	3.67	4.40	5.14	5.87
.37	$4\frac{7}{16}$	.75	1.49	2.23	2.98	3.72	4.46	5.21	5.95
.38	$4\frac{9}{16}$	.75	1.51	2.26	3.02	3.77	4.52	5.28	6.03
.39	$4\frac{11}{16}$	.76	1.53	2.29	3.05	3.82	4.58	5.35	6.11
.40	$4\frac{13}{16}$	.77	1.55	2.32	3.09	3.87	4.64	5.42	6.19
.41	$4\frac{15}{16}$	.78	1.57	2.35	3.12	3.92	4.70	5.48	6.27
.42	$5\frac{1}{16}$	.79	1.59	2.38	3.17	3.96	4.75	5.55	6.34
.43	$5\frac{1}{8}$	.80	1.60	2.41	3.21	4.01	4.81	5.61	6.42
.44	$5\frac{1}{4}$	.81	1.62	2.43	3.24	4.06	4.87	5.68	6.49
.45	$5\frac{3}{8}$	.82	1.64	2.46	3.28	4.10	4.92	5.74	6.56
.46	$5\frac{1}{2}$	.83	1.66	2.49	3.32	4.15	4.98	5.81	6.64
.47	$5\frac{5}{8}$	.84	1.68	2.52	3.36	4.20	5.04	5.87	6.71
.48	$5\frac{3}{4}$	.85	1.70	2.54	3.39	4.24	5.08	5.93	6.78
.49	$5\frac{7}{8}$	.86	1.71	2.57	3.42	4.28	5.14	5.99	6.85
.50	6	.87	1.73	2.59	3.46	4.32	5.19	6.05	6.92
.51	$6\frac{1}{8}$	.87	1.75	2.62	3.49	4.37	5.24	6.11	6.99
.52	$6\frac{1}{4}$	.88	1.76	2.65	3.53	4.41	5.29	6.17	7.05
.53	$6\frac{3}{8}$	.89	1.78	2.67	3.56	4.45	5.34	6.23	7.12
.54	$6\frac{1}{2}$	.90	1.80	2.70	3.59	4.49	5.39	6.29	7.19
.55	$6\frac{9}{16}$	.91	1.81	2.72	3.63	4.53	5.44	6.35	7.25
.56	$6\frac{11}{16}$	.92	1.83	2.75	3.66	4.58	5.49	6.41	7.32
.57	$6\frac{13}{16}$	.92	1.85	2.77	3.69	4.62	5.54	6.46	7.38
.58	$6\frac{15}{16}$	.93	1.86	2.79	3.73	4.66	5.59	6.52	7.45
.59	$7\frac{1}{16}$	.94	1.88	2.82	3.76	4.70	5.64	6.58	7.51

*Computed from the formula: $Q = 0.61A \sqrt{2gh}$.

Types of Orifices Used

Types of orifices used to measure irrigation water are: (1) submerged orifices with fixed dimensions, (2) adjustable submerged orifices, (3) miner's-inch boxes, and (4) calibrated gates.

Submerged orifices with fixed dimensions are used where the available head is insufficient for weirs. For best results they are usually rectangular with the horizontal dimension from two to six times the height. They are generally installed in sufficiently large channels so that the contractions[1] are complete or very nearly so. The coefficient of discharge will then be approximately 0.61. There is a lack of information regarding this coefficient for incomplete contractions. Submerged orifices should have a smooth vertical face of sufficient size, smooth sharp edges, accurate dimensions, and a provision for accurate head measurements.

There are several kinds of adjustable submerged orifices. Some resemble submerged orifices with fixed dimensions except that their height is adjustable to accommodate a wide range of flow without excessive head loss. Table 6.1 may be used to determine the flow through orifices in which the channel is sufficiently large to insure complete end contractions. The more usual type of adjustable submerged orifice is a combination head gate, or turnout, and measuring device. Such structures are usually made of wood and generally have one or two slide gates that may be held open in any desired position. Two gates can be used together to provide a constant head orifice with greater accuracy. One gate is used to control the water level. Single submerged orifice head gates are generally not accurate measuring devices and are gradually being replaced by better types of structures.

A development of recent years has been the calibrated commercial gates for water measurement (Calco Meter Gates for example). Tests have been conducted on gates of various size, and curves and tables prepared giving the flow in cfs for different gate openings measured on the rising stem. The head is the difference in elevation of the water surface in the supply canal and the outlet ditch. Each Calco Meter Gate is calibrated individually, and a discharge chart and table are furnished with the gate at the time of purchase.

Many ingenious miner's-inch boxes have been devised; most of them contain an orifice plate with adjustable opening and some auxiliary

---

[1] Full contraction is secured on an orifice or weir whenever the sides, bottom, and water surface are far enough from the opening that no change in discharge would occur if they were further removed, other conditions remaining constant.

means of regulating the required pressure head. The accuracy of measurement through any of these structures depends primarily upon (1) the ratio of head to height of orifice, (2) the accuracy with which the pressure head can be regulated or maintained, (3) the velocity of approach, (4) the accuracy with which the area of the orifice can be determined, (5) the conditions affecting the contraction of the jet, and (6) freedom from submergence.

For accurate measurements with free-flowing orifices, the height of the orifice should not be greater than the head. The pressure head is the most difficult to regulate. The means generally employed is either an adjustable regulating gate or a spill crest at the desired level allowing excess water to overflow. Some of the miner's-inch boxes employ both principles to insure accurate measurement. With complete contraction, velocity of approach is usually negligible.

6.3 FLOW THROUGH A WEIR

Suppose that the water surface is lowered until it drops below the upper edge of the orifice, Fig. 6.3. Then the pressure head in feet, which causes the average velocity, as represented by h of Fig. 6.3 is one-half the total depth of water over the bottom edge of the orifice. Representing the length of orifice by the symbol L, the area becomes:

$$A = 2hL$$

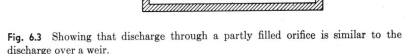

Fig. 6.3 Showing that discharge through a partly filled orifice is similar to the discharge over a weir.

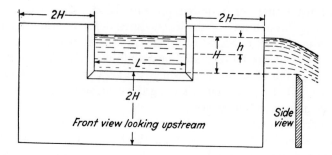

Fig. 6.4 A rectangular weir with complete end contractions.

Substituting this value of A in Equation 6.4

$$Q = 2CLh\sqrt{2gh} = 2CLh^{\frac{3}{2}}\sqrt{2g} \qquad (6.6)$$

Since the acceleration due to gravity g is nearly constant, it is convenient to represent the product $2C\sqrt{2g}$ by a single symbol, say C'. Then it follows that

$$Q = C'Lh^{\frac{3}{2}} \qquad (6.7)$$

Equation 6.7 gives the theoretical discharge of an orifice when the top edge of the orifice is above the water surface.

The term weir, as used in measurement of water, is defined as a notch in a wall built across a stream. The notch may be rectangular, trapezoidal, or triangular in shape. The orifice in Fig. 6.3, when flowing partly full, is a weir according to the above definition. In measuring the flow of water over weirs, it is convenient and customary to measure the total depth; i.e., $2h$ of Fig. 6.3. Although the total depth, as represented in Fig. 6.4 by the symbol H, does not represent the point in the stream of average velocity, it can be used in Equation 6.7 by changing only the coefficient C'. Substituting for h its equivalent $H/2$ in Equation 6.7 there results:

$$Q = C'L\left(\frac{H}{2}\right)^{\frac{3}{2}} = \frac{C'}{2^{\frac{3}{2}}} LH^{\frac{3}{2}}$$

or

$$Q = C''LH^{\frac{3}{2}} \qquad (6.8)$$

in which

$$C'' = \frac{C'}{2^{\frac{3}{2}}}$$

Equation 6.8 is the general form for discharge of rectangular and trapezoidal weirs. Although, as shown above, it is based on the fundamental equation $Q = AV$, the only measurements essential in using Equation 6.8 are those of length of weir crest, L, and depth of water flowing over it, H. The velocity need not be measured directly. The coefficient C'', ordinarily represented by C, has been determined by experiment by many workers. For rectangular weirs it was early found by Francis to be 3.33 and hence, the widely used sharp-crested weir discharge equation

$$Q = 3.33LH^{3/2} \qquad (6.9)$$

Equation 6.9 without modification applies accurately only to rectangular weirs in which the length of weir is the same as the width of the rectangular channel immediately above the weir; i.e., weirs having suppressed end contractions. For weirs having complete end contractions, such as represented in Fig. 6.4, the effective length of weir crest, L, is found from the relation

$$L = L' - 0.2H \qquad (6.10)$$

in which $L' =$ the measured length of weir crest. In actual use of Equation 6.9 and other discharge equations, it is customary to compute tables from the equation, using $L = 1$, and for many values of H. Table 6.2 gives discharge per foot of length of weir crest based on Equation 6. 9 for values of H from 0.20 up to 1.24 feet. For example, columns 1 and 3 show that for a head H of 0.45 foot the discharge is 1.005 cfs per foot length of weir having suppressed end contractions. The effective length for a 1 foot weir having complete end contractions, according to Equation 6.10 is:

$$L = 1.00 - 0.2 \times 0.45 = 0.91 \text{ ft}$$

and hence, the discharge per foot of measured length is $1.005 \times 0.91 =$ 0.915 cfs.

The Italian engineer, Cipolletti, long ago designed a trapezoidal weir with complete contractions in which the discharge is believed to be directly proportional to the length of weir crest, making it unnecessary to correct for end contraction. It has been widely used because of its advantages. The equation giving the discharge is:

$$Q = 3.37LH^{3/2} \qquad (6.11)$$

In this trapezoidal weir the sides have a slope of 1 horizontal to 4 vertical, Fig. 6.5. Aside from the small correction necessary due to the fact that the sides of the weir slope outward, Equation 6.11 may be arrived at in the same way as Equation 6.9. Table 6.3 gives dis-

TABLE 6.2

Discharge in Cubic Feet per Second (Second-feet) per Foot of Length of Weir Crest by the Francis Formula: $Q = 3.33H^{3/2}$

1	2	3	1a	2a	3a	1b	2b	3b
Depth of Water or Head, H		Discharge in Second-feet (q)	Depth of Water or Head, H		Discharge in Second-feet (q)	Depth of Water or Head, H		Discharge in Second-feet (q)
Feet	Inches		Feet	Inches		Feet	Inches	
0.20	$2\frac{3}{8}$	0.298	0.55	$6\frac{5}{8}$	1.358	0.90	$10\frac{13}{16}$	2.843
.21	$2\frac{1}{2}$	.320	.56	$6\frac{3}{4}$	1.395	.91	$10\frac{15}{16}$	2.890
.22	$2\frac{5}{8}$	.344	.57	$6\frac{13}{16}$	1.433	.92	$11\frac{1}{16}$	2.938
.23	$2\frac{3}{4}$	.367	.58	$6\frac{15}{16}$	1.470	.93	$11\frac{3}{16}$	2.986
.24	$2\frac{7}{8}$	.392	.59	$7\frac{1}{16}$	1.509	.94	$11\frac{1}{4}$	3.035
.25	3	.416	.60	$7\frac{3}{16}$	1.547	.95	$11\frac{3}{8}$	3.083
.26	$3\frac{1}{8}$	.442	.61	$7\frac{5}{16}$	1.586	.96	$11\frac{1}{2}$	3.132
.27	$3\frac{1}{4}$	.467	.62	$7\frac{7}{16}$	1.626	.97	$11\frac{5}{8}$	3.181
.28	$3\frac{3}{8}$	.493	.63	$7\frac{9}{16}$	1.665	.98	$11\frac{3}{4}$	3.230
.29	$3\frac{1}{2}$	.520	.64	$7\frac{11}{16}$	1.705	.99	$11\frac{7}{8}$	3.280
.30	$3\frac{5}{8}$	.547	.65	$7\frac{13}{16}$	1.745	1.00	12	3.300
.31	$3\frac{3}{4}$	.575	.66	$7\frac{15}{16}$	1.785	1.01	$12\frac{1}{8}$	3.380
.32	$3\frac{13}{16}$	.603	.67	$8\frac{1}{16}$	1.826	1.02	$12\frac{1}{4}$	3.430
.33	$3\frac{15}{16}$	.631	.68	$8\frac{3}{16}$	1.867	1.03	$12\frac{3}{8}$	3.481
.34	$4\frac{1}{16}$	.660	.69	$8\frac{1}{4}$	1.908	1.04	$12\frac{1}{2}$	3.532
.35	$4\frac{3}{16}$	.689	.70	$8\frac{3}{8}$	1.950	1.05	$12\frac{5}{8}$	3.583
.36	$4\frac{5}{16}$	.719	.71	$8\frac{1}{2}$	1.992	1.06	$12\frac{3}{4}$	3.634
.37	$4\frac{7}{16}$	.749	.72	$8\frac{5}{8}$	2.034	1.07	$12\frac{13}{16}$	3.686
.38	$4\frac{9}{16}$	.780	.73	$8\frac{3}{4}$	2.070	1.08	$12\frac{15}{16}$	3.737
.39	$4\frac{11}{16}$	.811	.74	$8\frac{7}{8}$	2.120	1.09	$13\frac{1}{16}$	3.789
.40	$4\frac{13}{16}$	.842	.75	9	2.163	1.10	$13\frac{3}{16}$	3.842
.41	$4\frac{15}{16}$	.874	.76	$9\frac{1}{8}$	2.206	1.11	$13\frac{5}{16}$	3.894
.42	$5\frac{1}{16}$	.906	.77	$9\frac{1}{4}$	2.250	1.12	$13\frac{7}{16}$	3.947
.43	$5\frac{3}{16}$	.939	.78	$9\frac{3}{8}$	2.294	1.13	$13\frac{9}{16}$	4.000
.44	$5\frac{1}{4}$	.972	.79	$9\frac{1}{2}$	2.340	1.14	$13\frac{11}{16}$	4.053
.45	$5\frac{3}{8}$	1.005	.80	$9\frac{5}{8}$	2.383	1.15	$13\frac{13}{16}$	4.107
.46	$5\frac{1}{2}$	1.039	.81	$9\frac{3}{4}$	2.428	1.16	$13\frac{15}{16}$	4.160
.47	$5\frac{5}{8}$	1.073	.82	$9\frac{13}{16}$	2.473	1.17	$14\frac{1}{16}$	4.214
.48	$5\frac{3}{4}$	1.107	.83	$9\frac{15}{16}$	2.520	1.18	$14\frac{3}{16}$	4.268
.49	$5\frac{7}{8}$	1.142	.84	$10\frac{1}{16}$	2.564	1.19	$14\frac{1}{4}$	4.323
.50	6	1.177	.85	$10\frac{3}{16}$	2.610	1.20	$14\frac{3}{8}$	4.377
.51	$6\frac{1}{8}$	1.213	.86	$10\frac{5}{16}$	2.660	1.21	$14\frac{1}{2}$	4.432
.52	$6\frac{1}{4}$	1.249	.87	$10\frac{7}{16}$	2.702	1.22	$14\frac{5}{8}$	4.487
.53	$6\frac{3}{8}$	1.285	.88	$10\frac{9}{16}$	2.749	1.23	$14\frac{3}{4}$	4.543
.54	$6\frac{1}{2}$	1.321	.89	$10\frac{11}{16}$	2.796	1.24	$14\frac{7}{8}$	4.598

charges for the trapezoidal weir, as computed from Equation 6.11 for length of crest from 1 to 18 feet.

For the 90° triangular weir shown in Fig. 6.6, the water cross-sectional area is $H \times H$, or H^2, and therefore, from Equation 6.4,

$$Q = CH^2\sqrt{2gh} = C'H^{5/2}$$

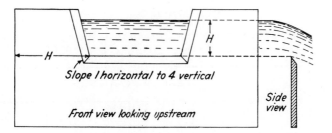

Fig. 6.5 A trapezoidal or Cipolletti weir.

is the theoretical discharge. The actual discharge has been found by experiment to be approximately

$$Q = 2.49H^{5/2} \tag{6.12}$$

Table 6.4 gives discharges for 90° triangular weirs.

Summary of Weir and Orifice Formulas

To show the similarity between the discharge equations and to summarize the material of the preceding section, Table 6.5 has been prepared. When the orifice is no longer flowing full with the result that h becomes equivalent to H then $H \times H^{1/2} = H^{3/2}$, or the form of Equation 6.9 is obtained to describe the rectangular weir without side contraction. However, when the sides are contracted, the effective width of the crest L is reduced on each side by $0.1H$. The 1 to 4 side slopes of a trapezoidal weir nearly produce the same flow as a rectangular weir, the only difference being a slight change in coefficient of

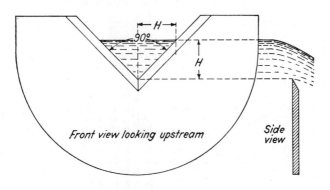

Fig. 6.6 A 90° triangular notch weir.

TABLE 6.3

DISCHARGE OVER CIPOLLETTI'S TRAPEZOIDAL WEIR. FOR VARIOUS
LENGTHS AND HEADS. FORMULA: $Q = 3.37LH^{3/2}$

Head H on Crest		Length of Weir Crest in Feet											
In Feet	In Inches	1	1½	2	2½	3	3½	4	5	7½	10	12½	15
		Discharge in Cubic Feet per Second											
0.21	2½	0.324	0.49	0.65	0.81	0.97	1.13	1.30	1.62	2.43	3.24	4.05	4.86
.22	⅝	.347	.52	.69	.87	1.04	1.22	1.39	1.74	2.61	3.47	4.34	5.21
.23	¾	.371	.56	.74	.93	1.11	1.30	1.49	1.86	2.79	3.71	4.64	5.57
.24	⅞	.396	.59	.79	.99	1.19	1.39	1.58	1.98	2.97	3.96	4.95	5.94
.25	3	.421	.63	.84	1.05	1.26	1.47	1.68	2.10	3.16	4.21	5.26	6.31
.26	3⅛	.446	.67	.89	1.12	1.34	1.56	1.79	2.23	3.35	4.46	5.58	6.70
.27	¼	.472	.71	.94	1.18	1.42	1.65	1.89	2.36	3.54	4.72	5.90	7.09
.28	⅜	.499	.75	1.00	1.25	1.50	1.75	2.00	2.49	3.74	4.99	6.24	7.48
.29	½	.526	.79	1.05	1.31	1.58	1.84	2.10	2.63	3.94	5.26	6.57	7.89
.30	⅝	.553	.83	1.11	1.38	1.66	1.94	2.21	2.77	4.15	5.53	6.92	8.30
.31	3¾	....	.87	1.16	1.45	1.74	2.03	2.32	2.91	4.36	5.81	7.26	8.72
.32	⅞	....	.91	1.22	1.52	1.83	2.13	2.44	3.05	4.57	6.09	7.62	9.14
.33	4	....	.96	1.28	1.60	1.91	2.23	2.55	3.19	4.79	6.38	7.98	9.57
.34	⅛	....	1.00	1.33	1.67	2.00	2.34	2.67	3.34	5.01	6.67	8.34	10.01
.35	¼	....	1.05	1.39	1.74	2.09	2.44	2.79	3.49	5.23	6.97	8.71	10.46
.36	4⅜	....	1.09	1.45	1.82	2.18	2.56	2.91	3.64	5.45	7.27	9.09	10.91
.37	½	....	1.14	1.52	1.89	2.27	2.65	3.03	3.79	5.68	7.58	9.47	11.37
.38	9/16	....	1.18	1.58	1.97	2.37	2.76	3.15	3.94	5.91	7.89	9.86	11.83
.39	⅝	....	1.23	1.64	2.05	2.46	2.87	3.28	4.10	6.15	8.20	10.25	12.30
.40	¾	....	1.28	1.70	2.13	2.56	2.98	3.41	4.26	6.39	8.52	10.65	12.78
.41	4⅞	....	1.33	1.77	2.21	2.65	3.09	3.54	4.42	6.63	8.84	11.05	13.26
.42	5	....	1.37	1.83	2.29	2.75	3.21	3.67	4.58	6.87	9.16	11.46	13.75
.43	⅛	....	1.42	1.90	2.37	2.85	3.32	3.80	4.75	7.12	9.49	11.87	14.24
.44	¼	....	1.47	1.97	2.46	2.95	3.44	3.93	4.91	7.37	9.83	12.28	14.74
.45	⅜	....	1.52	2.03	2.55	3.05	3.56	4.07	5.08	7.62	10.16	12.70	15.24
.46	5½	....	1.58	2.10	2.63	3.15	3.68	4.20	5.25	7.88	10.50	13.13	15.76
.47	⅝	....	1.63	2.17	2.71	3.25	3.80	4.34	5.42	8.14	10.85	13.56	16.27
.48	¾	....	1.68	2.24	2.80	3.36	3.92	4.48	5.60	8.40	11.20	14.00	16.79
.49	⅞	....	1.73	2.31	2.89	3.46	4.04	4.62	5.77	8.66	11.55	14.43	17.32
.50	6	....	1.79	2.38	2.98	3.57	4.17	4.76	5.95	8.93	11.90	14.88	17.85
.51	6⅛	....	1.84	2.45	3.07	3.68	4.29	4.90	6.13	9.20	12.26	15.33	18.39
.52	¼	....	1.89	2.52	3.16	3.79	4.42	5.05	6.31	9.47	12.62	15.78	18.94
.53	⅜	....	1.95	2.60	3.25	3.90	4.55	5.20	6.50	9.74	12.99	16.24	19.49
.54	½	....	2.00	2.67	3.34	4.01	4.68	5.34	6.68	10.02	13.36	16.70	20.04
.55	⅝	....	2.06	2.75	3.43	4.12	4.81	5.49	6.87	10.30	13.73	17.17	20.60
.56	6¾	....	2.12	2.82	3.53	4.23	4.94	5.64	7.05	10.58	14.11	17.64	21.16
.57	⅞	....	2.17	2.90	3.62	4.35	5.07	5.80	7.24	10.87	14.49	18.11	21.73
.58	7	....	2.23	2.97	3.72	4.46	5.20	5.95	7.44	11.15	14.87	18.59	22.31
.59	⅛	....	2.29	3.05	3.81	4.58	5.34	6.10	7.63	11.44	15.26	19.07	22.89
.60	¼	....	2.35	3.13	3.91	4.69	5.48	6.26	7.82	11.74	15.65	19.56	23.47

TABLE 6.3 (*Concluded*)

In Feet	In Inches	2	2½	3	3½	4	5	7½	10	12½	15	18
						Discharge in Cubic Feet per Second						
0.61	7⅜	3.21	4.01	4.81	5.61	6.42	8.02	12.03	16.04	20.05	24.06	28.87
.62	½	3.29	4.11	4.93	5.75	6.57	8.22	12.33	16.44	20.54	24.65	29.58
.63	⁹⁄₁₆	3.37	4.21	5.05	5.89	6.73	8.42	12.63	16.83	21.04	25.25	30.30
.64	⅝	3.45	4.31	5.17	6.03	6.89	8.62	12.93	17.24	21.55	25.86	31.03
.65	¾	3.53	4.41	5.29	6.18	7.06	8.82	13.23	17.64	22.05	26.46	31.76
.66	7⅞	3.61	4.51	5.42	6.32	7.22	9.03	13.54	18.05	22.56	27.08	32.49
.67	8	3.69	4.62	5.54	6.46	7.39	9.23	13.85	18.46	23.08	27.70	33.23
.68	⅛	3.78	4.72	5.66	6.61	7.55	9.44	14.16	18.88	23.60	28.32	33.98
.69	¼	3.86	4.82	5.79	6.75	7.72	9.65	14.47	19.30	24.12	28.94	34.73
.70	⅜	3.94	4.93	5.92	6.90	7.89	9.86	14.79	19.72	24.65	29.58	35.49
.71	8½	4.03	5.04	6.04	7.05	8.06	10.07	15.11	20.14	25.18	30.21	36.25
.72	⅝	4.11	5.14	6.17	7.20	8.23	10.28	15.43	20.57	25.71	30.85	37.03
.73	¾	4.20	5.25	6.30	7.35	8.40	10.50	15.75	21.00	26.25	31.50	37.80
.74	⅞	4.29	5.36	6.43	7.50	8.57	10.72	16.07	21.43	26.79	32.15	38.58
.75	9	4.37	5.47	6.56	7.65	8.75	10.93	16.40	21.87	27.33	32.80	39.36
.76	9⅛	4.46	5.58	6.69	7.81	8.92	11.15	16.73	22.31	27.88	33.46	40.15
.77	¼	4.55	5.69	6.82	7.96	9.10	11.37	17.06	22.75	28.43	34.12	40.95
.78	⅜	4.64	5.80	6.96	8.12	9.28	11.60	17.39	23.19	28.99	34.79	41.75
.79	½	4.73	5.91	7.09	8.27	9.46	11.82	17.73	23.64	29.55	35.46	42.55
.80	⅝	4.82	6.02	7.23	8.43	9.64	12.05	18.07	24.09	30.11	36.13	43.36
.81	9¾	4.91	6.14	7.36	8.59	9.82	12.27	18.41	24.54	30.68	36.81	44.18
.82	⅞	5.00	6.25	7.50	8.75	10.00	12.50	18.75	25.00	31.25	37.50	45.00
.83	10	5.09	6.36	7.64	8.91	10.18	12.73	19.09	25.46	31.82	38.19	45.82
.84	⅛	5.18	6.48	7.78	9.07	10.37	12.96	19.44	25.92	32.40	38.88	46.65
.85	¼	5.28	6.60	7.92	9.23	10.55	13.19	19.79	26.38	32.98	39.57	47.49
.86	10⅜	5.37	6.71	8.06	9.40	10.74	13.43	20.14	26.85	33.56	40.28	48.33
.87	½	5.46	6.83	8.20	9.56	10.93	13.66	20.49	27.32	34.15	40.97	49.18
.88	⁹⁄₁₆	5.56	6.95	8.34	9.73	11.12	13.90	20.84	27.79	34.74	41.69	50.03
.89	⅝	5.65	7.07	8.48	9.89	11.31	14.13	21.20	28.27	35.33	42.40	50.88
.90	¾	5.75	7.19	8.62	10.06	11.50	14.37	21.56	28.75	35.93	43.12	51.74
.91	10⅞	....	7.31	8.77	10.23	11.69	14.61	21.92	29.23	36.53	43.84	52.61
.92	11	....	7.43	8.91	10.40	11.88	14.85	22.28	29.71	37.14	44.56	53.48
.93	⅛	....	7.55	9.06	10.57	12.08	15.10	22.65	30.19	37.74	45.29	54.35
.94	¼	....	7.67	9.20	10.74	12.27	15.34	23.01	30.68	38.35	46.02	55.23
.95	⅜	....	7.79	9.35	10.91	12.47	15.59	23.38	31.17	38.97	46.76	56.11
.96	11½	....	7.92	9.50	11.08	12.67	15.83	23.75	31.67	39.58	47.50	57.00
.97	⅝	....	8.04	9.65	11.26	12.87	16.08	24.12	32.16	40.20	48.24	57.89
.98	¾	....	8.17	9.80	11.43	13.06	16.33	24.49	32.66	40.83	48.99	58.79
.99	⅞	....	8.29	9.95	11.61	13.27	16.58	24.87	33.16	41.45	49.74	59.69
1.00	12	....	8.42	10.10	11.78	13.47	16.83	25.25	33.67	42.08	50.50	60.60

Head *H* on Crest — Length of Weir Crest in Feet

TABLE 6.4
Discharge Table for 90° Triangular Notch Weir*

Head in Feet	Head in Inches	Discharge in Second-feet (q)	Head in Feet	Head in Inches	Discharge in Second-feet (q)	Head in Feet	Head in Inches	Discharge in Second-feet (q)
0.20	$2\frac{3}{8}$	0.046	0.55	$6\frac{5}{8}$	0.564	0.90	$10\frac{13}{16}$	1.92
.21	$2\frac{1}{2}$	.052	.56	$6\frac{3}{4}$	.590	.91	$10\frac{15}{16}$	1.97
.22	$2\frac{5}{8}$	.058	.57	$6\frac{13}{16}$	.617	.92	$11\frac{1}{16}$	2.02
.23	$2\frac{3}{4}$	.065	.58	$6\frac{15}{16}$	.644	.93	$11\frac{3}{16}$	2.08
.24	$2\frac{7}{8}$	.072	.59	$7\frac{1}{16}$	.672	.94	$11\frac{1}{4}$	2.13
.25	3	.080	.60	$7\frac{3}{16}$	.700	.95	$11\frac{3}{8}$	2.19
.26	$3\frac{1}{8}$	.088	.61	$7\frac{5}{16}$	.730	.96	$11\frac{1}{2}$	2.25
.27	$3\frac{1}{4}$	.096	.62	$7\frac{7}{16}$	.760	.97	$11\frac{5}{8}$	2.31
.28	$3\frac{3}{8}$	.106	.63	$7\frac{9}{16}$	.790	.98	$11\frac{3}{4}$	2.37
.29	$3\frac{1}{2}$	.115	.64	$7\frac{11}{16}$	.822	.99	$11\frac{7}{8}$	2.43
.30	$3\frac{5}{8}$	.125	.65	$7\frac{13}{16}$	.854	1.00	12	2.49
.31	$3\frac{3}{4}$	.136	.66	$7\frac{15}{16}$	.887	1.01	$12\frac{1}{8}$	2.55
.32	$3\frac{13}{16}$	.147	.67	$8\frac{1}{16}$	.921	1.02	$12\frac{1}{4}$	2.61
.33	$3\frac{15}{16}$	.159	.68	$8\frac{3}{16}$	.955	1.03	$12\frac{3}{8}$	2.68
.34	$4\frac{1}{16}$	.171	.69	$8\frac{1}{4}$	.991	1.04	$12\frac{1}{2}$	2.74
.35	$4\frac{3}{16}$	.184	.70	$8\frac{3}{8}$	1.03	1.05	$12\frac{5}{8}$	2.81
.36	$4\frac{5}{16}$	.197	.71	$8\frac{1}{2}$	1.06	1.06	$12\frac{3}{4}$	2.87
.37	$4\frac{7}{16}$	.211	.72	$8\frac{5}{8}$	1.10	1.07	$12\frac{13}{16}$	2.94
.38	$4\frac{9}{16}$	.226	.73	$8\frac{3}{4}$	1.14	1.08	$12\frac{15}{16}$	3.01
.39	$4\frac{11}{16}$	.240	.74	$8\frac{7}{8}$	1.18	1.09	$13\frac{1}{16}$	3.08
.40	$4\frac{13}{16}$	.256	.75	9	1.22	1.10	$13\frac{3}{16}$	3.15
.41	$4\frac{15}{16}$	.272	.76	$9\frac{1}{8}$	1.26	1.11	$13\frac{5}{16}$	3.22
.42	$5\frac{1}{16}$	.289	.77	$9\frac{1}{4}$	1.30	1.12	$13\frac{7}{16}$	3.30
.43	$5\frac{3}{16}$	.306	.78	$9\frac{3}{8}$	1.34	1.13	$13\frac{9}{16}$	3.37
.44	$5\frac{1}{4}$	.324	.79	$9\frac{1}{2}$	1.39	1.14	$13\frac{11}{16}$	3.44
.45	$5\frac{3}{8}$	.343	.80	$9\frac{5}{8}$	1.43	1.15	$13\frac{13}{16}$	3.52
.46	$5\frac{1}{2}$	.362	.81	$9\frac{3}{4}$	1.48	1.16	$13\frac{15}{16}$	3.59
.47	$5\frac{5}{8}$	.382	.82	$9\frac{13}{16}$	1.52	1.17	$14\frac{1}{16}$	3.67
.48	$5\frac{3}{4}$	.403	.83	$9\frac{15}{16}$	1.57	1.18	$14\frac{3}{16}$	3.75
.49	$5\frac{7}{8}$	.424	.84	$10\frac{1}{16}$	1.61	1.19	$14\frac{1}{4}$	3.83
.50	6	.445	.85	$10\frac{3}{16}$	1.66	1.20	$14\frac{3}{8}$	3.91
.51	$6\frac{1}{8}$	.468	.86	$10\frac{5}{16}$	1.71	1.21	$14\frac{1}{2}$	3.99
.52	$6\frac{1}{4}$	.491	.87	$10\frac{7}{16}$	1.76	1.22	$14\frac{5}{8}$	4.07
.53	$6\frac{3}{8}$	.515	.88	$10\frac{9}{16}$	1.81	1.23	$14\frac{3}{4}$	4.16
.54	$6\frac{1}{2}$	.539	.89	$10\frac{11}{16}$	1.86	1.24	$14\frac{7}{8}$	4.24
						1.25	15	4.33

* Computed from the formula: $Q = 2.49H^{2.48}$.

discharge from 3.33 to 3.37. A 90° triangular weir has a flow proportional to $H^{5/2}$ because the length also increases directly as the head increases; and L becomes equal to H, and $H \times H^{3/2} = H^{5/2}$. The coefficient of discharge must be varied slightly to take care of changing contractions and losses when boundaries are changed.

Effect of Boundary Form on Coefficient of Discharge

When silt deposits in front of an orifice or weir, the flow is changed. This causes one of the major sources of error in these measuring

TABLE 6.5

SUMMARY OF ORIFICE AND WEIR FORMULAS

Measuring Device (all sharp crested)	Views	Formula	Equation Number
Orifice	Front view / Side view	$Q = 0.61 A \sqrt{2gh}$	(6.5)
Rectangular Weir (without contraction)	Top view	$Q = 3.33 L H^{3/2}$	(6.9)
Rectangular Weir (with contraction)	Top view ($L - 0.2H$) / Side view	$Q = 3.33 (L - 0.2H) H^{3/2}$	(6.9) and (6.10)
Trapezoidal Weir	End view	$Q = 3.37 L H^{3/2}$	(6.11)
90° Triangular Weir	End view	$Q = 2.49 H^{5/2}$	(6.12)

devices. Orifice and weir formulas presented in the preceding sections are for average conditions and do not reflect changes in boundary form. Figure 6.7 can be used to determine a more accurate discharge coefficient for orifice and weir formulas when the installations do not conform to standard conditions.

For orifices the coefficient of discharge C_d in the following formula is a function of the height of opening b and the depth of water h over the center of the opening:

$$Q = C_d A \sqrt{2gh} \qquad (6.13)$$

From Fig. 6.7 a coefficient of 0.61 corresponds to conditions where $b/h = 0$, which is not always the case.

For weirs, the coefficient 3.33 in Equation 6.9 corresponds to a value

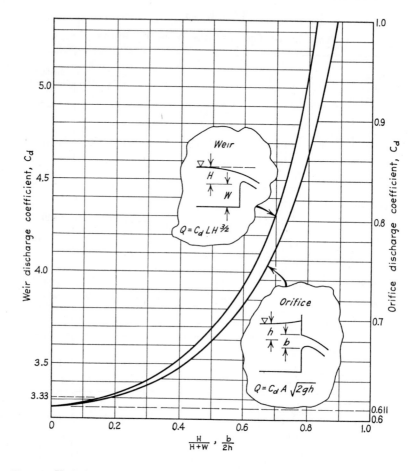

Fig. 6.7 Variation of discharge coefficients with boundary form for a rectangular weir and for an orifice.

of $h/(h + w)$ of 0.22—essentially the conditions shown in Fig. 6.4, where the depth over the crest is one-third the distance from the bottom to the crest. Hence, when the bottom is other than $3H$ below the crest, the coefficient of discharge given in Fig. 6.7 should be used in the following formula for more accurate results:

$$Q = C_d L H^{3/2} \qquad (6.14)$$

The same coefficient can be used for a contracted weir or a trapezoidal weir with very little error.

Often silt settles behind a weir changing the geometry of approach. The discharge for a silted weir is always more than the flow calculated from the measured head. Figure 6.8 shows the necessary dimensions for determining from Table 6.6 the increase in discharge occurring as

TABLE 6.6

PERCENT INCREASE IN DISCHARGE CAUSED BY SILTING BEHIND A WEIR

$\dfrac{P}{W}$	$\dfrac{X}{W}$					
	0	0.5	1.0	1.5	2.0	2.5[1]
0	0	10	13	15	16	16
0.25	0	5	8	10	10	10
0.50	0	3	4	5	6	6
0.75	0	1	2	2	3	3
1.00	0	0	0	0	0	0

[1] For values of X/W greater than about 3.0 the increase in discharge will reduce until for large values of X/W, the increase reduces to zero.

a result of the silting. For example, note that if $P/W = 0.25$ and $X/W = 1.0$, an increase of 8 percent in the flow will result. Note that this is the increase over the discharge given by use of the revised coefficient obtained from Fig. 6.7, not the discharge calculated from Equation 6.9 using a coefficient of 3.33.

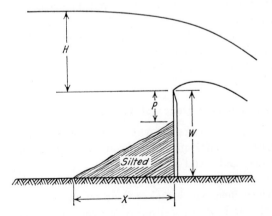

Fig. 6.8 Definition sketch for silting behind a weir to be used with Table 6.6.

The error in orifice and weir measurement is generally not greater than 5 percent. However, it should be clearly understood that boundary form does have an influence on discharge. Therefore, when accuracy greater than 5 percent is necessary, measuring devices should be carefully constructed to the specifications of the researchers who have developed the formula to be used. Many formulas are available, and each one applies to a rather narrow range of boundary conditions. Only the simpler formulas which are sufficiently accurate for most irrigation work have been presented herein.

Properties of Weirs and Orifices

It is important to note the influence that change of depth of water H has on the discharge Q for the various weirs and orifices. For example, when the stream over a rectangular weir increases until the depth H is doubled, the cross-sectional area A is doubled, and the discharge Q increased by 2.8 times; whereas doubling H over a trapezoidal weir slightly more than doubles the area, and increases the discharge 2.88 times. When the H on a triangular weir is doubled, area is increased 4 times and discharge is increased nearly 5.66 times. When the depth of water H causing the discharge through a submerged orifice is doubled, the area remains unchanged, and the discharge Q is made only 1.4 times the original discharge. Comparisons similar to the above may readily be made for any change in depth by remembering that discharge Q varies with the three-halves power of depth H for the rectangular and trapezoidal weirs, with the five-halves power for the triangular weir, and with the one-half power of the head h for the submerged orifice. The variations of discharge with depth of water have a very important bearing on irrigation practice and should be clearly understood.

In order to illustrate the influence of change of depth on the discharge further, the curves of Fig. 6.9 are presented. The above relations may be confirmed by examination of these curves and of Tables 6.1 to 6.4.

Advantages and Disadvantages of Weirs

The advantages of weirs for water measurement are: (1) accuracy, (2) simplicity and ease of construction, (3) non-obstruction by moss or floating materials, and (4) durability.

The disadvantages of weirs are: (1) requirement of considerable fall of water surface, which makes their use in areas having level land impracticable, and (2) deposition of gravel, sand, and silt above the weir prevents accurate measurements.

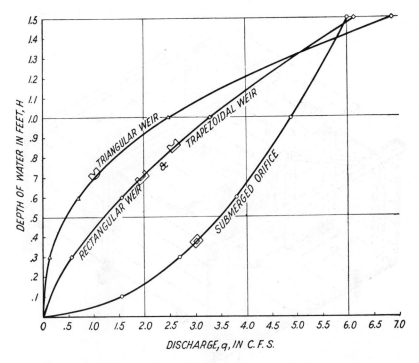

Fig. 6.9 Curves showing the relation between the discharge of water in cubic feet per second and the depth of water on the weir crest in feet for a 1-foot rectangular weir with suppressed end contractions, a 1-foot trapezoidal weir, a 90° triangular notch weir, and a submerged orifice of 1 square foot cross-sectional area.

The principal advantage for the submerged orifice is the small loss of head, or difference in elevation of the water surfaces on the upstream and the downstream sides of the orifice, making it suitable for use in canals and ditches having very small slopes, where it is difficult to obtain fall enough to use weirs. Orifices have in addition most of the advantages enumerated for weirs.

More serious disadvantages of orifices are: (1) collecting of floating debris, and (2) collecting of sand and silt above the orifice, thus preventing accurate measurement.

Weir Box and Pond

When using weirs, the ditch or canal must be made wider and deeper than average for some distance upstream from the weir. This is to

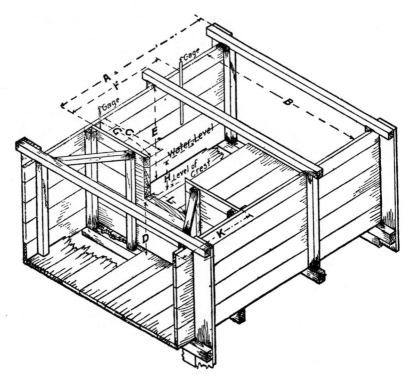

Fig. 6.10   Plan of weir box.   (*U.S.D.A. Farmers' Bul.* 813.)

make the water approach the weir at a low velocity (usually less than 0.5 foot per second) by flowing through a relatively large channel. The enlarged section of the ditch should be gradually tapered to natural size.   Cross currents upstream from and near the weir must be prevented.   This may be done by placing baffle boards across the weir channel.   The weir may be placed in a weir box built of lumber or concrete, Fig. 6.10, or it may simply be placed in an enlargement of the ditch, Fig. 6.11.

Less room is required when a box is used, but cleaning is made more difficult.   For temporary use, the placing of a weir in an open ditch as in Fig. 6.11 is the more economical method.

Cleaning is also less expensive in an open ditch, since a scraper may be used.   The ditch downstream must be protected with loose rock or other material to prevent erosion by the falling water.

Table 6.7 gives the sizes of weirs best adapted to measuring streams

TABLE 6.7

Weir-box Dimensions for Rectangular, Cipolletti, and 90° Triangular Notch Weirs

(All dimensions are in feet. The letters at the heads of the columns in this table refer to Fig. 6.18.)

Rectangular and Trapezoidal Weirs with End Contractions

Flow (Second-feet)	*H* Maximum Head	*L* Length of Weir Crest	*A* Length of Box Above Weir Notch	*K* Length of Box Below Weir Notch	*B* Total Width of Box	*E* Total Depth of Box	*C* End of Crest to Side	*D* Crest to Bottom	*F* Hook Gage Distance Upstream	*G* Hook Gage Distance Across Stream
½ to 3	1.0	1	6	2	5½	3½	2¼	2	4	2
2 to 5	1.1	1½	7	3	7	4	2¾	2½	4½	2
4 to 8	1.2	2	8	4	8½	4½	3¼	2¾	5	2½
6 to 14	1.3	3	9	5	12	5	4½	3¼	5½	3
10 to 22	1.5	4	10	6	14	5½	5	3½	6	3
90° Triangular Notch Weir										
½ to 2½	1.00	..	6	2	5	3	2½	1½	4	2
2 to 4⅓	1.25	..	6½	8½	6½	3¼	3¼	1½	5	2½

Fig. 6.11 Weir notch and bulkhead in weir pond. (*U.S.D.A. Farmers' Bul.* 813.)

of water varying from ½ to 22 cfs, and also the proper dimensions for each size of rectangular, trapezoidal, and 90° triangular notch weirs.

Weir dimensions, listed in Table 6.7 and illustrated in Fig. 6.10, are a little smaller than would be necessary to obtain rigid accuracy, but boxes of these sizes generally will give results within 5 percent of the correct values.

For temporary wooden weirs, the wood of which the weir is constructed may well be used also to form the weir crest and sides. However, since wood warps easily and its sharp edges become worn and splintered, its use on permanent weirs for crests and sides is seldom desirable.

Measurement of Head or Depth on Weir Crest

Measurement of head or depth of water on a weir crest is obtained with a specially constructed scale or a carpenter's rule. A scale, called a weir gage, must be set upstream from the weir a distance no less than four times the depth of water H flowing over the crest, or upstream in the corner near the bank where the velocity is essentially zero. This is made necessary by the downward curvature of the water surface near the crest. The zero point on the scale must be set level with the crest of the rectangular or trapezoidal weir, or with the vertex of the triangular weir. The scale, or a lug upon which to place a rule, may be fastened to the bulkhead at a lateral distance from the end of the notch of not less than twice the greatest depth of water H over the crest. To get the zero point of the scale or the lug level with the crest, a carpenter's level may be used. Allowing the water to flow into the pond and slowly rise until it flows over the weir crest is inaccurate since the water surface will rise appreciably above the crest before flow over the crest begins. Small errors in reading H cause relatively large errors in discharge determination. Table 6.8 shows the error in measurement caused by an error in reading of only 0.01 foot (less than ⅛ inch). The scale or gage may be marked to read cubic feet per second directly, and thus avoid the necessity of having to refer to a weir table to find the flow each time a reading is made.

Hook gages are widely used and considered the most accurate for determining water depths or stages. They consist of two essential parts; a moveable scale on which is fastened a hook; a fixed part containing an index mark and usually a vernier scale. The moveable part is raised until the point produces a slight pimple on the water surface, and the gage height is read opposite the index. A blunt point is preferable to a sharp point.

Recording gages, called water-level recorders, are used to obtain

TABLE 6.8

PERCENTAGE OF ERROR IN DISCHARGE OVER WEIRS CAUSED
BY 0.01 FOOT ERROR IN READING THE HEAD

Head			Length of Weir Crest					90° Notch
Feet	Feet	Inches	Per cent 1 Foot	Per cent 1.5 Feet	Per cent 2 Feet	Per cent 3 Feet	Per cent 4 Feet	Per cent
0.20	0	$2\frac{3}{8}$	7.2	7.5	7.5	7.6	7.6	..
.30	0	$3\frac{5}{8}$	5.0	5.1	5.1	5.6	4.8	8.5
.50	0	6	3.5	3.2	3.0	2.9	2.9	5.0
.70	0	$8\frac{3}{8}$	2.1	1.9	2.1	2.2	2.2	3.9
.90	0	$10\frac{13}{16}$	1.8	1.8	1.8	1.7	1.7	2.9
1.10	1	$1\frac{13}{16}$	...	1.4	1.3	1.3	1.3	2.2
1.25	1	3	...	...	...	1.1	1.1	2.1
1.50	1	6	...	...	...	0.9	1.0	...

a continuous graph of gage height. Essential parts of a recording gage are: (1) float or pressure-indicating device, (2) recording mechanism, and (3) clock. Several different kinds of recording gages are available.

Stilling wells for measuring water elevation are essential if accuracy is desired. A box or large-diameter pipe set vertically at one side of a connected stream or channel can be used as a stilling well. Stilling wells are used to eliminate wave action and provide a still water surface. To function properly, the cross-sectional area of a stilling well should be about 100 times the area of the inlet pipe or opening. Care should be taken to prevent the inlet pipe from clogging, and a convenient means of cleaning both the inlet pipe and stilling well should be provided. Flow of high velocity past a protruding stilling-well pipe should be avoided.

Portable Weirs

It is sometimes desirable to measure small streams at points where the cost of installing permanent weirs would not be warranted. For example, the occasional measurement of surface runoff from various fields, though desirable, would hardly warrant the installation of a permanent weir. In situations like this, a small steel plate cut like a half circle and having a weir notch serves well, such as shown in Fig. 6.6. The notch may be cut as a rectangle, trapezoid, or triangle, depending on the type of weir desired. Bulkheads of plastic or canvas attached to weirs or orifices make excellent portable measuring devices.

Weirs Without End Contractions

A standard rectangular weir without end contractions consists of a bulkhead having a sharp crest, built across a rectangular channel, high enough to cause a complete deflection of water filaments as the stream passes over the weir. The conditions for accuracy are the same as for the standard rectangular weir with contractions, except for those relating to side contractions. This type of weir can be used only in channels having a uniform rectangular cross section. Air holes must be made through the weir box just below the weir crest to fully admit air under the sheet of over-falling water. Lack of an adequate air supply at atmospheric pressure under the sheet of falling water will cause a suction, drawing down the water surface and increasing the discharge. Equation 6.9 or 6.14 can be used to determine the discharge.

Rules for Setting and Operating Weirs

1. The weir should be set at the lower end of a long pool sufficiently wide and deep to give an even, smooth current with a velocity of approach of not over 0.5 foot per second, which means practically still water.

2. The center line of the weir box should be parallel to the direction of the flow.

3. The face of the weir should be vertical, i.e., leaning neither upstream nor downstream.

4. The crest of the weir should be level, so the water passing over it will be of the same depth at all points along the crest.

5. The upstream edge should be sharp so that the over-falling water touches the crest at only one point.

6. The distance of the crest above the bottom of the pool should be about three times the depth of water flowing over the weir crest.

7. The sides of the pool should be at a distance from the sides of the crest not less than twice the depth of the water passing over the crest.

8. For accurate measurements the depth over the crest should be no more than one-third the length of the crest.

9. The depth of water over the crest should be no less than 2 inches, as it is difficult with smaller depths to get sufficiently accurate gage readings to give close results.

10. The crest should be placed high enough so water will fall freely below the weir, leaving an air space under the over-falling sheet of water. If the water below the weir rises above the crest, this free fall

is not possible, and the weir is then said to be submerged. Unless complicated corrections are made, measurements on submerged weirs are unreliable.

11. The gage or weir scale may be placed on the upstream face of the weir structure and far enough to one side so that it will be in comparatively still water, Fig. 6.10, or it may be placed at any point in the weir pond or box, so long as it is a sufficient distance from the weir notch as to be essentially beyond the downward curve of the water as it flows over the weir crest. The zero of the weir scale or gage should be placed level with the weir crest. This may be done with a carpenter's level, or where greater refinement is desired, with an engineer's level.

12. To prevent erosion by the falling water, the ditch downstream from the weir should be protected by loose rock or by other material.

There are notable differences in opinion among irrigation authorities concerning the accuracy of different weir formulas and the suitability of different water measuring devices. The reader who desires further information concerning weirs, especially for precise measurements of water, should consult the references given at the end of the text.

6.4 PARSHALL FLUMES

Parshall has designed a measuring device with which the discharge is obtained by measuring the loss in head caused by forcing a stream of water through a throat or converged section of a flume with a depressed bottom. The disadvantages of weirs and submerged orifices are largely overcome by the Parshall flume. Since the head H is small on which measurement is based, care must be exercised in determining the differences in water level to get accurate measurements. The Parshall flume is illustrated in Figs. 6.12, 6.13, and 6.14.

The Parshall measuring flume is a product of many years of painstaking research. It was first known as the venturi flume, being similar to the venturi tube or meter early designed to measure the flow of water in pipes.

The accuracy of the Parshall measuring flume is within limits that are allowable in irrigation practice, ordinarily within 5 percent. Flumes ranging from 1 inch to 10 feet in throat width are used to measure flows from 5 gpm to 200 cfs or larger. Small flumes are well suited to the requirement of measuring farm water deliveries and flows in furrows. The Parshall flume operates successfully with less loss of head than required for weirs.

Silt will not deposit in the structure where it will affect the accuracy,

Fig. 6.12 Finished 9-inch Parshall flume with staff gage. One-inch angle irons at the upstream end, crest, and downstream end of the structure serve as guides for striking off the floors at exact elevations. (Courtesy Soil Conservation Service.)

Fig. 6.13 Parshall measuring flume constructed with reinforced concrete, near Longmont, Colorado. (Courtesy Soil Conservation Service.)

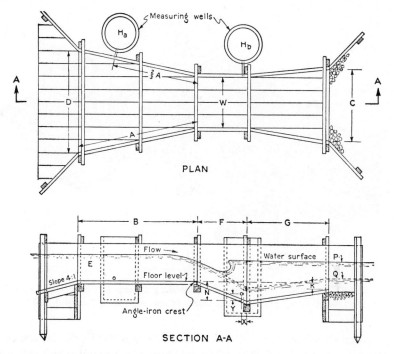

Fig. 614(A) Plan and longitudinal section of Parshall measuring flume. Dimensions are given in Table 9. (*Calif. Agr. Exp. Sta. Bul.* 588.)

because velocity is higher in the flume than in the channel. Ordinary velocities of approach have little or no effect on the measurement. The flume may be used with recording or registering instruments when continuous records of flow are desired, or with an indicating gage graduated to give the flow in any unit desired.

The Parshall flume cannot readily be combined with a turnout. For free-flow conditions the exit velocity is relatively high, and channel protection is generally necessary downstream from the flume to prevent erosion.

Only a single head need be measured for free-flow conditions which exist when the head at the lower gage is less than about 60 percent of that at the upper gage. Free-flow is determined from a measurement of the head at the upper gage by use of Table 6.9. When the head at the lower gage is greater than 70 percent of that of the upper gage, the upper gage reading is affected and submerged flow results. Fairly accurate measurements can be made with a submergence of 90 percent,

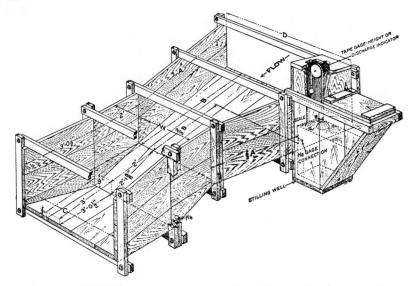

Fig. 6.14(B) Parshall measuring flume, including stilling-well equipment with indicating tape device. Staff gage in well. (*Colo. Agr. Exp. Sta. Bul.* 336.)

provided that the heads at both places are measured and the amount determined from Fig. 6.15 is subtracted from the flow given in Table 6.9. The correction for larger flumes is obtained by multiplying the correction for a 1-foot flume, Fig. 6.15, by the appropriate factor in Table 6.10.

For example, consider a 2-foot flume in which the upper head H_a is 1.6 feet and the lower head H_b is 1.2 feet. The ratio 1.2/1.6 = 0.75, which shows 75 percent submergence and also shows that a correction is required. It is not necessary to compute the percentage of submergence, except to determine whether a correction is necessary—often answered by inspection. On the left margin of the diagram Fig. 6.15, for a 1-foot flume, take a point about one-fifth of the distance between the lines for H_a, 1.5 and 2.0, respectively, and follow horizontally to the right until the imaginary line intersects the curved line for $H_b = 1.2$. Then follow an imaginary vertical line downward to the bottom of the diagram and read the correction, which is approximately 0.5 cfs. This amount is now multiplied by the factor 1.8 for a 2-foot flume, Table 6.10, and the product 0.9 is subtracted from the free flow, 16.6 cfs, Table 6.9, to obtain 15.7 cfs, the correct flow under these conditions.

The successful operation of the Parshall flume depends largely upon the correct selection of size and proper setting of the flume. The probable maximum and minimum flow to be measured is estimated,

TABLE 6.9

Free Flow Through Parshall Measuring Flume

Feet	Inches (Approx.)	1 in.	2 in.	3 in.	6 in.	9 in.	1 ft	2 ft	3 ft	4 ft	5 ft	6 ft	8 ft	10 ft
0.10	1 3/16	0.009	0.019	0.03	0.05	0.09								
0.12	1 5/16	0.013	0.025	0.04	0.07	0.12								
0.14	1 11/16	0.016	0.032	0.05	0.09	0.15								
0.16	1 15/16	0.020	0.039	0.06	0.11	0.19								
0.18	2 3/16	0.024	0.047	0.07	0.14	0.22								
0.20	3/8	0.028	0.055	0.08	0.16	0.26	0.35	0.66	0.97	1.26				
0.22	5/8	0.032	0.065	0.10	0.19	0.30	0.40	0.77	1.12	1.47				
0.24	7/8	0.037	0.074	0.11	0.22	0.35	0.46	0.88	1.28	1.69				
0.26	3 1/8	0.042	0.084	0.12	0.25	0.39	0.51	0.99	1.46	1.91	2.36	2.80		
0.28	3/8	0.047	0.094	0.14	0.28	0.44	0.58	1.11	1.64	2.15	2.65	3.15		
0.30	5/8	0.052	0.105	0.15	0.31	0.49	0.64	1.24	1.82	2.39	2.96	3.52	4.62	
0.32	13/16	0.058	0.116	0.17	0.34	0.54	0.71	1.37	2.02	2.65	3.28	3.90	5.13	
0.34	4 1/16	0.064	0.127	0.19	0.38	0.59	0.77	1.50	2.22	2.92	3.61	4.30	5.66	
0.36	5/16	0.069	0.139	0.21	0.41	0.64	0.84	1.64	2.42	3.19	3.95	4.71	6.20	
0.38	9/16	0.075	0.151	0.22	0.45	0.70	0.92	1.79	2.64	3.48	4.31	5.13	6.76	
0.40	13/16	0.082	0.163	0.24	0.48	0.76	0.99	1.93	2.86	3.77	4.68	5.57	7.34	9.1
0.42	5 1/16	0.088	0.176	0.26	0.52	0.81	1.07	2.09	3.08	4.07	5.05	6.02	7.94	9.8
0.44	1/4	0.095	0.189	0.28	0.56	0.87	1.15	2.24	3.32	4.38	5.43	6.48	8.55	10.6
0.46	1/2	0.101	0.203	0.30	0.61	0.94	1.23	2.40	3.56	4.70	5.83	6.96	9.19	11.4
0.48	3/4	0.108	0.217	0.32	0.65	1.00	1.31	2.57	3.80	5.03	6.24	7.44	9.8	12.2
0.50	6	0.115	0.230	0.34	0.69	1.06	1.39	2.73	4.05	5.36	6.66	7.94	10.5	13.1
0.52	1/4	0.123	0.245	0.36	0.73	1.13	1.48	2.90	4.31	5.70	7.09	8.46	11.2	13.9
0.54	1/2	0.130	0.260	0.38	0.78	1.20	1.57	3.08	4.57	6.05	7.52	8.98	11.9	14.8
0.56	3/4	0.138	0.275	0.40	0.82	1.26	1.66	3.26	4.84	6.41	7.97	9.52	12.6	15.7
0.58	15/16	0.145	0.290	0.43	0.87	1.33	1.75	3.44	5.11	6.77	8.43	10.1	13.3	16.6
0.60	7 3/16	0.153	0.306	0.45	0.92	1.40	1.84	3.62	5.39	7.15	8.89	10.6	14.1	17.5
0.62	7/16	0.161	0.322	0.47	0.97	1.48	1.93	3.81	5.68	7.53	9.37	11.2	14.8	18.5
0.64	11/16	0.169	0.338	0.50	1.02	1.55	2.03	4.01	5.97	7.91	9.85	11.8	15.6	19.5
0.66	15/16	0.177	0.355	0.52	1.07	1.63	2.13	4.20	6.26	8.31	10.3	12.4	16.4	20.4
0.68	8 3/16	0.186	0.372	0.55	1.12	1.70	2.23	4.40	6.56	8.71	10.6	13.0	17.2	21.5
0.70	3/8		0.389	0.57	1.17	1.78	2.33	4.60	6.86	9.11	11.4	13.6	18.0	22.5
0.72	5/8		0.406	0.60	1.23	1.86	2.43	4.81	7.17	9.53	11.9	14.2	18.9	23.5
0.74	7/8		0.424	0.62	1.28	1.94	2.53	5.02	7.49	9.95	12.4	14.9	19.7	24.6
0.76	9 1/8		0.442	0.65	1.34	2.02	2.63	5.23	7.81	10.1	12.9	15.5	20.6	25.7
0.78	3/8		0.459	0.68	1.39	2.10	2.74	5.44	8.13	10.8	13.5	16.2	21.5	26.8
0.80	5/8			0.70	1.45	2.18	2.85	5.66	8.46	11.3	14.0	16.8	22.4	27.9
0.82	13/16			0.73	1.50	2.27	2.96	5.88	8.79	11.7	14.6	17.5	23.3	29.0
0.84	10 1/16			0.76	1.56	2.35	3.07	6.11	9.13	12.2	15.2	18.2	24.2	30.2
0.86	5/16			0.79	1.62	2.44	3.18	6.33	9.48	12.6	15.8	18.9	25.1	31.4
0.88	9/16			0.81	1.68	2.52	3.29	6.56	9.82	13.1	16.3	19.6	26.1	32.5
0.90	13/16			0.84	1.74	2.61	3.41	6.80	10.2	13.6	16.9	20.3	27.0	33.7
0.92	11 1/16			0.87	1.81	2.70	3.52	7.03	10.5	14.0	17.5	21.0	28.0	35.0
0.94	1/4			0.90	1.87	2.79	3.64	7.27	10.9	14.5	18.1	21.8	29.0	36.2
0.96	1/2			0.93	1.93	2.88	3.76	7.51	11.3	15.0	18.8	22.5	30.0	37.5
0.98	3/4			0.96	2.00	2.98	3.88	7.75	11.6	15.5	19.4	23.2	31.0	38.7
1.00	12			0.99	2.06	3.07	4.00	8.00	12.0	16.0	20.0	24.0	32.0	40.0
1.02	1/4			1.02	2.12	3.17	4.12	8.25	12.4	16.5	20.6	24.8	33.0	41.3
1.04	1/2			1.05	2.19	3.26	4.25	8.50	12.8	17.0	21.3	25.6	34.1	42.6
1.06	3/4			1.09	2.26	3.36	4.37	8.76	13.2	17.5	21.9	26.3	35.1	44.0
1.08	15/16			1.12	2.32	3.45	4.50	9.01	13.5	18.1	22.6	27.1	36.2	45.3
1.10	13 3/16				2.40	3.55	4.62	9.27	13.9	18.6	23.3	27.9	37.3	46.7
1.12	7/16				2.46	3.65	4.75	9.54	14.3	19.1	23.9	28.8	38.4	48.0
1.14	11/16				2.53	3.75	4.88	9.80	14.7	19.7	24.6	29.6	39.5	49.4

TABLE 6.9 (*Concluded*)

Upper Head, H_a		Throat Widths												
Feet	Inches (Approx.)	1 in.	2 in.	3 in.	6 in.	9 in.	1 ft	2 ft	3 ft	4 ft	5 ft	6 ft	8 ft	10 ft
							Flow in Cubic Feet per Second							
1.16	13 15/16	…	…	…	2.60	3.85	5.01	10.1	15.1	20.2	25.3	30.4	40.6	50.8
1.18	14 3/16	…	…	…	2.68	3.95	5.15	10.3	15.6	20.8	26.0	31.3	41.8	52.3
1.20	3/8	…	…	…	2.75	4.06	5.28	10.6	16.0	21.3	26.7	32.1	42.9	53.7
1.22	5/8	…	…	…	2.82	4.16	5.41	10.9	16.4	21.9	27.4	33.0	44.1	55.2
1.24	7/8	…	…	…	2.89	4.27	5.55	11.2	16.8	22.5	28.1	33.8	45.2	56.6
1.26	15 1/8	…	…	…	…	4.37	5.69	11.5	17.2	23.0	28.9	34.7	46.4	58.1
1.28	3/8	…	…	…	…	4.48	5.82	11.7	17.7	23.6	29.6	35.6	47.6	59.6
1.30	5/8	…	…	…	…	4.59	5.96	12.0	18.1	24.2	30.3	36.5	48.8	61.1
1.32	13/16	…	…	…	…	4.69	6.10	12.3	18.5	24.8	31.1	37.4	50.0	62.7
1.34	16 1/16	…	…	…	…	4.80	6.25	12.6	19.0	25.4	31.8	38.3	51.2	64.2
1.36	5/16	…	…	…	…	4.92	6.39	12.9	19.4	26.0	32.6	39.2	52.5	65.7
1.38	9/16	…	…	…	…	5.03	6.53	13.2	19.9	26.6	33.3	40.1	53.7	67.3
1.40	13/16	…	…	…	…	…	6.68	13.5	20.3	27.2	34.1	41.1	55.0	68.9
1.42	17 1/16	…	…	…	…	…	6.82	13.8	20.8	27.8	34.9	42.0	56.2	70.5
1.44	1/4	…	…	…	…	…	6.97	14.1	21.2	28.5	35.7	42.9	57.5	72.1
1.46	1/2	…	…	…	…	…	7.12	14.4	21.7	29.1	36.5	43.9	58.8	73.7
1.48	3/4	…	…	…	…	…	7.26	14.7	22.2	29.7	37.3	44.9	60.1	75.4
1.50	18	…	…	…	…	…	7.41	15.0	22.6	30.3	38.1	45.8	61.4	77.0
1.52	1/4	…	…	…	…	…	7.57	15.3	23.1	31.0	38.9	46.8	62.7	78.7
1.54	1/2	…	…	…	…	…	7.72	15.6	23.6	31.6	39.7	47.8	64.0	80.4
1.56	3/4	…	…	…	…	…	7.87	15.9	24.1	32.3	40.5	48.8	65.4	82.1
1.58	15/16	…	…	…	…	…	8.02	16.3	24.6	32.9	41.4	49.8	66.7	83.8
1.60	19 3/16	…	…	…	…	…	8.18	16.6	25.1	33.6	42.2	50.8	68.1	85.5
1.62	7/16	…	…	…	…	…	8.34	16.9	25.5	34.3	43.0	51.8	69.5	87.2
1.64	11/16	…	…	…	…	…	8.49	17.2	26.0	34.9	43.9	52.8	70.9	89.0
1.66	15/16	…	…	…	…	…	8.65	17.6	26.5	35.6	44.7	53.9	72.3	90.7
1.68	20 3/16	…	…	…	…	…	8.81	17.9	27.0	36.3	45.6	54.9	73.7	92.5
1.70	3/8	…	…	…	…	…	8.97	18.2	27.6	37.0	46.4	56.0	75.1	94.3
1.72	5/8	…	…	…	…	…	9.13	18.5	28.1	37.7	47.3	57.0	76.5	96.1
1.74	7/8	…	…	…	…	…	9.29	18.9	28.6	38.3	48.2	58.1	77.9	97.9
1.76	21 1/8	…	…	…	…	…	9.46	19.2	29.1	39.0	49.1	59.1	79.4	99.7
1.78	3/8	…	…	…	…	…	9.62	19.6	29.6	39.7	49.9	60.2	80.8	101.4
1.80	5/8	…	…	…	…	…	9.79	19.9	30.1	40.5	50.8	61.3	82.3	103.4
1.82	13/16	…	…	…	…	…	9.95	20.2	30.7	41.2	51.7	62.4	83.8	105.3
1.84	22 1/16	…	…	…	…	…	10.1	20.6	31.2	41.9	52.6	63.5	85.3	107.1
1.86	5/16	…	…	…	…	…	10.3	20.9	31.7	42.6	53.6	64.6	86.8	109.0
1.88	9/16	…	…	…	…	…	10.5	21.3	32.3	43.3	54.5	65.7	88.3	110.9
1.90	13/16	…	…	…	…	…	10.6	21.6	32.8	44.1	55.4	66.8	89.8	112.9
1.92	23 1/16	…	…	…	…	…	10.8	22.0	33.3	44.8	56.3	67.9	91.3	114.8
1.94	1/4	…	…	…	…	…	11.0	22.4	33.9	45.5	57.3	69.1	92.8	116.7
1.96	1/2	…	…	…	…	…	11.1	22.7	34.4	46.3	58.2	70.2	94.4	118.7
1.98	3/4	…	…	…	…	…	11.3	23.1	35.0	47.0	59.1	71.4	95.9	120.6
2.00	24	…	…	…	…	…	11.5	23.4	35.5	47.8	60.1	72.5	97.5	122.6
2.05	5/8	…	…	…	…	…	11.9	24.3	36.9	49.7	62.5	75.4	101.4	127.6
2.10	25 3/16	…	…	…	…	…	12.4	25.3	38.4	51.6	64.9	78.4	105.4	132.7
2.15	13/16	…	…	…	…	…	12.8	26.2	39.8	53.5	67.4	81.4	109.5	137.8
2.20	26 3/8	…	…	…	…	…	13.3	27.2	41.3	55.5	69.9	84.4	113.6	143.0
2.25	27	…	…	…	…	…	13.7	28.1	42.7	57.5	72.4	87.5	117.8	148.3
2.30	5/8	…	…	…	…	…	14.2	29.1	44.2	59.6	75.0	90.6	122.0	153.7
2.35	28 3/16	…	…	…	…	…	14.7	30.1	45.7	61.6	77.6	93.8	126.0	159.1
2.40	13/16	…	…	…	…	…	15.2	31.1	47.3	63.7	80.3	97.0	130.7	164.6
2.45	29 3/8	…	…	…	…	…	15.6	32.1	48.8	65.8	82.9	100.2	135.1	170.2
2.50	30	…	…	…	…	…	16.1	33.1	50.4	67.9	85.6	103.5	139.5	175.8

and maximum allowable head is determined. The maximum allowable head will depend on the grade of the channel and the freeboard (distance from normal water surface to top of banks) at the place where the flume is to be installed. Whenever possible, the selection should be made so that free-flow will always result. For economy the smallest flume that will satisfy the conditions should be selected.

For example, suppose that a flume is to be installed in a ditch on a moderate grade and that the stream flow to be measured varies from 1 to 15 cfs. Assume that for the maximum flow the depth of water in the ditch is 2.5 feet and the freeboard is 6 inches, but that the banks could be raised slightly for a sufficient distance upstream from the flume and that the water level could be raised 6 inches with safety. The maximum allowable loss of head is therefore 6 inches with safety. Table 6.9 indicates that flumes with a throat width of 1, 2, or 3 feet could measure the entire range of flow. For a flow of 15 cfs, the head, H_a, would be 2.38 feet for a 1-foot flume, 1.50 for a 2-foot flume, and 1.16 for a 3-foot flume.

For free-flow, submergence should not exceed 60 percent, so that the loss of head should not be less than 40 percent of head, H_a. The required loss for a 1-foot flume would be $0.4 \times 2.38 = 0.95$ ft; for a 2-foot flume $0.4 \times 1.50 = 0.60$ ft; and for a 3-foot flume, $0.4 \times 1.16 = 0.46$ ft. A 3-foot flume is, therefore, the smallest size for which the maximum loss of head will be less than 6 inches.

The required depth upstream for a 3-foot flume is $2.50 + 0.46 = 2.96$ ft; and the head H_a for 15 cfs is 1.16 ft. The crest should be set $2.96 - 1.16 = 1.8$ ft above the bottom of the ditch. If a 2-foot flume

TABLE 6.10

FACTORS M TO BE USED IN CONNECTION WITH FIG. 6.15
FOR DETERMINING SUBMERGED DISCHARGES FOR
PARSHALL MEASURING FLUMES LARGER
THAN 1-FOOT THROAT WIDTH*

Throat Width, W, in Feet	Factor, M	Throat Width, W, in Feet	Factor, M
1	1.0	5	3.7
2	1.8	6	4.3
3	2.4	7	4.9
4	3.1	8	5.4

* These factors are to be multiplied by the correction obtained from Fig. 6.15 and subtracted from the free flow for the same upper head, H_a, Table 6.8, to determine flow for submerged conditions. Computed from the expression $M = W^{0.815}$.

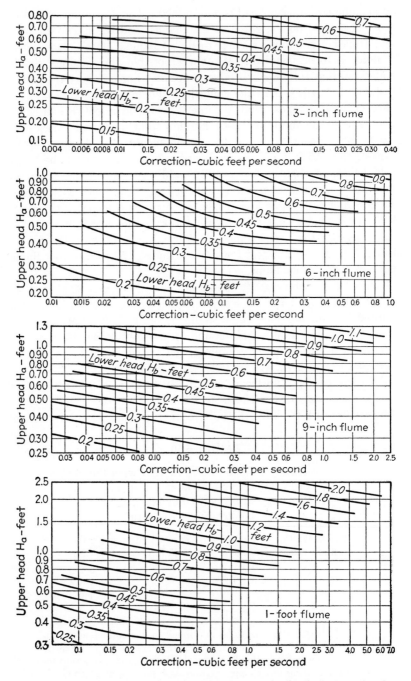

Fig. 6.15 Reduction in flow caused by submergence of a Parshall measuring flume.

is selected, the depth upstream will be $2.50 + 0.60 = 3.10$ ft; and since the head H_a in this case is 1.50, the elevation of the crest should be $3.10 - 1.50 = 1.60$ ft above the bottom of the ditch. In order to use a 2-foot flume, one would have to raise the ditch banks higher than assumed or permit a maximum submergence of about 67 percent, in which case the crest could be set 1.5 feet above the bottom of the ditch. If the available loss of head is sufficient to permit the use of a 1-foot flume, the upstream depth will be 3.45 feet. The crest will then be set $3.45 - 2.38 = 1.07$ ft above the bottom of the ditch. The greater the throat width, the higher the crest must be set to insure free-flow operation.

Parshall flumes may be built of wood, concrete, concrete blocks, concrete and brick, or, in the smaller sizes, of heavy sheet metal. The dimensions of flumes ranging from 1 inch to 10 feet in throat width are given in Tables 6.11 and 6.12.

To secure accuracy in measurement, these flumes must be built to exact dimensions, especially the converging and throat sections. The flow of the upstream converging section, especially the crest, must be level. Where the flume is more than 6 inches above the channel bottom, a short inclined floor should be provided as a reasonably uniform approach.

TABLE 6.11

Standard Dimensions of Parshall Measuring Flumes
from 3 to 9 Inches Throat Width

Dimension Letter*	Dimensions in Feet and Inches for Throat Widths (W) of				
	1 in.	2 in.	3 in.	6 in.	9 in.
A	$1'\,2\frac{9}{32}''$	$1'\,4\frac{5}{16}''$	$1'\,6\frac{3}{8}''$	$2'\,\frac{7}{16}''$	$2'\,10\frac{5}{8}''$
2/3 A	$0'\,9\frac{17}{32}''$	$0'\,10\frac{7}{8}''$	$1'\,\frac{1}{4}''$	$1'\,4\frac{5}{16}''$	$1'\,11\frac{1}{8}''$
B	$1'\,2''$	$1'\,4''$	$1'\,6''$	$2'\,0''$	$2'\,10''$
2/3 B	$0'\,9\frac{5}{16}''$	$0'\,10\frac{5}{8}''$	$1'\,0''$	$1'\,4''$	$1'\,10\frac{5}{8}''$
C	$0'\,3\frac{21}{32}''$	$0'\,5\frac{3}{16}''$	$0'\,7''$	$1'\,3\frac{1}{2}''$	$1'\,3''$
D	$0'\,6\frac{19}{32}''$	$0'\,8\frac{3}{8}''$	$0'\,10\frac{3}{16}''$	$1'\,3\frac{1}{2}''$	$1'\,10\frac{5}{8}''$
E	$0'\,9''$	$1'\,0''$	$1'\,3''$	$1'\,6''$	$2'\,0''$
F	$0'\,3''$	$0'\,4\frac{1}{2}''$	$0'\,6''$	$1'\,0''$	$1'\,0''$
G	$0'\,8''$	$0'\,10''$	$1'\,0''$	$2'\,0''$	$1'\,6''$
K	$0'\,\frac{3}{4}''$	$0'\,\frac{7}{8}''$	$0'\,1''$	$0'\,3''$	$0'\,3'$
N	$0'\,\frac{1}{8}''$	$0'\,1\frac{3}{16}''$	$0'\,2\frac{1}{4}''$	$0'\,4\frac{1}{2}''$	$0'\,4\frac{1}{2}''$
X	$0'\,\frac{5}{16}''$	$0'\,\frac{5}{8}''$	$0'\,1''$	$0'\,2''$	$0'\,2''$
Y	$0'\,\frac{1}{2}''$	$0'\,1''$	$0'\,1\frac{1}{2}''$	$0'\,3''$	$0'\,3''$

* Letters refer to Fig. 6.14(A).

TABLE 6.12

STANDARD DIMENSIONS OF PARSHALL MEASURING FLUMES FROM 1 TO 10 FEET THROAT WIDTH

Throat Width, W, in feet	Dimensions in Feet and Inches*					
	A	2/3 A	B	2/3 B	C	D
1.0	4′ 6″	3′ 0″	4′ 4$\frac{7}{8}$″	2′ 11$\frac{1}{4}$″	2′ 0″	2′ 9$\frac{1}{4}$″
2.0	5′ 0″	3′ 4″	4′ 10$\frac{7}{8}$″	3′ 3$\frac{1}{4}$″	3′ 0″	3′ 11$\frac{1}{2}$″
3.0	5′ 6″	3′ 8″	5′ 4$\frac{3}{4}$″	3′ 7$\frac{1}{8}$″	4′ 0″	5′ 1$\frac{7}{8}$″
4.0	6′ 0″	4′ 0″	5′ 10$\frac{5}{8}$″	3′ 11$\frac{1}{8}$″	5′ 0″	6′ 4$\frac{1}{4}$″
5.0	6′ 6″	4′ 4″	6′ 4$\frac{1}{2}$″	4′ 3″	6′ 0″	7′ 6$\frac{5}{8}$″
6.0	7′ 0″	4′ 8″	6′ 10$\frac{3}{8}$″	4′ 6$\frac{7}{8}$″	7′ 0″	8′ 9″
7.0	7′ 6″	5′ 0″	7′ 4$\frac{1}{4}$″	4′ 10$\frac{7}{8}$″	8′ 0″	9′ 11$\frac{3}{8}$″
8.0	8′ 0″	5′ 4″	7′ 10$\frac{1}{8}$″	5′ 2$\frac{3}{4}$″	9′ 0″	11′ 1$\frac{3}{4}$″
10.0	9′ 0″	6′ 0″	8′ 9$\frac{7}{8}$″	5′ 10$\frac{5}{8}$″	11′ 0″	13′ 6$\frac{3}{8}$″

* Letters refer to Fig. 6.14(A), in which other dimensions for these flumes are shown.

The diverging section can be omitted without affecting the discharge. The function of the diverging section is to reduce the head loss through the flume and to minimize erosion downstream.

6.5 TRAPEZOIDAL FLUMES

Trapezoidal flumes somewhat similar to Parshall's rectangular flumes have been developed, Fig. 6.16. The head-discharge relationships are shown in Fig. 6.17. Based upon the study of Robinson and Chamberlain (1958), the characteristics are listed as follows: (1) extreme approach conditions seem to have a minor effect upon head-discharge relationships; (2) material deposited in the approach did not change the head-discharge relationships noticeably; (3) a large range of flows can be measured through the structure with a comparatively small change in head; (4) the flumes will operate under greater submergence than rectangular-shaped ones without corrections being necessary to determine the exact discharge; (5) the trapezoidal shape fits the common canal section more closely than a rectangular one; (6) construction details such as transitions and form work are simplified; (7) the relationship between head and discharge is not as easily expressed in the form of an equation as is the rectangular-shaped flume; (8) since a small change in head results in a comparatively large change in discharge, the sensitivity of the flume to changes in discharge is less than that for the rectangular shaped ones.

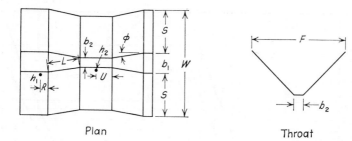

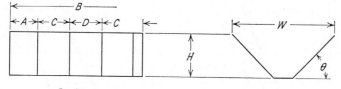

Flume no.	Description	b_1	b_2	A	L	C	D	E	F	H	B	R	S	U	W	θ	φ
1	Large 60°-v	2	0	7	7	$6\frac{15}{16}$	7	3	8	$6\frac{3}{4}$	$30\frac{7}{8}$	$1\frac{1}{2}$	4	$3\frac{1}{2}$	10	60°	8.25°
2	Small 60°-v	2	0	5	$4\frac{1}{8}$	$4\frac{3}{64}$	5	2	$4\frac{3}{4}$	4	$20\frac{3}{32}$	1	$2\frac{3}{8}$	$2\frac{1}{2}$	$6\frac{3}{4}$	60°	11.20°
3	2"-60° wsc	$4\frac{7}{8}$	2	8	$8\frac{1}{2}$	$8\frac{13}{32}$	$8\frac{1}{2}$	3	14	$13\frac{1}{2}$	$36\frac{5}{16}$	$1\frac{1}{2}$	6	$4\frac{1}{4}$	$16\frac{7}{8}$	60°	9.22°
4	2"-45° wsc	$4\frac{7}{8}$	2	8	$8\frac{1}{2}$	$8\frac{3}{8}$	$8\frac{1}{2}$	3	$22\frac{3}{16}$	$10\frac{19}{32}$	$36\frac{1}{4}$	$1\frac{1}{2}$	$10\frac{19}{32}$	$4\frac{1}{4}$	$26\frac{1}{16}$	45°	9.91°
5	2"-30° wsc	$4\frac{7}{8}$	2	8	$8\frac{1}{2}$	$8\frac{3}{8}$	$8\frac{1}{2}$	3	$36\frac{41}{64}$	10	$36\frac{1}{4}$	$1\frac{1}{2}$	$17\frac{5}{16}$	$4\frac{1}{4}$	$39\frac{33}{64}$	30°	9.80°
6	4"-60° wsc	8	4	9	10	$9\frac{13}{16}$	10	3	20	$13\frac{7}{8}$	$41\frac{5}{8}$	$1\frac{1}{2}$	8	5	24	60°	11.20°
7	2"-30° csu	10	2	10	$10\frac{3}{4}$	10	10	3	$35\frac{1}{2}$	$9\frac{21}{32}$	43	$1\frac{1}{2}$	$16\frac{3}{4}$	5	$43\frac{1}{2}$	30°	21.80°

Note: Dimensions in inches except as shown. See Fig. 6-16 for the relationship of Q to depth of each of the seven flumes.

Fig. 6.16 Details of designs of trapezoidal flumes. (After Robinson and Chamberlain.)

6.6 THE CURRENT METER

A device widely used by engineers for measuring flowing water is the current meter. Meters of two general types are used: the cup type with vertical axis, shown in Fig. 6.18, and the propeller type with horizontal axis, shown in Fig. 6.19. Cup meters are generally used by the U.S. Bureau of Reclamation and the U.S. Geological Survey.

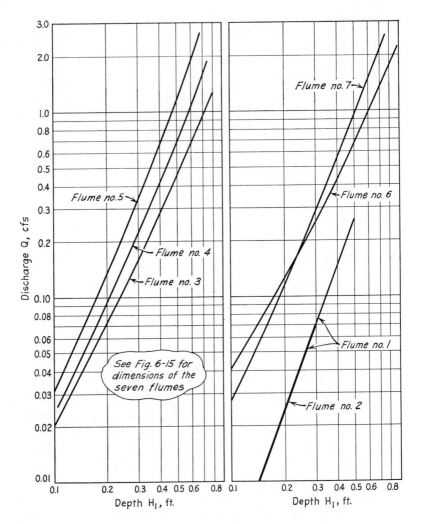

Fig. 6.17 Calibration curves for trapezoidal flumes under free-flow conditions.

Propeller meters are favored in European countries. Both have their advantages. The cup meter is more sensitive to disturbances such as being brought close to the bank and not being held stationary. The cup meter also offers greater resistance to the flow of water and is thus swept farther downstream than is the propeller meter. One distinct advantage of the cup meter, and one of the principal reasons why it is used extensively in the United States, is that it is a more rugged

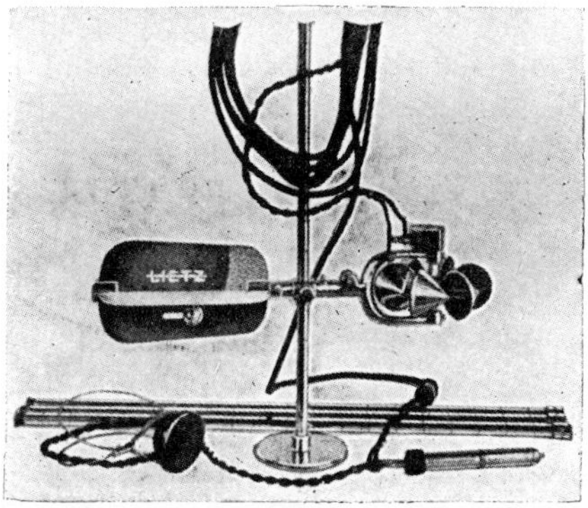

Fig. 6.18 Current meter showing rod suspension with double-end hanger and round wading base. (The A. Leitz Company.)

instrument than the propeller meter and thus can be used by relatively unskilled technicians. The propeller meter has been used for higher ranges of velocities than the cup meter (up to 20 to 30 ft per sec as against 10 to 15 ft per sec for the cup meter). The propeller meter can also be made much smaller, down to 2½ inches, whereas the minimum size for the cup meter is 5 inches. The small size is sometimes an advantage when measurements must be taken near the walls of pipes and the edges of channels where there is rapid change of velocity. Another major advantage of the screw-type propeller meter is that it is less likely to be fouled by floating weeds and debris. Hence, the choice of meter depends entirely upon conditions under which the meter will be used. However, for field observations the cup-type meter has generally been found superior; whereas for pipe measurements and laboratory observations the propeller meter has been found to be more useful.

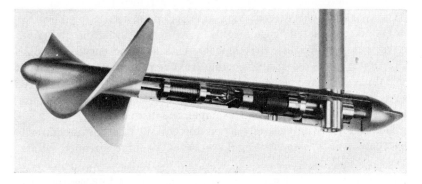

Fig. 6.19 Propeller type current meter. (Courtesy A. Ott.)

The accuracy of current meter measurements will depend largely upon the skill of the operator and conditions at the location of the measurement. Under ideal conditions the error may be less than 2 percent and is usually not more than 5 percent.

The meter is calibrated by passing it through still water at a known speed and noting the number of revolutions per second. When the calibrated meter is held still in running water at the proper depth, it is thus possible to determine the average velocity of the water by observing the number of revolutions per second. The average velocity in streams not over 1.5 feet in depth is at about 0.6 of the depth from the surface;[2] in streams over 1.5 feet in depth the average velocity is represented by the average of the velocities at 0.2 and 0.8 of the depth. In the measurement of flowing water it is essential that the current meter be placed at the point or points of average velocity.

Another method of determining the average velocity in a stream is the integration method, in which the current meter is raised and lowered slowly and at a constant rate from the bottom to the top of the stream. On practically all the larger canals and on rivers discharge measurements are computed from current-meter readings of velocity and measured cross-sectional areas. The cross section of a stream should be divided into twenty or more parts, except for a very small stream. Not more than 10 percent, and preferably not more than 5 percent, of the discharge should occur in any one section. The velocity is then measured in each section and multiplied by the area to obtain the discharge of the section. The flow in the channel is obtained by adding the flow in each section.

By measuring the discharge of a canal or river at several different stages (or depths) the engineer obtains data from which he determines a relationship between the depth of the water and the discharge of the stream. The changes in depth are usually referred to a permanent bench mark, or elevation datum, and distances vertically above the datum are designated "gage heights." After measuring the discharges at various gage heights, the engineer plots a rating curve, Fig. 6.20. This figure shows discharges ranging from zero cfs at zero gage height to 100 cfs at a gage height of 2.60 feet. At any gage heights between these limits the reader can determine the discharge from the figure. At a gage height of 1 foot, for example, the discharge is 13 cfs.

The major advantages of current meters are that they require no obstruction of stream flow and are suited to large streams. Water

---

[2] Some authorities have found that velocities measured at 0.6 of the depth from the surface in shallow streams usually range from 4 to 6 percent higher than the true average velocities.

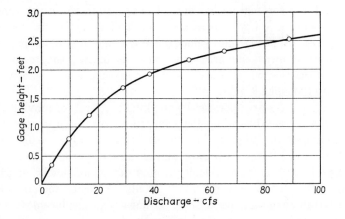

Fig. 6.20 Typical rating curve.

commissioners whose responsibility it is to distribute the public waters to those entitled to their use depend very largely on rating curves for their measurements. The gage height may be read by non-technical men, but the actual use of the meter and the making of rating tables and rating curves are tasks for a trained and experienced hydrographer or for an engineer.

6.7 MECHANICAL MEASURING AND RECORDING DEVICES

A number of mechanical devices for water measurement have been designed, most of which measure the rate of flow and also automatically register the total amount of water passing in any given period of time. Canal companies that base water charges on the number of acre-feet delivered to individual irrigators find self-recording devices serviceable and convenient. It is likely that more extended use of mechanical-automatic registering devices will be made as water increases in value and more irrigation companies, in order to stimulate economy among their irrigators, find it necessary to base water charges on actual amounts of water delivered.

Meters, although desirable under certain conditions, involve numerous practical difficulties. Therefore, one should understand the limitations of any meter before purchasing it for a particular purpose. The most common mechanical meters operate by the flow of water and may be classified as displacement, velocity, or by-pass type.

The displacement-type meter measures volumetrically. Water passing through the meter displaces a vane or disk, which in turn operates the recording mechanism. The operation is positive, and the measurement is fairly accurate. Such meters generally require a greater loss of head than velocity meters and are more expensive. Their use for measuring irrigation water is limited to pressure pipe lines in localities where water is rather expensive and stream flows less than 100 gpm.

Velocity meters are operated by the kinetic energy of moving water. They usually contain an impeller vane, turned by the water; and their operation is similar to that of a current meter. For high velocities, the impeller vane rotates at a rate almost directly proportional to the velocity of water; for low velocity it may turn more slowly, and below certain velocities it may not move at all. Velocity meters are less expensive and can accommodate larger flows than displacement meters.

The by-pass meter operates differently from either the velocity or displacement meters. Only a small part of the flow passes through the registering element, but the meter is calibrated to register the volume that passes. The percentage of water passing through it is not constant but usually varies slightly with the flow.

Mechanical meters have the advantage of eliminating computations in volume determination. This convenience often justifies the expense involved. Meters which are subject to clogging should never be used on pipe lines receiving water from open ditches unless the entrances are adequately screened to keep out debris.

6.8 VENTURI TUBES AND SIMILAR DEVICES

The venturi tube is a convergent-divergent tube which has been used extensively for measuring flow in pipe lines of diameters ranging from a few inches to several feet, but it has been used only to a limited extent for irrigation water. The standard venturi tube has been considered too expensive for irrigation purposes.

Modified venturi sections have been developed and used to measure irrigation water in open channels. They are used primarily where insufficient slope exists to permit accurate measurement by Parshall flume or weir. Losses of energy through the measuring section are minimized by using a moveable diverging section over the top of the channel to recover most of the kinetic energy of the flow after it passes through the restricted section. The recovery of the kinetic energy is accomplished by increasing the cross-sectional area and thereby reducing the velocity. When this can be accomplished without excessive turbulence, most of the kinetic energy is transformed into potential energy.

The sides and bottom of the channel are straight while the top is a plate with a typical venturi shape. This top confining plate is moved vertically to accommodate the flow and thereby give a variable head loss and a variable accuracy of flow, greater head loss insuring greater accuracy. When greater accuracy is desired, the throat section is made smaller with a resulting increase in velocity and head loss.

These modified venturi tubes are calibrated individually, and discharge charts and tables are provided for each tube. They have been in use for several years and have proved satisfactory. With them less head is lost than with most other practical devices. Their principal disadvantages are: the lack of standardized sizes and shapes, the lack of information regarding the coefficients of discharge, and that they are relatively expensive except when made in large quantities.

Flow nozzles, the thin plate orifice in pipes and the thin plate orifice on the end of a pipe, are being used for measuring irrigation water. These methods are particularly useful for testing pumps and measuring flow from wells.

6.9 COLLINS FLOW GAGE

The Collins flow gage is used to measure the flow of water in pipe lines, especially from pumping plants. This device consists essentially of two parts: an impact tube (a special form of Pitot tube) and a water-air manometer. The impact tube, a straight small-diameter brass tube inserted through a pipe, is divided by a partition at the center into two compartments, each containing a small orifice—an impact orifice on the upstream side and a trailing orifice on the downstream side. The differential head is twice the velocity head. Hose connections are made from the ends of the tube to the manometer, the scale of which is so graduated that it indicates the velocity in the pipe directly in feet per second. Diagrams are furnished for converting to any desired unit of measurement for various pipe diameters.

The velocity of flow through pipes is sometimes determined by measuring the time required for a dye or salt to pass through a measured length of pipe.

6.10 THE DIVISION OF IRRIGATION WATER

Irrigation companies in the western states divide their streams according to the number of shares of stock owned by individuals or groups of individuals. On small streams a single company owns the entire flow, or it is divided among two or three companies, each company owning a share of the total stream. Some users are less interested in the measurement of water than in the division of the stream. For example, one company may be entitled to five-twelfths of the stream and another company to seven-twelfths. Many times a division must be made where it is impracticable to make a measurement.

For satisfactory division, a few principles must be observed. The water must approach the divider in parallel paths; i.e., there must be no cross currents. To secure this condition the divider box must be placed at the lower end of a long flume or straight open channel. The floor of the channel immediately above the divider should be level transversely. If the water is reasonably free from silt, it is desirable to have it approach the divider at a low velocity. For streams carrying considerable silt and gravel there should be no obstruction in the channel in the form of a bulkhead, and the velocity with which the water enters should be maintained through the structure. It is very important that divider structures have a long, straight channel of approach. Any gravel or debris allowed to collect in the channel of approach will cause cross currents and interfere with proper division.

The flow over a weir can be easily divided by placing a sharp-edged partition below the weir to divide the stream as it falls over the crest. The crest of this partition should be placed a sufficient distance below the weir crest to permit a free circulation of air between the divider and the sheet of water falling over the weir.

The discharge over a weir is not exactly proportional to the length of the crest; the error in considering it so is slight for low heads. Trapezoidal weirs are the most desirable type for dividers. The flow over this weir is very nearly proportional to length of crest. If it is desired to divide the stream into two parts, one taking five-sixths and the other one-sixth of the flow, the divider should be placed one-sixth of the distance from the end of the weir. Figure 6.21 shows a trapezoidal weir divider fixed to divide a stream into three parts.

If it is desired to divide a stream into two equal parts, the rectangular weir, either with or without end contractions, is entirely satisfactory. In localities where water is distributed to several parts of a farm in underground pipes, it is essential to provide special boxes for making an appropriate division of the water. A typical concrete proportional division box connected to a pipe line is illustrated in Fig. 6.22, and a concrete division box with flashboards is shown in Fig. 6.23.

6.11   FLOAT MEASUREMENTS

Quick, inexpensive measurements of discharge are frequently made by timing the velocity of an object on the surface of the water. The observed surface velocity must be reduced, since it is greater than the average velocity. The correction factor is influenced by roughness, shape of channel, and depth of flow. The observed surface

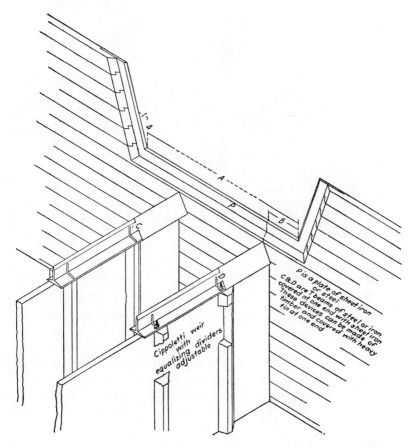

Fig. 6.21 Adjustable divider. (*Utah Agr. Exp. Sta. Circ.* 6.)

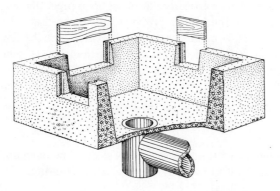

Fig. 6.22 Proportional division box. (*U.S.D.A. Farmers' Bul.* 348.)

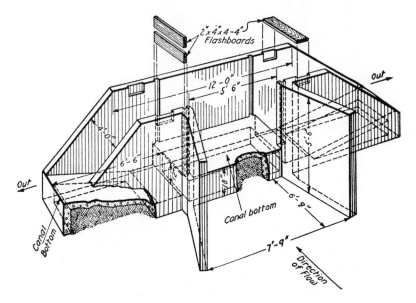

Fig. 6.23   Concrete division box.   (*U.S.D.A. Farmers' Bul.* 1243.)

velocity should be multiplied by about 0.85 to obtain the mean velocity. The correction factor may be as low as 0.8 and as high as 0.95. It increases as depth increases and diminishes as roughness of channel bed increases. Float measurements can be expected to involve errors of at least 10 percent.

6.12   FLOW FROM A VERTICAL PIPE

When water flows vertically out of an open pipe, the height to which it will rise above the pipe is proportional to the flow. Lawrence and Braunworth (1906) made careful measurements of the flow and found that when the height of the jet was less than $0.37D$, where D is the inside diameter of the pipe, the flow is similar to the flow over a weir and can be expressed by the following empirical equation, when the lower nappe springs free from the pipe:

$$Q = 8.8D^{2.5}H^{3.5} \qquad (6.15)$$

When the height of the jet exceeds $1.4D$ the flow is comparable to jet flow and can be expressed by the empirical equation:

$$Q = 5.57D^{1.99}H^{0.53} \qquad (6.16)$$

In both equations Q is the flow in cfs, D is the inside diameter, and H is the average height of rise of the center of the jet, with both D and H measured in feet. For heights between $0.37D$ and $1.4D$, the flow is somewhat less than the flow given by either equation. Using these equations, accuracies of 10 percent can be expected; however, the accuracy increases as H increases.

6.13   FLOW FROM A HORIZONTAL PIPE

The trajectory of a stream of water from a horizontal pipe can be used to estimate the discharge. Such a procedure is rapid, inexpensive, and convenient. Measurements are made of the X and Y coordinates, where X is measured parallel to the pipe and Y is measured vertically. The formula is obtained by combining the following three equations:

$$Y = \tfrac{1}{2}gt^2 \qquad X = Vt \qquad Q = AV$$

$$Q = C'\sqrt{\frac{g}{2}}\,\frac{AX}{\sqrt{Y}} = CA\,\frac{X}{\sqrt{Y}} \tag{6.17}$$

where Q = discharge in cfs
  C = coefficient of discharge
  g = acceleration of gravity, 32.2 ft per sec^2
  A = cross-sectional area of water at end of pipe in sq ft
  X = coordinate of the point on the surface measured in ft parallel to the pipe
  Y = vertical coordinate measured in ft

When A, X, and Y are measured in inches, Q in gpm is given by:

$$Q = 3.6C\,\frac{AX}{\sqrt{Y}} \tag{6.18}$$

Note that the coordinates are measured from the surface of the water as it emerges from the pipe. The coefficient of discharge C varies greatly depending primarily upon the depth of flow in the pipe and the coordinates X and Y. Variation for a partially full pipe is shown in Table 6.13, where the depth of flow at the end of the pipe is designated by d and the inside diameter of pipe by D. The table indicates that the coefficient of discharge has the value of 1.33 for $d/D = 0.8$ and $X/D = 1.00$. Hence, this coefficient is extremely important, if accuracy is to be obtained.

The error involved with the coordinate method will seldom be less than 5 percent. Errors of 10 percent can be expected, even when the coefficient is used, if the jet is turbulent and the free surface poorly defined. Wherever practical, X/D should not be less than 3.

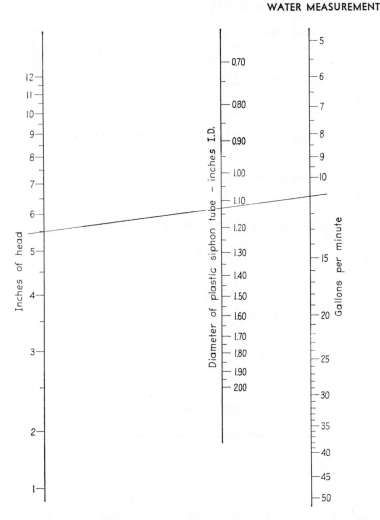

Fig. 6.24 Discharge vs. head for plastic siphon tubes.

6.14 DISCHARGE FROM SIPHON TUBES

When water is diverted from a ditch using a siphon tube, the head and diameter can be used to estimate the discharge, Fig. 6.24, with an accuracy of 5 to 10 percent. In essence, the orifice-discharge equation applies to this type of flow with the discharge coefficient depending upon the length of siphon and entrance and exit conditions of flow. The effective head causing flow is the difference between the water level in the ditch and the level of water in the furrow for a submerged

TABLE 6.13

APPROXIMATE COEFFICIENTS OF DISCHARGE FOR USE IN THE
COORDINATE METHOD FOR MEASURING THE RATE OF
WATER FLOW FROM HORIZONTAL PIPES

A. Flowing partially full at the end

d/D	X/D							
	1.00	1.50	2.00	2.50	3.00	4.00	5.00	8.00
0.2	1.02	1.01	1.00					
0.3	1.11	1.06	1.03	1.02	1.01	1.00		
0.4	1.17	1.10	1.06	1.03	1.02	1.01	1.00	
0.5	1.22	1.13	1.07	1.04	1.03	1.01	1.01	1.00
0.6	1.26	1.15	1.09	1.05	1.03	1.02	1.01	1.00
0.7	1.30	1.17	1.10	1.06	1.04	1.02	1.01	1.00
0.8	1.33	1.18	1.11	1.07	1.04	1.02	1.01	1.00

B. Flowing completely full at the end

Y/D	X/D							
	1.00	1.50	2.00	2.50	3.00	4.00	5.00	8.00
0.5	1.44	1.28	1.18	1.13	1.10	1.06	1.03	1.00
1.0	1.37	1.24	1.17	1.12	1.09	1.06	1.03	1.00
2.0		1.11	1.09	1.08	1.07	1.05	1.03	1.00
3.0			1.04	1.04	1.04	1.04	1.03	1.00
4.0			1.01	1.01	1.02	1.03	1.02	1.00
5.0			0.97	0.99	1.00	1.01	1.01	1.00

outlet. When the outlet is not submerged, the effective head is the height of the water in the ditch above the center of the outlet end.

REFERENCES

Addison, Herbert, "Hydraulic Measurements," John Wiley and Sons, New York, second edition, 1946.

Benson, J. R., "Canal Seepage Control at Low Cost," *Western Construction News*, Vol. 25, No. 9, p. 72, 1949.

Brown, Floyd E., "Measuring Irrigation Water," *Colorado State University Ext. Ser. Bul.* 448-A, November 1957.

Byrne, W. S., "Canal Lined with Stabilized Earth," *Eng. News-Record*, Vol. 136, pp. 410–412, March 1946.

Christiansen, J. E., "Measuring Water for Irrigation," *California Agr. Exp. Sta. Bul.* 588 (reprinted January 1947).

Code, W. E., "Linings for Small Irrigation Ditches," *Colo. Farm Bul.*, Vol. 8, pp. 2, 14–15, March-April 1946.

Corbett, Don M., et al., "Stream Gaging Procedure, A Manual Describing Methods and Practices of the Geological Survey," *U.S. Geol. Survey Water Supply Paper* 888, 1943.

Criddle, Wayne D. and Eldon M. Stock, "Water Measurement," *Utah State Eng. Exp. Sta. Bul.* 2, June 1941.

Dai Ly, "Coefficient of Discharge for a Silted Weir," Master's Thesis, Utah State University, Logan, Utah, 1962.

Debler, E. B., "Conveyance Losses in Canals," *Civil Eng.*, Vol. 11, pp. 584–585, October 1941.

Elevatorski, E. A., "Discharge Characteristics of Irrigation Delivery Gates," *University of Arizona Eng. Exp. Sta. No.* 1, March 1958.

Ferguson, J. E. and J. E. Garton, "A Modified Venturi Section for Measuring Irrigation Water in Open Channels," *Agr. Eng.*, Vol. 30, No. 12, pp. 548–585, December 1949.

Golze, Alfred R., "Lower-Cost Canal Lining Program," *Reclamation Era*, Vol. 32, pp. 165–167, August 1946.

Grover, Nathan Clifford and A. W. Harrington, "Stream Flow Measurements and their Uses," John Wiley and Sons, New York, 1943.

Hale, C. S., "Coachella Canal, Excavating and Lining 141-Mile Canal," *Western Construction News*, Vol. 21, No. 12, pp. 73–77, 1946.

Hamman, A. J., "Measuring Irrigation Water on the Farm," *Colo. Agr. and Mech. College Agr. Ext. Ser. and Agr. Exp. Sta. Circ.* 172-A, May 1952.

Johnston, C. N., "A New Portable Field Water Meter and New Furrow Water Meter," *Agr. Eng.*, Vol. 27, p. 29, January 1946.

King, H. W., "Handbook of Hydraulics," third edition, McGraw-Hill Book Co., New York, 1939.

Lauritzen, C. W. and O. W. Israelsen, "West's Canal Linings Studied," *Western Construction News*, Vol. 22, No. 5, pp. 85–87, 1947.

Lauritzen, C. W. and O. W. Israelsen, "Canal Linings Tested in Field," *Soil Cons.*, Vol. XV, No. 4, p. 80, 1949.

Lawrence, F. E. and Braunworth, P. L., "Fountain Flow of Water in Vertical Pipes," *Trans. Am. Soc. Civil Eng.*, 1906.

Lewis, M. R., "Design of Small Irrigation Pipe Lines," *Oregon Agr. Exp. Sta. Circ.* 142, June 1941.

Parshall, R. L., "Measuring Water in Irrigation Channels with Parshall Flumes and Small Weirs," *U.S.D.A. Soil Conservation Service Circ.* 843, May 1950.

Robinson, A. R. and A. R. Chamberlain, "Trapezoidal Flumes for Open Channel Flow Measurement," *Trans. Am. Soc. Agr. Eng.*, Gen. Ed., Vol. 3, No. 2, pp. 120–124, 1960.

Rouse, Hunter, "Elementary Mechanics of Fluids," John Wiley and Sons, New York, 1945.

Scott, Verne H. and Clyde E. Houston, "Measuring Irrigation Water," *California Agr. Exp. Sta. Ext. Ser. Bul.* 5, June 1955.

Stock, Eldon M., "Measurement of Irrigation Water," *Utah State Eng. Exp. Sta. and Utah Cooperative Ext. Ser. Bul.* 5, June 1955.

Thomas, Charles W., "Common Errors in Measurement of Irrigation Water," *Proc. Am. Soc. Civil Eng.*, Paper 1362, 24 pp., Sept. 1957.

U.S. Dept. of Interior Bur. of Reclamation, "Water Measurement Manual," Denver, May 1953.

Vennard, John K., "Elementary Fluid Mechanics," John Wiley and Sons, New York, 1940.

CHAPTER 7 BASIC SOIL—WATER RELATIONS

Knowledge of soil and water relationships is valuable to all who have the opportunity to improve irrigation practices, including the irrigators who desire to obtain the best use of water available for their farms.

Chapters 7, 8, and 9 are devoted to a consideration of the relationships of soils and water, with special reference to their influence on irrigation and drainage practices. Soil and water relationships of special importance in irrigated regions include the capacities of well-drained soils in the field to retain water available for plants, and the flow, or movement, of water in soils. The salinity and alkali conditions, together with translocation and concentration of soluble salts due to the movement and evaporation of soil water, also constitute important soil and water relationships.

The functions of soil moisture in plant growth are very important. Excessive volumes of water in soils retard or inhibit plant growth and make drainage essential. Sterility of arid-region soils is usually caused by deficient amounts of water. Irrigation is an artificial means of preventing deficiencies in the moisture content of soils. Intelligent irrigation practices, based on knowledge of soil and moisture, is also a means of preventing, or at least retarding, the occurrence of water-logging of soils. Students of irrigation are interested in the basic physical laws that influence the distribution, storage, flow, and consumptive use of water in unsaturated soils.

Intermolecular forces and tensions in unsaturated soils are influenced by soil texture and structure. These forces give rise to "capillary phenomena" in unsaturated soils which are of special interest in the study of soil and water relations. The close contact of soil particles and particles having thin water films results in attractive forces of large magnitude.

7.1 SOIL TEXTURE

The sizes of particles making up a soil determine its texture. These particles range in size from fine gravel to clay. Particles larger than 1.00 mm in diameter are gravel, particles from 0.05 to 1.00 mm are sand, particles from 0.002 to 0.05 mm are silt, and smaller than 0.002 mm are clay. Most soils contain a mixture of sand, silt, and clay. If sand particles dominate, the soil is called sand. If clay particles dominate, it is called clay. Silts fall between clays and sands. Loams are medium-textured soils having about equal amounts of clay, silt, and sand particles.

Sand grains feel gritty to the fingers and can be distinguished without difficulty by the unaided eye. Silt, barely visible to the naked eye, has the appearance and feel of flour. The individual particles of the clay fraction, many of which are inorganic colloids, are not distinguishable by the eye, and a large portion of them are too small to be seen under a microscope. It is this fraction that makes soils swell and become sticky when wet, and shrink and become brittle on drying.

Where the main soil-forming process is due to mechanical disintegration, relatively coarse-grained particles are found. The size of particles is designated by the word texture. Sandy soils are classed as coarse-textured, loam soils as medium-textured, and clay soils as fine-textured. The texture of a soil has a very important influence on the flow of soil water, the circulation of air, and the rate of chemical transformations which are of importance to plant life. The farmer is unable to modify the texture of his soil by any practical means. The size of soil particles has a great influence on crop production the world over, but to the irrigation farmer it is particularly important, because it determines in a large measure the depth of water he can store in a given depth of soil.

7.2 SOIL STRUCTURE

Soils in which the particles are relatively uniform in size have comparatively large spaces between the particles, whereas those soils in which the size of the grains varies greatly may become more closely packed, and thus the volume of spaces between the grains is restricted. Fine-grained irrigated soils, if properly managed, function as granules, each consisting of many soil particles, whereas in coarse-textured soils each particle functions separately. The existence of granules assures a desirable soil structure. Excessive irrigation, plowing, or otherwise working fine-textured soils, when either too wet or too dry, tends to break down these granules. A soil worked too wet is said to be puddled and has a poor structure. Favorable structure in fine-textured soils is essential to the satisfactory movement of water and air.

Favorable soil structure is recognized by soil scientists as the key to soil fertility. Adequate amounts of chemical nutrients in soils, though essential to crop production, do not assure satisfactory plant growth and crop yields. The permeability of soils to water, air, and roots, provided by favorable soil structure, is equally important to crop growth as are adequate supplies of nutrients.

Soil structure is fundamentally built by wetting and drying, freezing and thawing, or combinations of these conditions. Penetrating roots build soil structure by removing water from the soil, which causes drying and permits subsequent wetting. The primary functions of organic matter and of humus in soil is to add stability to soil aggregates, serving as a cushion against the shock of tillage.

Some of the more important practices to be followed to maintain and improve the structure of irrigated soils are as follows: (1) plow below compacted layers, not at the same depth each year; (2) allow soil to air as long as practical after plowing before giving pre-planting irrigation or before preparing seed bed; (3) return all possible organic matter to the soil; (4) follow a good crop rotation of legumes, cash crops, and fibrous-rooted crops; (5) reduce cultivation and tillage operations to a minimum.

Remember that modern implements such as rubber-tired wheels and disks are used to compact soils for roads, airports, and race tracks. Likewise, compaction will occur on farm soils when these implements are used. Hence, they should be used to a minimum, consistent with the needs for seed-bed preparation and weed control.

7.3   REAL SPECIFIC GRAVITY

The real specific gravity of a soil is a dimensionless quantity and is defined as the ratio of the weight of a single soil particle to the weight of a volume of water equal to the volume of the particle of soil. The specific gravity of the common soil-forming minerals varies from 2.5 to more than 5. A few minerals, such as quartz and feldspar, usually make up the bulk of a soil. The real specific gravity of soils which have a low percentage of organic matter varies but little, and approaches closely an average of 2.65. Some irrigated soils, which are formed largely of organic matter, have a real specific gravity of 1.5 to 2.0, depending on the amount of mineral matter present.

7.4 APPARENT SPECIFIC GRAVITY

The apparent specific gravity of a soil is defined as the ratio of the weight of a given volume of dry soil, air space included, to the weight of an equal volume of water. This ratio is known also as the "volume weight" or "bulk density." Whereas apparent specific gravity is a dimensionless quantity, being weight of soil per weight of water, bulk density is grams per cubic centimeter or mass per unit volume. Therefore, the dimensions are not equal. However, since 1 gram of water fills a volume of 1 cubic centimeter at normal temperatures, the two terms do have equal numerical values.

The apparent specific gravity is influenced by structure; i.e., the arrangement of soil particles, and by its texture and compactness. Apparent specific gravity is a soil property of great importance to the irrigation farmer, as will be shown more fully in connection with the capacity of soils to retain irrigation water.

Compacting a soil of fixed real specific gravity will increase the apparent specific gravity, because it reduces the space between soil particles and thereby reduces the volume of pore space. When working with irrigated soils, it is necessary to know their apparent specific gravity in order to account for the water applied in irrigation, since it is impractical to measure, by direct means, the volume of water which exists in the form of soil moisture in a given volume of soil. It is necessary to measure the weight of water in a given weight of soil by observing the loss of weight in drying, and then convert the weight percentage so obtained to a volume percentage by use of the apparent specific gravity; thus, the volume of water in a given volume of soil may be determined.

The usual method for determining apparent specific gravity is to obtain a soil sample of known volume. This is usually done by driving a tube with a cutting edge into the soil and obtaining an uncompacted core within the tube. Sometimes pits are dug and blocks of soil are obtained directly. Also, a hole can be bored with an auger and all the soil that has been removed can be dried and weighed. The volume is determined by measuring the average size of the hole or by placing a flexible rubber or plastic tube in the hole and determining the volume of water required to fill it.

Another method of measuring apparent specific gravity is by the gamma-ray absorption technique. Absorption of gamma rays has been found to be independent of the chemical composition of soils for the energy used and to depend almost entirely upon the density of soil. Cobalt 60, a relatively inexpensive and stable radio isotope, is an ideal source of gamma radiation. A portable survey meter is satisfactory as a counting device, and small, stable detectors are readily available. These have been combined into commercially available units. The source and detector are lowered into a prepared hole to the depth to be tested; the counts per minute are calibrated to yield apparent specific gravity readings.

7.5 PORE SPACE

The volume of a 1-inch diameter marble is 0.524 cu in. If the marble is placed in a cubical box having a capacity of 1 cu in., there will remain an air space of 0.476 cu in., or 47.6 percent of the total volume. The same air space will be left by any number of marbles of any diameter if arranged in vertical columns. If, however, the marbles are arranged in oblique order, there will remain only 25.9 percent air space. These facts show that considerable variation of pore space between spherical particles may result from change in arrangement, but they do not show the maximum range or variation that may occur in soil. There is great variation in the size and shape of soil particles, and these differences in particles influence the closeness of contact and the interpacking of small particles between large ones, thus determining the total percentage of pore space, which for convenience is represented by the symbol n. In general, coarse-textured, gravelly, and sandy soils have a smaller percentage of total pore space, and fine-textured clay loams and clays have a greater percentage. It is not unusual in irrigated soils for the pore space to vary from 35 percent to 55 percent. To compute the percentage pore space n, it is necessary to know the real and the apparent specific gravity of the soil. The ratio of the apparent to the real specific gravity gives the proportionate space occupied by the soil, and this ratio subtracted from unity gives the pore space. The pore space on a percentage basis is given by the equation:

$$n = 100 \left(1 - \frac{A_s}{R_s}\right) \tag{7.1}$$

where n = the percentage pore space
 A_s = the apparent specific gravity
 R_s = the real specific gravity, approximately 2.65 for most agricultural soils

The term porosity, equivalent to pore space, is much used in soil mechanics. It is defined as the ratio of the volume of voids (air- and water-filled space) to the total volume of soil plus water and air.

Pore space has a direct bearing upon productive value of soils because of its influence upon water-holding capacity and upon the movement of air, water, and roots through the soil. When the pore space of a productive soil is reduced 10 percent, movement of air, water, and roots is greatly restricted and growth is very seriously impeded.

7.6 INFILTRATION

A property of soils, of great importance to irrigators, is the time rate at which water will percolate into soil, or rate of infiltration. Usually, the infiltration rate is much higher at the beginning of a rain or irrigation than it is several hours later. It is influenced by soil properties and also by moisture gradient. Moisture tension, explained in the following chapters, may be zero in a surface inch of soil shortly after wetting and may be very high a few inches below, thus causing a large downward force (in addition to gravity) pulling the water into the unsaturated soil. Several hours after wetting, these differences in tension may be very small, and gravity then becomes the dominant force causing infiltration. The decrease of infiltration with time after wetting a soil is of importance in rainfall-runoff studies and in irrigation. Water standing on gravelly or coarse sandy soils percolates into the soil so rapidly that the water surface may be lowered several inches an hour. On fine-textured clay soils, water may collect and stand on soil, seemingly with very little infiltration, for many days. Desirable infiltration rates are between these two extremes. A convenient means of expressing infiltration is in terms of inches lowering of water surface per hour. For example, if an acre of level land at 9 o'clock is covered with water to a depth of 2 inches and at 10 o'clock the water is but 1 inch deep, the infiltration rate is 1 inch per hour, neglecting evaporation losses.

7.7 INTAKE

The rate of infiltration from a furrow into the soil is referred to as the intake rate. This term indicates that infiltration occurs under a particular soil surface configuration. Intake rate is therefore influenced by furrow size and shape, whereas infiltration rate applies to a level surface covered with water. Hence, whenever the configuration of a soil surface influences the rate of water entry, the term intake rate should be used rather than the term infiltration rate to refer to the rate of entry of water into soil.

7.8 PERMEABILITY

One of the most important properties of soils is the velocity of water flow through the pore spaces caused by a given force. The permeability[1] of soil is defined as the velocity of flow caused by a unit hydraulic gradient in which the driving force is 1 lb per lb of water. Permeability is not influenced by the hydraulic slope, and this is an important point of difference between permeability and infiltration. Also the term permeability is used for designating flow through soils in any direction. Permeability is influenced most by the physical properties of the soil. Changes in water temperature influence permeability slightly. The meaning of the term soil permeability is explained by the symbols and equations in Chapter 9, where typical measurements of permeability are reported. In saturated field soils permeability varies between wide limits: from less than 1 foot per year in compact clay soils, up to several thousand feet per year in gravel formations. For unsaturated soils the moisture content is one of the dominant factors influencing permeability. Permeability is a velocity having the physical dimensions of length divided by time.

Infiltration, intake, and permeability are discussed in greater detail in Chapter 9.

7.9 DEPTH OF SOIL

The importance of having an adequate depth of soil in which to store satisfactory amounts of irrigation water at each irrigation should be emphasized. Soils of arid regions are relatively deep compared to humid-region soils. There are many areas of productive shallow soils in irrigated regions underlain at depths of 1 to 3 feet by coarse gravel, hardpan, or other formations in which plants obtain little or no sustenance. Shallow soils require frequent irrigations to keep crops growing. Excessive deep percolation losses usually occur when shallow soils overlying coarse-textured, highly permeable sands and gravels are irrigated. Deep soils of medium texture and loose structure permit plants to root deeply, provide for storage of large volumes of irrigation water in the soil, and consequently sustain satisfactory plant growth during relatively long periods between irrigations. The volume of water actually absorbed by some plant roots and consumed to produce a crop may be practically the same for shallow and deep soils. Yet nearly all practical irrigators recognize that more water is required during the crop-growing season to irrigate a given crop on a shallow soil than is required for the same crop on a deep soil. The larger

[1] Permeability and conductivity are frequently used interchangeably.

number of irrigations required for shallow soils and greater unavoidable water losses at each irrigation on shallow soils account for
differences in practical water requirements for different soils during the
season. The relationship of depth of soils to their water capacities
is considered in more detail in Chapter 8.

7.10 PLANT FOOD COMPOUNDS

In order to produce large and satisfactory yields of crops, all soils
must have adequate supplies of available plant nutrient elements.
Many chemical elements are essential to plant growth. Calcium,
carbon, hydrogen, iron, magnesium, nitrogen, oxygen, potassium,
phosphorus, and sulfur are of major importance because they are
required in considerable amounts. Because of the sparse or scanty
growth of native vegetation on the virgin soils of arid regions, nitrogen is relatively deficient. Plants absorb nitrogen in the form of
soluble nitrates which are dissolved in the soil water. In order to
assure an adequate supply of available nitrogen irrigated soils must
contain sufficient amounts of nitrogenous matter which may be supplied from barnyard manure, from the growing of legume crops as
green manures, or from commercial fertilizers. Soil moisture content,
soil structure, and soil aeration should be made favorable to the bacterial activity which is essential to the formation of nitrates.

The low rainfall of arid regions results in a comparatively small
amount of leaching, and hence, arid soils are usually high in the
more important mineral plant-food elements, particularly calcium
and potassium. The open structure of most arid soils permits very
favorable aeration to great depths, and consequently favorable bacterial activity occurs at much greater depths in arid- than in humid-
climate soils.

7.11 EXCESS SOLUBLE SALTS

The arid-region conditions of low precipitation, high evaporation,
and relatively small amounts of soil leaching, although favorable to
the occurrence of satisfactory quantities of calcium, phosphorus, and
potassium in soils, result in the accumulation of excessive quantities
of soluble salts that retard or inhibit plant growth. Soils having excess
soluble salts are designated saline soils, and those having excessive
exchangeable sodium are designated alkali or sodic soils. Even such
compounds as sodium nitrate and potassium nitrate which normally
function as plant nutrients, if accumulated in the soil in excessive
amounts, become toxic to plants. There are many areas of highly
productive soils in arid regions which, because of good drainage, are
entirely free from salinity and alkali troubles. Moreover, some of

these areas probably never will be adversely influenced by the occurrence of the accumulation of excess soluble salts or of exchangeable sodium. In other areas these accumulations are the cause of widespread sterility and barrenness of arid-region soils. A large amount of research has been directed toward the solution of the salinity and alkali problem in arid regions. The waters of some arid-region streams and rivers contain appreciable amounts of soluble salts, the actual amounts being influenced by the soil from which the waters flow into streams and rivers. Because of the importance of salinity and alkali in arid-region soils and their relations to irrigation practices, Chapter 10 is devoted to a consideration of these topics.

7.12 SURFACE TENSION

Surface tension is due to unbalanced molecular forces. In any body of water the particles in the interior of the liquid are attracted equally in all directions by the other particles of the liquid, as illustrated by the particle at point A in Fig. 7.1. A particle on the water surface, on the contrary, is not attracted equally on all sides, since the molecules of the air surrounding the particle exert less attraction upon the water-surface particle than is exerted by the interior particles of the liquid. There is consequently a resultant inward attraction along a line perpendicular to the surface of the liquid as illustrated in points B and C of Fig. 7.1.

7.13 TENSION HEADS[2]

The water in the capillary tube illustrated in Fig. 7.2 is held in the position shown at a height h_t above the water surface by an upward force due to surface tension in the water.

[2] Since the surface tension of water causes a suction on water within the soil, the term "suction head" is being used by many researchers instead of "tension head." Suction head is generally expressed in terms of an equivalent length of vertical water column.

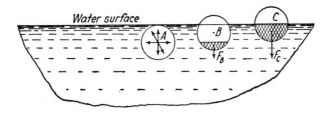

Fig. 7.1 Illustrating the fact that surface tension is due to unbalanced molecular forces. (From *Mechanics, Molecular Physics and Head*, by Millikan, Ginn and Company, New York.)

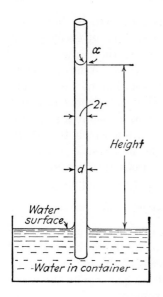

Fig. 7.2 Showing that water rises in a capillary tube, owing to surface tension, to a height well above the surface of the water in the container.

Let $\qquad$ F_u = the total upward force

$\qquad\qquad\qquad$ F_d = the total downward force

Then $F_u = F_d$ because the water is at rest. For the purpose of this analysis, the angle of contact between the water and the glass tube α in Fig. 7.2 is considered zero. Then the upward force is equal to the circumference of the tube times the surface tension of the water, and the downward force is equal to the volume of water times its specific weight w. Using symbols and equations,

$$F_d = \pi r^2 h_t w = F_u = 2\pi r T$$

It follows that

$$h_t = \frac{2T}{rw} \qquad (7.2)$$

The length h_t in Equation 7.2 is the height to which water will rise in a capillary tube of radius r as the result of the surface tension of the water T. This is here designated the tension head. Unsaturated soil pores, like the capillary tube of Fig. 7.2, develop concave water surfaces and suction forces or tensions that tend to pull water in soils from points of high moisture percentages to points of lower moisture percentages.

Multiply each side of Equation 7.2 by w and there results a negative pressure or suction p

$$p = wh_t = \frac{2T}{r} \tag{7.3}$$

Each side of Equation 7.3 is a force intensity (F/L^2) due to tension.

Using metric units for T and w of Equation 7.2, since the surface tension of water is 75.6 dynes per cm and its specific weight is 980 dynes per cc, it follows that

$$h_t = \frac{2 \times 75.6}{980r} = \frac{0.15}{r} \tag{7.4}$$

Equation 7.4 gives the height in centimeters to which water at 4°C will rise in a capillary tube of radius r centimeters.

Applying essentially the same reasoning as underlies Equation 7.4 to ideal soils in which the capillary tubes are triangular in cross section, Keen has shown that the maximum height of rise of water in centimeters is given approximately by the equation:

$$h_t = 0.75/r, \quad \text{and} \quad h = 1.5/d \tag{7.5}$$

in which r is the radius and d the diameter in millimeters of the soil particles.

In soils the pores are irregular in size. Furthermore, the liquid-vapor interface, where the surface tension forces act, is not only irregular in shape but usually inclined to the horizontal. Thus, the actual heights that water can rise in a soil by capillary action is usually less than the theoretical heights computed from Equation 7.5. Under average conditions capillarity usually acts freely to 4 or 5 feet, frequently to 10 feet, and rarely to 30 or more feet.

Equation 7.5 shows that film tension of soil moisture increases as the radius of a capillary water film decreases. Soils of a given texture and structure having low percentages of water have, therefore, high moisture tensions, whereas the same soils having high percentages of water have low moisture tensions. These facts explain observed soil and water relationships such as, for example, the decrease in capillary water with increase in distance above a water table when at equilibrium.

7.14 SOIL-MOISTURE TENSION

The tension force of water in unsaturated soils has been described by several different expressions, such as soil-pull, the force of suction, and capillary tension.

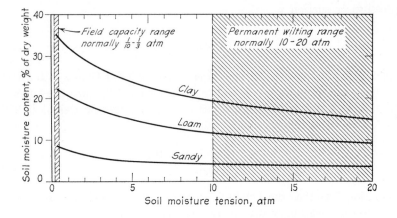

Fig. 7.3 Typical curves of soil moisture variation with tension.

The moisture content of the soil corresponding to a particular tension force is influenced by soil texture, soil solution, and temperature. Soil texture influences particularly the curvatures of capillary water films, and the other two factors of surface tension of the liquid; thus, all influence tensions.

As shown in the previous section, the tension head $h_t = 2T/wr$. The surface tension T and specific weight w can be measured, but it is not feasible to measure the radius r in soils; hence, h_t is measured by means of the tensiometer.

The relationships of soil-moisture content to tension for typical sandy, loam, and clay soils are presented Fig. 7.3.

7.15 SOIL-MOISTURE CONTENT

Direct methods of measuring moisture content of soils, although tedious and costly, have substantial value. The practice is to bore or drive to desired depths with a soil auger or a soil tube, to remove samples of the moist soil and place it in cans provided with covers, and to take the soil samples to a laboratory for weighing and drying. Samples of 100 or more grams of moist soils are kept in an oven having a temperature of 105° to 110°C, until the soil is free from moisture. The loss of weight in drying, divided by the weight of the water-free soil, yields the moisture percentage on the dry-weight basis, represented by the symbol P_w. For example:

Weight of moist soil	100 gm
Weight of water-free soil	80 gm
Loss of weight in drying	20 gm

Then P_w (i.e., the moisture content on the dry-weight basis) = 100 (20/80) = 25%.

A short cut for computing moisture percentage on a dry weight basis is to divide the wet weight by the dry weight, subtract 1.0, and multiply by 100 to obtain the percentage. In the above example dividing the wet weight of 100 grams by the dry weight of 80 grams gives 1.25, subtracting 1.0, the result is 0.25, or 25 percent moisture. This procedure is satisfactory when a calculator is used, but may not be sufficiently accurate when a 10-inch slide rule is used to make the calculation.

Measurements of soil-moisture content are sometimes reported on the wet-weight basis; that is, in the above example, the moisture percentage on the wet-weight basis is 20. The apparent advantage of the wet-weight basis is the simplicity of computation if 100 gm of moist soil is used. In reality, however, the wet-weight basis is irrational because the reference base for the percentage computation, i.e., the weight of a wet sample of soil, varies according to the moisture content of the soil.

Interpretations of the significance and influence of different quantities of water in soil, both in relation to water storage and to plant growth, are facilitated by converting the moisture percentage on the dry-weight basis to the volume basis. The percentage on a volume basis is defined as the volume of water per unit volume of space within the body of soil. For example, if a cubic foot of space within the soil contains ¼ cu ft of air, ¼ cu ft of water, and ½ cu ft of solid soil particles, the percentage of moisture on the volume basis, represented by the symbol P_v, is 25. Drying the soil in an oven, as a means of extracting the water from it, results in a loss of water in the form of vapor. It is therefore desirable to convert dry-weight basis moisture percentages P_w to volume percentages P_v. The apparent specific gravity of the soil, represented by a symbol A_s, is defined as the ratio of the weight of a given volume of soil to the weight of an equal volume of water, and:

$$P_v = P_w A_s \tag{7.6}$$

Based on Equation 7.6, Table 7.1 gives moisture percentages on a volume basis that are equivalent to different percentages on a dry-

TABLE 7.1

Moisture Percents on the Volume Basis, P_v, Equivalent to Various Percents on the Dry-Weight Basis, P_w, for Soils of Different Apparent Specific Gravity, A_s, Based on Equation 7.6, $P_v = P_w A_s$

Moisture Percent on Dry-Weight Basis (P_w)	Apparent Specific Gravity (A_s)							
	1.1	1.2	1.3	1.4	1.5	1.6	1.7	1.8
1.0	1.10	1.20	1.30	1.40	1.50	1.60	1.70	1.80
1.2	1.32	1.44	1.56	1.68	1.80	1.92	2.04	2.16
1.4	1.54	1.68	1.82	1.96	2.10	2.24	2.38	2.52
1.6	1.76	1.92	2.08	2.24	2.40	2.56	2.72	2.88
1.8	1.98	2.16	2.34	2.52	2.70	2.88	3.06	3.24
2.0	2.20	2.40	2.60	2.80	3.00	3.20	3.40	3.60
2.2	2.42	2.64	2.86	3.08	3.30	3.52	3.74	3.96
2.4	2.64	2.88	3.12	3.36	3.60	3.84	4.08	4.32
2.6	2.86	3.12	3.38	3.64	3.90	4.16	4.42	4.68
2.8	3.08	3.36	3.64	3.92	4.20	4.48	4.76	5.04
3.0	3.30	3.60	3.90	4.20	4.50	4.80	5.10	5.40
3.2	3.52	3.84	4.16	4.48	4.80	5.12	5.44	5.76
3.4	3.74	4.08	4.42	4.76	5.10	5.44	5.78	6.12
3.6	3.96	4.32	4.68	5.04	5.40	5.76	6.12	6.48
3.8	4.18	4.56	4.94	5.32	5.70	6.08	6.46	6.84
4.0	4.40	4.80	5.20	5.60	6.00	6.40	6.80	7.20
4.2	4.62	5.04	5.46	5.88	6.30	6.72	7.14	7.56
4.4	4.84	5.28	5.72	6.16	6.60	7.04	7.48	7.92
4.6	5.06	5.52	5.98	6.44	6.90	7.36	7.82	8.28
4.8	5.28	5.76	6.24	6.72	7.20	7.68	8.16	8.64
5.0	5.50	6.00	6.50	7.00	7.50	8.00	8.50	9.00
5.2	5.72	6.24	6.76	7.28	7.80	8.32	8.84	9.36
5.4	5.94	6.48	7.02	7.56	8.10	8.64	9.18	9.72
5.6	6.16	6.72	7.28	7.84	8.40	8.96	9.52	10.08
5.8	6.38	6.96	7.54	8.12	8.70	9.28	9.86	10.44
6.0	6.60	7.20	7.80	8.40	9.00	9.60	10.20	10.80
6.2	6.82	7.44	8.06	8.68	9.30	9.92	10.54	11.16
6.4	7.04	7.68	8.32	8.96	9.60	10.24	10.88	11.52
6.6	7.26	7.92	8.58	9.24	9.90	10.56	11.22	11.88
6.8	7.48	8.16	8.84	9.52	10.20	10.88	11.56	12.24
7.0	7.70	8.40	9.10	9.80	10.50	11.20	11.90	12.60
7.2	7.92	8.64	9.36	10.08	10.80	11.52	12.24	12.96
7.4	8.14	8.88	9.62	10.36	11.10	11.84	12.58	13.32
7.6	8.36	9.12	9.88	10.64	11.40	12.16	12.92	13.68
7.8	8.58	9.36	10.14	10.92	11.70	12.48	13.26	14.04
8.0	8.80	9.60	10.40	11.20	12.00	12.80	13.60	14.40
8.2	9.02	9.84	10.66	11.48	12.30	13.12	13.94	14.76
8.4	9.24	10.08	10.92	11.76	12.60	13.44	14.28	15.12
8.6	9.46	10.32	11.18	12.04	12.90	13.76	14.62	15.48
8.8	9.68	10.56	11.44	12.32	13.20	14.08	14.96	15.84
9.0	9.90	10.80	11.70	12.60	13.50	14.40	15.30	16.20
9.2	10.12	11.04	11.96	12.88	13.80	14.72	15.64	16.56
9.4	10.34	11.28	12.22	13.16	14.10	15.04	15.98	16.92
9.6	10.56	11.52	12.48	13.44	14.40	15.36	16.32	17.28
9.8	10.78	11.76	12.74	13.72	14.70	15.68	16.66	17.64

weight basis. Moisture-content of soil can also be represented as a depth d obtained by multiplying the percentage volume P_v by the depth of soil D:

$$d = \frac{P_v}{100} D \qquad (7.7)$$

or combining Equations 7.6 and 7.7:

$$d = \frac{P_w}{100} A_s D \qquad (7.8)$$

Since A_s and P_w are dimensionless quantities, the dimension of D will also be the dimension of d. Hence, if the depth of the soil D is measured in inches, d will be in inches; and if D is in centimeters, then d also will be in centimeters. Table 7.2, based upon Equation 7.8,

TABLE 7.2

DEPTH OF IRRIGATION WATER IN INCHES REQUIRED TO ADD DIFFERENT AMOUNTS OF MOISTURE TO 1 FOOT OF SOIL FOR SOILS HAVING DIFFERENT APPARENT SPECIFIC GRAVITIES, BASED ON EQUATION 7.8,

$$d = \frac{P_w A_s D}{100}$$

Available Field Moisture Capacities (P_w)	Apparent Specific Gravity (A_s)							
	1.2	1.3	1.4	1.5	1.6	1.7	1.8	1.9
4.0	0.58	0.62	0.67	0.72	0.77	0.82	0.86	0.91
4.2	.60	.65	.71	.76	.81	.86	.91	.96
4.4	.63	.68	.74	.79	.84	.90	.95	1.01
4.6	.66	.72	.77	.82	.88	.94	.99	1.05
4.8	.69	.75	.81	.86	.92	.98	1.04	1.09
5.0	.72	.78	.84	.90	.96	1.02	1.08	1.14
5.2	.75	.81	.87	.94	1.00	1.06	1.12	1.19
5.4	.78	.84	.91	.97	1.04	1.10	1.16	1.23
5.6	.81	.87	.94	1.01	1.08	1.14	1.21	1.28
5.8	.83	.90	.97	1.04	1.11	1.18	1.25	1.32
6.0	.86	.93	1.01	1.08	1.15	1.22	1.30	1.37
6.2	.89	.97	1.04	1.12	1.19	1.26	1.34	1.41
6.4	.92	1.00	1.08	1.15	1.23	1.31	1.38	1.46
6.6	.95	1.03	1.11	1.19	1.27	1.35	1.43	1.50
6.8	.98	1.06	1.14	1.22	1.31	1.39	1.47	1.55
7.0	1.01	1.09	1.18	1.26	1.34	1.43	1.51	1.60
7.2	1.04	1.12	1.21	1.30	1.38	1.47	1.56	1.64
7.4	1.07	1.15	1.24	1.33	1.42	1.51	1.60	1.69
7.6	1.09	1.19	1.28	1.37	1.46	1.55	1.64	1.73
7.8	1.12	1.22	1.31	1.40	1.50	1.59	1.68	1.78
8.0	1.15	1.25	1.34	1.44	1.54	1.63	1.73	1.82

shows the depth of water in inches required to add different amounts of moisture to 1 foot of soil for soils having different apparent specific gravities.

7.16 CLASSES AND AVAILABILITY OF SOIL WATER

For many years soil water has been classified as hygroscopic, capillary, and gravitational. Hygroscopic water is on the surface of the soil grains and is not capable of movement by the action of gravity or capillary forces. Capillary water is that part in excess of the hygroscopic water which exists in the pore space of the soil and is retained against the force of gravity in a soil that permits unobstructed drainage. Gravitational water is that part in excess of hygroscopic and capillary water which will move out of the soil if favorable drainage is provided. There is no precise boundary or line of demarcation between these three classes of soil water. The proportion of each class depends on soil texture, structure, organic matter content, temperature, and depth of soil column considered.

Water may also be classified as unavailable, available, and gravitational or superfluous. Such a grouping refers to the availability of soil water to plants. Gravitational water drains quickly from the root zone under normal drainage conditions. Unavailable water is held too tightly by capillary forces and is generally not accessible to plant roots. Available water is the difference between gravitational and unavailable water.

Water drains from the soil under the constant pull of gravity. Sandy soils drain readily, while clay soils drain very slowly. Hence, one day after irrigating a sandy soil most of the gravitational water has drained out of the soil, whereas clay may require 4 or more days for gravitational water to drain. The rate of drainage is most rapid immediately after irrigating and decreases constantly, nevertheless it continues to drain at a relatively slow rate, even after gravitational water has been removed. On the average, 2 days are required before the rate of drainage decreases rather sharply and gravitational water has been removed from the root zone. This implies, of course, that no restricting layers are in the root zone to impede the downward flow of water.

Relationships between the various terms used in identifying classes and availability of soil water are shown in Fig. 7.4.

Field Capacity

When gravitational water has been removed, the moisture content of soil is called field capacity. Field capacity cannot be determined

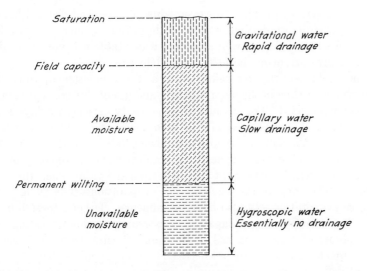

Fig. 7.4 Classes of soil-water availability to plants and drainage characteristics.

accurately because there is no discontinuity in the curve of moisture content versus time. Nevertheless, the concept of field capacity is extremely useful in arriving at the amount of water available in the soil for plant use. Most of the gravitational water drains through the soil before it can be used consumptively by plants.

In practice, field capacity is usually determined 2 days after an irrigation. Therefore, field capacity defines a specific point on the moisture-content time curve. Specifying the time also makes it possible to calculate the water used consumptively by plants while gravitational water is draining from the soil. A soil will come to field capacity more quickly when an active crop is growing than when there are no roots removing water from the soil. Special precaution should be taken not to overlook the amount of water consumptively used by the crop between the time of irrigation and the time at which field capacity is determined. The amount of water thus used when light irrigations are given is very important, and the principle should never be overlooked in irrigation practice.

Field capacity can be measured by determining moisture content of soil after an irrigation which is sufficiently heavy to insure thorough wetting of the soil to be tested. Observing the decrease in moisture by making moisture determinations at different times after irrigation is valuable in understanding and properly interpreting the field capacity characteristics of soil. However, soils must be well drained before reliable field determinations can be made in this manner. Restricting layers of silt and clay as well as a high water table will impede drainage and give erroneous indications of field capacity.

Since field capacity by definition is a direct function of time, the importance of time in the determination and use of this concept cannot be overemphasized. For example, field capacity measured 2 days after an irrigation cannot be used without modification to estimate the water remaining in the soil in the spring following fall irrigation. Furthermore, the value of field capacity determined when a crop is growing on the land will be somewhat different from the value when the land is bare. The effect of excessive surface evaporation versus little or no surface evaporation will also have its influence. Despite these limitations the concept of field capacity is extremely useful for estimating the volume of water retained in the soil. Precise determinations of field capacity are generally not necessary for field applications.

Soil moisture tension is normally between $\frac{1}{10}$ and $\frac{1}{3}$ atmospheres[3] when the soil is at field capacity. The correct value depends upon the drainage characteristics of the soil and the time after irrigation at which the soil is assumed to reach field capacity. Sandy soils tend to be near $\frac{1}{10}$ atm at field capacity, while clays tend toward $\frac{1}{3}$ atm. Some soils have gone as high as 0.6 atm. For most agricultural soils, a tension of $\frac{1}{10}$ corresponds more closely than does $\frac{1}{3}$ atm to the generally accepted values of field capacity determined by moisture content. The large difference in moisture content and volume of water in the soil between $\frac{1}{10}$ and $\frac{1}{3}$ atm makes it important to define the tension correctly at which field capacity occurs. Considering field capacity to occur at $\frac{1}{3}$ atm tension when it is $\frac{1}{10}$ of an atm leads to gross errors in estimating the volume of water available to plants following irrigation.

Permanent Wilting Point

The soil-moisture content when plants permanently wilt is called the permanent wilting point or the wilting coefficient. The permanent wilting point is at the lower end of the available moisture range. A plant will wilt when it is no longer able to extract sufficient moisture from the soil to meet its water needs. Temporary wilting will occur in many crops on a hot windy day, but the plants recover in the cooler portion of the day. Permanent wilting, as well as temporary wilting, depends upon the rate of water used by the plant, the depth of the

[3] One atmosphere equals 14.7 pounds per square inch.

root zone, and the water holding capacity of the soil. Permanent wilting will occur at a higher moisture content in a hot climate than in a cool climate. A plant is considered to be permanently wilted when it will not recover after being placed in a saturated atmosphere where little or no consumptive water use occurs.

Field estimates of wilting point can often be made by determining the moisture content of soils in which plants have permanently wilted. This method is subject to more error and requires more judgment than field determinations of field capacity. Allowance must be made for depth and nature of rooting. Furthermore, it is often difficult to find plants in the condition of permanent wilt. For a plant to reach permanent wilting following an irrigation, provided that the plant is using considerable water, will require 1 week in sands to perhaps 4 weeks in clays, and even longer if the plant is deeply rooted.

The tension at which permanent wilting occurs can vary from 7 to as high as 40 atm, depending upon rate of consumptive use, crop, salt content of soil, and soil texture. As the temperature and rates of consumptive use increase, permanent wilting will occur at significantly lower tensions and higher moisture contents. The tension in the soil moisture when the soil is at permanent wilting is generally considered to be 15 atm. Whether in reality it is 10 or 20 atm makes very little difference, since the change of moisture is slight with rather large changes of moisture tension.

As an approximation, the permanent wilting percentage can be estimated by dividing the field capacity by a factor varying from 2.0 to 2.4, depending upon the amount of silt in the soil. For soils of high silt content 2.4 should be used.

Available Moisture

The difference in moisture content of the soil between field capacity and permanent wilting is termed the available moisture. This represents the moisture which can be stored in the soil for subsequent use by plants. Available moisture can be expressed as percentage moisture P_w, as percentage volume P_v, or as depth d, whichever is most convenient.

Often it becomes desirable to refer to the amount of available moisture remaining in or extracted from the soil. This can also be expressed by P_w, P_v, or d, when it is desired to know the quantity of water in the soil. However, when correlating the moisture condition of the soil to crop response, the amount of moisture remaining, or extracted, can be represented as a percentage of the available moisture to obtain a more significant expression of moisture condition in the soil.

Readily Available Moisture

Soil moisture content near the wilting point is not readily available to the plant. Hence, the term readily available moisture has been used to refer to that portion of the available moisture that is most easily extracted by plants, approximately 75 percent of the available moisture.

7.17 FILLING THE AVAILABLE SOIL WATER RESERVOIR

Having determined the amount of water to be added to the soil, the irrigator needs to know how long it will be necessary to apply a given stream of water. The relationship between size of stream, time of application, area to be irrigated, and depth of water to be applied is as follows:

$$qt = ad \tag{7.9}$$

where q = the size of stream in cu ft per sec (or acre-inches per hr)
    t = the time in hours required to irrigate the area
    a = the area, in acres
    d = the depth in inches that the volume of water used would cover the land irrigated, if quickly spread uniformly over its surface

The quantity in cubic feet per second (or acre-inches per hour) multiplied by the time in hours equals the acre-inches applied. Also the number of acres covered times the depth in inches equals the acre-inches applied.

A stream of 1.9 cfs running 3 hours should uniformly cover 1 acre to a depth of 5.7 inches. Also consider that a loam soil 4 feet deep underlain by coarse sand and gravel is to be irrigated. If the irrigator finds that it is taking 4 hours to irrigate one acre using a stream of 2.8 cfs, he may determine from Equation 7.9 that he has applied enough water to cover the land to a depth of 11.2 inches. Since a loam soil will hold only about 2 inches of water per foot of depth, or no more than ·8 inches per 4 feet, even when the crop is wilting, the irrigator sustains a deep percolation loss of at least 3 inches. He will probably want to modify his method of application in order to apply water more efficiently.

When Equation 7.9 is combined with Equation 7.8 it is apparent that:

$$t = \frac{P_w A_s Da}{100q} \tag{7.10}$$

Provided the apparent specific gravity A_s is known, it is possible to

TABLE 7.3

TIME IN HOURS t REQUIRED WITH A STREAM q TO ADD VARIOUS PERCENTAGES OF AVAILABLE MOISTURE P_w TO 1 ACRE-FOOT OF SOIL, THE APPARENT SPECIFIC GRAVITY OF WHICH IS 1.4 BASED ON $t = \dfrac{P_w A_s Da}{100q}$

Col. No.		1	2	3	4	5	6	7	8	9	10
Line No.	Available Capacity (P_w)	Size of Stream, q cfs									
		0.5	1.0	1.5	2.0	2.5	3.0	3.5	4.0	4.5	5.0
1	4.0	1.34	0.67	0.45	0.34	0.27	0.22	0.19	0.17	0.15	0.13
2	4.2	1.41	.71	.47	.35	.28	.24	.20	.18	.16	.14
3	4.4	1.48	.74	.49	.37	.29	.25	.21	.18	.16	.15
4	4.6	1.55	.77	.52	.39	.31	.26	.22	.19	.17	.15
5	4.8	1.61	.81	.54	.40	.32	.27	.23	.20	.18	.16
6	5.0	1.68	.84	.56	.42	.34	.28	.24	.21	.18	.17
7	5.2	1.75	.87	.58	.44	.35	.29	.25	.22	.19	.17
8	5.4	1.81	.90	.60	.45	.36	.30	.26	.23	.20	.18
9	5.6	1.88	.94	.63	.47	.37	.31	.27	.23	.21	.19
10	5.8	1.95	.97	.65	.49	.39	.32	.28	.24	.22	.19
11	6.0	2.02	1.01	.67	.50	.40	.34	.29	.25	.22	.20
12	6.2	2.08	1.04	.69	.52	.42	.35	.30	.26	.23	.21
13	6.4	2.15	1.08	.72	.54	.43	.36	.31	.27	.24	.22
14	6.6	2.22	1.11	.74	.55	.44	.37	.32	.28	.24	.22
15	6.8	2.29	1.15	.76	.57	.46	.38	.33	.29	.25	.23
16	7.0	2.35	1.17	.78	.59	.47	.39	.34	.29	.26	.23
17	7.2	2.42	1.21	.81	.60	.48	.40	.35	.30	.27	.24
18	7.4	2.49	1.25	.83	.62	.50	.41	.36	.31	.28	.25
19	7.6	2.56	1.28	.85	.64	.51	.42	.36	.32	.28	.26
20	7.8	2.62	1.31	.87	.65	.52	.44	.37	.33	.29	.26
21	8.0	2.69	1.35	.90	.67	.54	.45	.38	.34	.30	.27

TABLE 7.4

Representative Physical Properties of Soils

Soil Texture	Infiltration[1] and Permeability Inches/hour I_f	Total Pore Space % N	Apparent Specific Gravity A_s	Field Capacity % FC	Permanent Wilting % PW	Total Available Moisture[2]		
						Dry Weight % $P_w = FC - PW$	Volume % $P_v = P_w A_s$	Inches per Foot $d = \dfrac{P_w}{100} A_s D$
Sandy	2 (1–10)	38 (32–42)	1.65 (1.55–1.80)	9 (6–12)	4 (2–6)	5 (4–6)	8 (6–10)	1.0 (0.8–1.2)
Sandy Loam	1 (0.5–3)	43 (40–47)	1.50 (1.40–1.60)	14 (10–18)	6 (4–8)	8 (6–10)	12 (9–15)	1.4 (1.1–1.8)
Loam	0.5 (0.3–0.8)	47 (43–49)	1.40 (1.35–1.50)	22 (18–26)	10 (8–12)	12 (10–14)	17 (14–20)	2.0 (1.7–2.3)
Clay Loam	0.3 (0.1–0.6)	49 (47–51)	1.35 (1.30–1.40)	27 (23–31)	13 (11–15)	14 (12–16)	19 (16–22)	2.3 (2.0–2.6)
Silty Clay	0.1 (0.01–0.2)	51 (49–53)	1.30 (1.25–1.35)	31 (27–35)	15 (13–17)	16 (14–18)	21 (18–23)	2.5 (2.2–2.8)
Clay	0.2 (0.05–0.4)	53 (51–55)	1.25 (1.20–1.30)	35 (31–39)	17 (15–19)	18 (16–20)	23 (20–25)	2.7 (2.4–3.0)

Note: Normal ranges are shown in parentheses.

[1] Intake rates vary greatly with soil structure and structural stability, even beyond the normal ranges shown above.

[2] Readily available moisture is approximately 75% of the total available moisture.

compute the hours required to add a given moisture percentage P_w to a field of given area a and a soil of certain depth D when using a stream of water of q cfs (acre-inches per hour).

It is impracticable to add small percentages of water in capillary form to great depths of soil, since the full capillary capacity of water must be satisfied for the surface soil before the water moves to lower depths. Likewise, it is very difficult to spread water uniformly over the land.

Keeping these factors in mind, Equation 7.10 has practical utility. To simplify the use of this equation, Table 7.3 may be used for soils having an apparent specific gravity of 1.4. For soils having higher or lower values of A_s, proportionate corrections must be made. The use of Table 7.3 is illustrated by the following example: an irrigator has at his disposal a stream of 3 cfs and he wants to apply enough water to increase the 4-foot depth root-zone soil moisture from 10 percent to 15 percent. How many hours will be required if the water is spread uniformly and losses are neglected? Column 6 of Table 7.3 shows that 0.28 hour is required to supply enough water with a 3 cfs stream to add 5 percent moisture to 1 acre-foot of soil. Therefore, in 1.12 hours, enough water is applied to add 5 percent moisture to 4 acre-feet of soil, and longer application is likely to result in deep percolation loss provided 5 percent satisfies the field capillary capacity for usable or available water or fills the capillary reservoir.

7.18 REPRESENTATIVE PHYSICAL PROPERTIES

Normal ranges in average values of several physical properties of irrigated soils described in the preceding section are summarized in Table 7.4. Extremes can and will occur. However, Table 7.4 will be applicable to most areas. It will be particularly helpful to estimate physical properties, and to get a clear understanding of the variation that occurs in the physical properties of normal irrigated soils.

REFERENCES

Bouyoucos, G. J. and A. H. Mick, "An Electrical Resilience Method for the Continuous Measurement of Soil Moisture under Field Conditions," *Mich. Agr. Exp. Sta. Tech. Bul.* 172, 1940.

Dreibelbis, F. R. and F. A. Post, "An Inventory of Soil-Water Relationships on Woodland, Pasture, and Cultivated Soils," *Proc. Soil Sci. Am.*, Vol. 6. pp. 462–473, 1941.

Edlefsen, N. E., A. B. C. Anderson, and W. B. Marcum, "Methods of Measuring Soil Moisture," *Ann. Am. Soc. Sugar Beet Tech.*, 1942.

Jamison, Vernon C., "Make Full Use of Stored Soil Moisture," reprinted from *J. of Soil and Water Cons.*, Vol. 12, No. 3, May 1959.

Jamison, Vernon C., "Pertinent Factors Governing the Availability of Soil Moisture to Plants," reprinted from *Soil Sci.*, Vol. 81, No. 6, June 1956.

Richards, L. A., "Hydraulics of Water in Unsaturated Soil," *Agr. Eng.*, Vol. 22, No. 9, pp. 325–326, September 1941.

Richards, L. A., "Retention and Transmission of Water in Soil," Yearbook Separate No. 2580, reprinted from pp. 144–154, U.S.D.A. Yearbook of Agriculture, 1955.

Richards, L. A. and L. R. Weaver, "Moisture Retention by Some Irrigated Soil as Related to Soil Moisture Tension," *J. Agr. Research*, Vol. 69, No. 6, pp. 215–235, September 1944.

Richards, L. A. and S. J. Richards, "Soil Moisture," Soil—U.S.D.A. Yearbook of Agriculture, 1957.

Richards, S. J. and R. M. Hagan, "Soil Moisture Tensiometer," *Calif. Agri. Exp. Sta. Ext. Ser.*, University of California, February 1958.

Rouse, Hunter, Fluid Mechanics for Hydraulic Engineers, McGraw-Hill Book Company, New York, 1938.

Russell, E. W., "Soil Structure," *Imp. Bur. Soil Sci. Tech. Commun.* 37, Harpenden, England, 1938.

Scofield, Carl S., "The Measurement of Soil Water," *J. Agr. Research*, Vol. 71, No. 9, pp. 375–402, November 1945.

Soil—U.S.D.A. Yearbook of Agriculture, 1957.

Vomocil, James A., "In Situ Measurement of Soil Bulk Density," *Agr. Eng.*, September, 1954.

CHAPTER 8 MEASUREMENT OF SOIL MOISTURE

Measurement of water stored in soils and capacity of soils to store water are important in humid and arid regions. That some humid-climate soils produce crops despite the lapse of many days, and sometimes weeks, between periods of rainfall is evidence of their capacity to store available water, since all growing plants require water continuously. In irrigated regions the capacity of soils to store available water for the use of growing crops is of special importance and interest, because the depth of water to apply in each irrigation and the interval between irrigations are both influenced by storage capacity of the soil. Irrigated soils of large water-storage capacity may produce profitable crops in places where, and at times when, the shortage of irrigation water makes it impossible to irrigate as frequently as would be desirable.

Knowledge of the capacity of soils to retain available irrigation water is also essential for efficient irrigation. If the irrigator applies more water than the root-zone soil reservoir can retain at a single irrigation, the excess is wasted. If he applies less than the soil will retain, the plants may wilt from lack of water before the next irrigation unless water is applied more frequently than otherwise would be necessary. Water losses which result from deep percolation below the root zone of crops cannot be seen. They can be measured approximately by subtracting the storage capacity of the various soils from the amount of water applied in single irrigations, less the runoff.

Determining the moisture in unsaturated soils that is available to plants is of major importance and is therefore given special attention here. In localities where the late-season water supply is always low and surface-storage reservoirs are impracticable or economically prohibitive, storage of water in saturated soils in gravitational form is sometimes practical and helpful. Excessive application of water for the purpose of water storage may cause the ground water to rise and become injurious to crops during the early part of the season.

It is important to find the available water capacity for different soils; i.e., the field capacity less the moisture content at the permanent wilting point. Some soils having high field capacities also have high wilting points, thus making the available water capacity rather low and requiring frequent irrigation. The amount of available water that can be stored increases with root-zone depth. Methods of estimating and measuring depths and volumes of water that may be stored and made available to crops are considered in this chapter.

8.1 DRILLING HOLES AND OBTAINING SOIL SAMPLES

Soil Auger

It is highly desirable that irrigation farmers observe by inspection, and sometimes by measurement, the quantities of moisture in their soils. Boring or drilling deeply into soils of arid regions to determine soil moisture conditions is essential to profitable farm operation.

Two types of soil augers are used: a spiral-shaped bit made from carpenter's standard wood-auger bit, usually 1½ inches in diameter, and a post-hole auger, usually 2 or 3 inches in diameter, made especially for boring into the soil.

The spiral-bit auger, well below the soil surface in Fig. 8.1*A*,

A   B

Fig. 8.1 A soil auger helps to show when water is needed, and the depth of water penetration from irrigation. (Courtesy Union Pacific Railroad Company.)

Fig. 8.2 Post-hole auger. Small-diameter augers of this type are used for studying the distribution of irrigation water in soil. (Courtesy Iwan Brothers.)

is illustrated in Fig. 8.1*B*. The screw-point and the side-cutting edges of a carpenter's auger are removed in preparing the bit for soil boring. The spiral auger is light in weight and convenient to carry around. When using it in compact soils, care is necessary to avoid boring too deeply at one time, thus causing the auger to lodge in the soil. The post-hole auger, illustrated in Fig. 8.2, does not involve this danger, because when it is filled with soil it does not easily advance further. However, because of the open sides of both augers shown in Figs. 8.1 and 8.2, soil obtained at a particular depth is contaminated somewhat with other soil as the auger is removed from the hole. For normal farm work this is not serious, but for research activity wherein soil moisture is to be determined, better results are obtained by use of a post-hole auger having the sides enclosed.

Another type of auger particularly useful to obtain samples of known volume for apparent specific gravity determinations is shown in Fig. 8.3.

Hand Soil Sampling Tubes and Probes

F. H. King, one of America's pioneer soil scientists, designed and used a steel tube for sampling soils. The "King soil tube" has been improved by Veihmeyer and associates of the California Agricultural Experiment Station. In soils containing fine gravel it is frequently difficult, and sometimes impossible, to obtain samples by means of a soil auger, whereas with a properly made sampling tube it is possible to cut through several layers of gravel and still obtain satisfactory samples.

As illustrated in Fig. 8.4, each tube consists of three parts: a tube of seamless steel of the desired length, a driving head, and a point. Each end of the tube is threaded to facilitate attachment of the head and point. A special hammer drives the tube into the soil.

In the design of the improved tube precautions have been taken, as reported by Veihmeyer, to reduce the difficulties of removing the

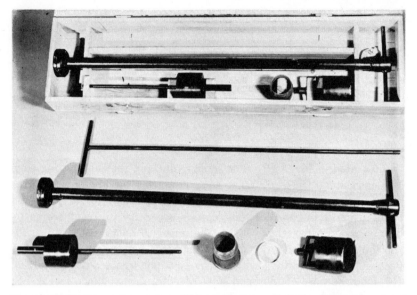

Fig. 8.3 Tube-type soil sampler. This particular unit is the Uhland hand soil sampler. (Courtesy Utah Scientific Research Foundation.)

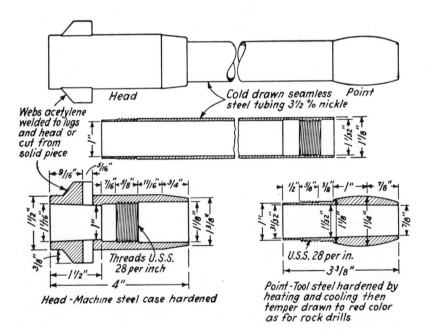

Fig. 8.4 Details of soil-sampling tube. (Williams and Wilkins Company, *Soil Sci.*, Vol. 27, No. 2.)

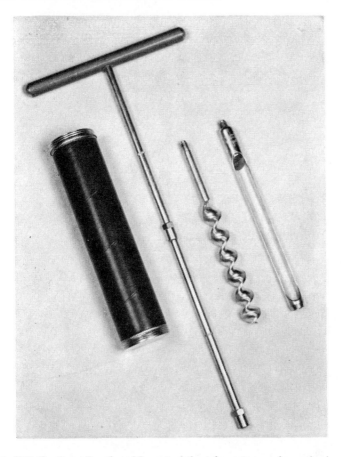

Fig. 8.5 Details of a soil probe with part of the tube cut away for easier insertion, inspection, and removal. A screw-type auger is also shown. (Courtesy Oakfield Apparatus Company.)

tube from the soil. Where samples need to be taken at great depths, 12 to 18 feet, a puller, made of two automobile jacks mounted on a base and connected at the top by a yoke, may be needed to withdraw the tube from the soil.

Short tubes, about 12 inches in length, have been developed with a design similar to the King tube except that the points are not removable. They are designed so that (1) the tube can be pushed into the soil with a minimum of effort, (2) the soil will enter the tube readily, and (3) the tube can be extracted easily from the soil. Part of the tube is cut out to inspect and remove the soil better. A soil probe of this type is shown in Fig. 8.5. Soil samples generally can be ob-

Fig. 8.6 Powered soil sampler capable of obtaining a 4-inch undisturbed core to a depth of 10 feet. The tube has just been removed from the soil. (Courtesy Utah Scientific Research Foundation.)

tained faster with an Oakfield probe shown in Fig. 8.5 than with the King tube shown in Fig. 8.4. Under ideal conditions both the tube and probe can be used to obtain cores of undisturbed soil from which apparent specific gravity of the soil can be determined.

Power Auger Tubes and Probes

Powered augers, tubes, and probes have been used to secure soil samples faster and to conserve man power. Usually these units are attached to the rear of a truck. A larger, self-contained sampler that can secure a 4-inch diameter undisturbed sample to a depth of 10 feet is shown in Fig. 8.6.

8.2 RESISTANCE OF SOIL TO PENETRATION

Irrigators have long been aware that the wetter the soil, the deeper one sinks into the mud. This observation has been used to indicate how well the soil is irrigated. True, it is only an index, but the man who knows his land soon learns to interpret the tendency to penetrate wet soil in terms of the adequacy of the irrigation.

Inserting a shovel into the soil gives a better indication of soil moisture than the tendency to sink into the wet ground. A still better tool is to push a steel rod, about ½ inch in diameter into the soil. In many soils the depth that the rod can be pushed into the soil will be a good indication of the depth of wetting. Welding a crossbar on the top of the rod will facilitate insertion and removal.

8.3 APPEARANCE AND FEEL OF SOIL INDICATE MOISTURE CONTENT

One of the oldest and most widely used methods of estimating soil moisture content is to look at the soil and to feel it. Using the soil auger, samples of soil usually can be obtained readily throughout the root zone of the soil. Table 8.1[1] can be used as a guide for judging how much available moisture has been removed from the soil and consequently, how much must be added during the irrigation period. Reasonably good estimates of irrigation needs can be obtained by using Table 8.1. In many applications, greater accuracy is not needed, nor is it justified economically. However, as technology improves, more precise and more usable tools become available. As the prices paid for land and its products increase, the feasibility of using more accurate methods of determining moisture content of soil also increases.

8.4 GRAVIMETRIC DETERMINATION OF MOISTURE

Determining soil moisture by weighing the soil before and after drying in an oven has been described in the previous chapter under the heading "Soil Moisture Content." The method is for primary measurement, chiefly limited in its usefulness by the time required to collect samples and dry them in an oven. Usually 24 hours are required for drying. Samples of about 200 grams are most often used, so that they will dry in a reasonable time and with reasonable uniformity. Such small samples naturally may not be, and usually are

[1] The original concept and outline using feel and appearance of soil as an indication of soil moisture was developed largely by U.S. Soil Conservation Service technicians.

TABLE 8.1
GUIDE FOR JUDGING HOW MUCH OF THE AVAILABLE MOISTURE HAS BEEN REMOVED FROM THE SOIL

Feel or Appearance of Soil and Moisture Deficiency in Inches of Water Per Foot of Soil

Soil Moisture Deficiency	Coarse Texture	Moderately Coarse Texture	Medium Texture	Fine and Very Fine Texture
0% (Field capacity)	Upon squeezing, no free water appears on soil but wet outline of ball is left on hand. 0.0	Upon squeezing, no free water appears on soil but wet outline of ball is left on hand. 0.0	Upon squeezing, no free water appears on soil but wet outline of ball is left on hand. 0.0	Upon squeezing, no free water appears on soil but wet outline of ball is left on hand. 0.0
0–25%	Tends to stick together slightly, sometimes forms a very weak ball under pressure. 0.0 to 0.2	Forms weak ball, breaks easily, will not slick. 0.0 to 0.4	Forms a ball, is very pliable, slicks readily if relatively high in clay. 0.0 to 0.5	Easily ribbons out between fingers, has slick feeling. 0.0 to 0.6
25–50%	Appears to be dry, will not form a ball with pressure. 0.2 to 0.5	Tends to ball under pressure but seldom holds together. 0.4 to 0.8	Forms a ball somewhat plastic, will sometimes slick slightly with pressure. 0.5 to 1.0	Forms a ball, ribbons out between thumb and forefinger. 0.6 to 1.2
50–75%	Appears to be dry, will not form a ball with pressure.[1] 0.5 to 0.8	Appears to be dry, will not form a ball.[1] 0.8 to 1.2	Somewhat crumbly but holds together from pressure. 1.0 to 1.5	Somewhat pliable, will ball under pressure.[1] 1.2 to 1.9
75–100% (100% is permanent wilting)	Dry, loose, single-grained, flows through fingers. 0.8 to 1.0	Dry, loose, flows through fingers. 1.2 to 1.5	Powdery, dry, sometimes slightly crusted but easily broken down into powdery condition. 1.5 to 2.0	Hard, baked, cracked, sometimes has loose crumbs on surface. 1.9 to 2.5

[1] Ball is formed by squeezing a handful of soil very firmly.

not, representative of large areas. Hence, several samples must be used to obtain a satisfactorily representative indication of moisture content. To save time and money, infrared heat has been used to dry the soil at a much more rapid rate than is done in a conventional soil oven. However, care must be exercised not to overdry the soil and burn out the organic matter.

Gravimetric absorption blocks have been used to a limited extent. A porous block is placed in a prepared hole in contact with the soil at a desired depth. Moisture moves into or out of the block until the soil and block have essentially equal amounts of moisture. The block is removed periodically from the hole and weighed to determine moisture content.

The principal disadvantages of the absorption-block method are the lag in approaching equilibrium, the high sensitivity required in weighing, and the accompanying fragile nature of the weighing device. Preparing the holes so that blocks can be placed in close contact with the soil and yet be readily removable for weighing is time-consuming and costly.

8.5 USING ELECTRICAL PROPERTIES OF A POROUS BLOCK

Electrical properties of resistance (or conductance), capacitance, and dielectric strength have been used to indicate moisture content. Changes of moisture affect all of these electrical properties. Porous blocks containing desired electrical elements are placed into the soil. As the moisture content of the blocks change, the electrical properties change. Only limited success has been achieved using capacitance or dielectric properties; but practical units have been developed using the resistance or conductance principle.

Dr. G. J. Bouyoucos developed a gypsum (plaster of Paris) block in which two electrodes of straight wire were embedded. Moisture in this block tends to come to equilibrium with the moisture in the soil. As the moisture increases, the amount of gypsum in solution increases and the resistance between the wires decreases. Other materials have been used for making blocks, such as nylon, fiber glass, and combinations of these materials with gypsum.

Many designs have been used successfully and others are now being developed. Not only have the shapes of the blocks been altered, but the size, shape, and configuration of the electrodes have also been varied. In general, nylon and fiber-glass units are more sensitive than gypsum blocks at high moisture content and low tension conditions. Nylon units are most sensitive at tensions less than two atmospheres, and gypsum blocks operate best at tensions between 1 to 15 atm.

The gypsum blocks are soluble and deteriorate in one to three seasons of use. However, gypsum is less sensitive to soil salts than nylon and fiber glass because of the concentration of soluble gypsum within the water of a gypsum block. Normally there is considerable variation between blocks and considerable change occurs in the calibration during the season, especially in gypsum blocks.

Resistance blocks and small transistorized meters for reading electrical properties are available commercially and are being used in many areas for moisture control. Some of the outstanding examples of their use are found on some of the large sugarcane plantations in Hawaii, where field crews spend their full time reading the blocks and scheduling irrigations. Figure 8.7 shows some typical units with a portable meter. These units are generally installed by digging a hole to the desired depth, placing a unit on its side, backfilling, and tamping the soil in place. In this manner several units can be placed in a single hole. To increase ease of placement and to facilitate removal at the end of the season, individual units have been cast into single rods and the rods inserted into the soil as a unit.

8.6 TENSIOMETERS

A tensiometer consists of a porous cup filled with water and attached to a vacuum gauge or mercury manometer. A hole is bored or dug in the soil to a desired depth, a handful of loose soil is placed into the hole, and the cup pushed firmly into the soil. Additional soil is packed around the cup and around the tube wherever necessary to insure firm contact with the soil. A temporary connection is soon established between the water inside the cup and the water in the soil outside. As water moves out of the cup because of the suction or tension existing in the soil water, the vacuum created in the cup is registered on the gauge. Conversely, an increase of water in the soil will lower the tension, water will move into the cup, and the gauge will read less tension. Fluctuations in soil moisture are registered by the tensiometer, as long as the tension does not exceed about 0.8 atm. At greater degrees of tension, air enters the closed system through the pores of the cup and the instrument is no longer accurate. The soil moisture tension must be lowered by irrigation or rainfall and the system filled with water before it will again operate properly.

Considerable progress has been made in the technology of tensiometer construction since the first units were made by Drs. L. A. Richards and Willard Gardner at Utah State University. Units that perform very well, such as those shown in Fig. 8.8, are now available commercially.

Fig. 8.7 Portable meter and resistance blocks used to measure soil moisture. (Courtesy Industrial Instrument, Inc.)

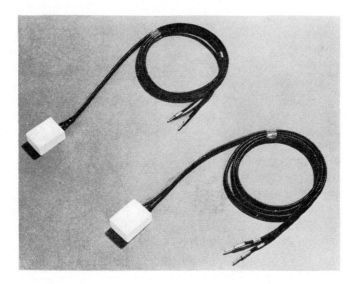

Fig. 8.8. Tensiometer in a recently harvested field of sugarcane measuring moisture at five depths. (Courtesy Hawaian Commercial and Sugar Company.)

Since these units operate satisfactorily only up to tensions of 0.8 atm, they are most useful in sandy soils, where this represents a major portion of the available water, or for crops that must be irrigated frequently. How tensiometers can be used to predict the time of less frequent irrigations is illustrated in the following example: If it takes 3 days following an irrigation for the tensiometer to become inoperative, it may be desirable when growing a particular crop to irrigate at an interval two times this period, or on a 6-day interval for sandy soil or perhaps three times this period, or on a 9-day interval for a silty loam soil. In this case the tensiometer is used to indicate the rate of depletion of soil moisture, but cannot be used for the full moisture range.

Because of its narrow range the tensiometer is used for moist and resistance blocks for dryer soil conditions. Sometimes the two instruments are used together, since the blocks become sensitive at about the same degree of moisture content that the tensiometer ceases to function. With both tensiometers and blocks it is desirable to place units within the zone of major root activity so that they will be most sensitive to changes in soil moisture. Yet the moisture available at deeper depths in the soil should not be overlooked when the instrument is placed at shallower depths.

8.7 NEUTRON METHOD OF MEASURING SOIL MOISTURE

Considerable potentialities exist in the neutron method of measuring soil moisture. In this method fast neutrons are emitted from a source into the surrounding soil. The fast neutrons are slowed down by the water. The resulting slow neutrons which reach the counting tube are recorded. Fast neutrons are not registered by the counter. The greater the water content of the soil, the greater the number of slow neutrons reaching the counting tube. Hydrogen is the principal element which absorbs fast neutrons. Since water is the principal source of hydrogen in the soil, there exists a very good correlation between the amount of water in the soil and the number of slow neutrons reaching the counter. In practice, a hole is dug with an auger, and a metal tube is driven into the hole to retain the soil. The neutron source and counting device are placed within the hole and lowered to the desired depth. The reading obtained in a unit of time is proportional to the moisture content of the soil surrounding the source and counter. Sufficient holes should be dug to obtain a representative indication of soil moisture.

Some minerals, as well as boron and chlorine, also absorb fast neutrons and thus influence the reading. Boundary geometry also has an influence on the readings. The number of slow neutrons recorded in a given time while the probe is near the surface will be different from a reading obtained at the same moisture content well beneath the surface, because of the geometrical configuration of the soil surrounding the probe. The zone of influence decreases as moisture content increases. Improvements are being made in reducing weight, making the units more dependable, and reducing the radiation exposure hazard. Since readings can be taken as soon as the instrument is placed in a hole, lag time is negligible, resulting in a fast method of measurement. Also repeated non-destructive sampling in the same location is feasible. The neutron method holds considerable promise for an effective practical method of measuring soil moisture.

8.8 USING THERMAL PROPERTIES

Heat conductivity of the soil can be used as an index of soil moisture, since the rate of conductivity depends upon moisture present in the soil. The thermistor, a thermally sensitive resistor, has given the best control, because of the influence of temperature on heat conductivity. A permanent porous material with proper pore size distribution must be developed, into which a thermally sensitive device can be embedded without damaging the heating and sensing element.

8.9 SAMPLING ERROR

Error of sampling has long plagued the investigator as he seeks to determine the amount of moisture in the soil. Obtaining representative samples is a major problem. Uneven growing of plants and non-uniform root penetration must be considered, since they cause variations of moisture content. Texture and structure variations of soils alter the intake, transmission, and retention of moisture. Variations in land-surface configuration change the opportunity for intake of rainfall, and furrowed shape and size alter the rate of intake of irrigation water. All of these factors cause the moisture content to vary from point to point in a field. To obtain a representative soil moisture sample requires that several samples be taken, unless the method of determining soil moisture inherently integrates a large volume of soil. The number of samples required to obtain a representative sample increases as the moisture variation increases.

Another complexity is that essentially all methods of moisture determination are based upon small samples. Individual samples can be expected to vary at least plus or minus 20 percent from the mean of a large number of samples. Concentrated research is needed to develop practical and faster methods of moisture determination which will give more representative values of moisture content of the soil.

REFERENCES

Colman, E. A., "A Laboratory Procedure for Determining the Field Capacity of Soils," *Soil Sci.*, Vol. 63, p. 277, 1947.

Garton, James E. and Frank R. Crow, "Rapid Methods of Determining Soil Moisture," *Agr. Eng.*, Vol. 35. pp. 486–487, July 1954.

Haise, Howard R., "How to Measure the Moisture in the Soil," Water—U.S.D.A. Yearbook of Agriculture, 1955.

Israelsen, O. W., Wayne D. Criddle, Dean K. Fuhriman, and Vaughn E. Hansen, "Water Application Efficiencies in Irrigation," *Utah Agr. Exp. Sta. Bul.* 311, March 1944.

Merriam, J. L., "Field Method of Approximating Soil Moisture for Irrigation," *Trans. Am. Soc. Agr. Eng.*, Paper, Vol. 3, No. 1, pp. 31–32, Special Soil and Water Edition, 1960.

Stewart, Gordon L. and Sterling A. Taylor, "Field Experience with the Neutron Scattering Method of Measuring Soil Moisture," reprinted from *Soil Sci.*, Vol. 33, No. 2, February 1957.

van Bavel, C. H. M., "Measurement of Soil Moisture Content by the Neutron Method," *U.S.D.A. Agr. Res. Ser.*, ARS 41–24, August 1958.

CHAPTER 9 FLOW OF WATER INTO AND THROUGH SOILS

Water characteristics considered in Chapters 7 and 8 dealt mainly with the capacity of the soil to hold water. Practical estimates must be made of the amount of water available in the soil for plant use following an irrigation. Out of this necessity for a volumetric expression of available water have grown the useful concepts of field capacity, permanent wilting, and resulting water-holding capacities discussed in the previous chapters. In reality, as was pointed out, these are not static quantities but are only points on a continuous dynamic curve of moisture content versus time.

The dynamic aspects are now considered in more detail as they affect the movement of water into, through, and out of the soil. The phenomenon of water entry is first considered and then the movement through the soil follows. Finally, consideration is given to unsaturated flow and vapor movement within the soil. However, as a basis for the subsequent discussions, the basic laws and equations governing flow are presented first.

9.1 ENERGY IN FLOWING WATER

The principles of mechanical work and energy are applied in the derivation of fundamental formulas for the flow of fluids, including the flow of water in soils. Energy is defined as capacity to do work.

In fluids, energy may be in two forms:

1. Kinetic energy.
2. Potential energy.
   a. Energy resulting from pressure differences.
   b. Energy resulting from elevation differences.

A pound of water flowing at a velocity of v feet per sec has a kinetic energy of $v^2/2g$ ft-lb per lb, where g is the acceleration of gravity.

A pound of water having potential energy has the energy only because of the existence of a difference of pressure or a difference of elevation. The potential energy of a pound of water resulting from a pressure difference can be represented by the ratio p/w where p is the pressure per unit area and w is the weight per unit volume. A pound of fluid having a potential energy due to elevation can be represented by y, representing elevation above some datum.

Actually, a pound of water may have both kinetic and potential energy, where the potential energy is composed of both pressure and elevation. The combined energy can be represented by the following form of the widely used Bernoulli equation[1] showing energy per unit weight (ft-lb/lb):

$$H = \frac{V^2}{2g} + \frac{p}{w} + y \qquad (9.1)$$

Note that this equation when applied to two points in the flow is identical to Equation 5.2. Energy per unit weight, having the dimension of length, is used extensively in engineering practice because of its simplicity, especially when applied to open-channel flow and flow through soils. Open-channel flow and the use of Equation 9.1 were discussed in Chapter 5. When applying this equation to flow through soils, a simplification can readily be made. Since the velocity through soils is very small, the kinetic energy $(V^2/2g)$ is even smaller and can usually be ignored. Hence, Bernoulli's equation applied to flow through soils reduces to:

$$h = \frac{p}{w} + y \qquad (9.2)$$

which has been referred to in Chapter 5 and will be used here as the piezometric head.[2] This quantity is measured in saturated soil by the simple piezometer and in unsaturated soil by a tensiometer.

Figure 9.1 illustrates flow of water in three directions and the use of piezometers to measure the direction of flow and the magnitude of the energy. Note that flow always occurs in the direction of decreasing piezometric head.

When flow similar to that shown in Fig. 9.1 occurs in a pipeline, the principles discussed in Chapter 5 can be used to compute the flow, loss of energy, etc. When the flow occurs in soils, the equations are

---

[1] See Chapter 5 for definitions of the terms.

[2] Piezometric head is also frequently referred to as hydraulic head, the two terms being synonymous.

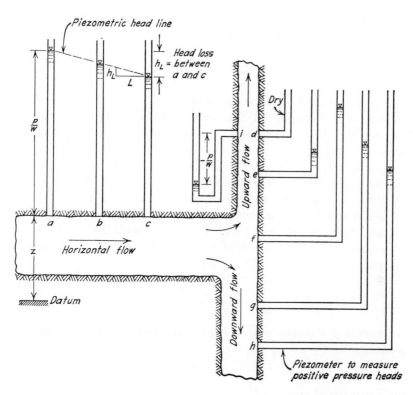

Fig. 9.1 Illustrating horizontal, upward, and downward flow through soil and the use of piezometers to indicate direction of flow and the magnitude of the piezometric gradient.

similar but are simplified because of the negligible kinetic energy. Hence, the Darcy-Weisbach equation, shown earlier as Equation 5.3,

$$h_L = f \frac{L}{D} \frac{V^2}{2g} \tag{5.3}$$

is used in the following form known as the Darcy equation:

$$V = k \frac{h_L}{L} \tag{9.3}$$

As long as the flow is not turbulent, the velocity V and flow through saturated soils is proportional to the soil permeability k and the slope of the piezometric headline h_L/L. The soil permeability k has the dimension of length per unit time L/t, and is a combined measure of soil and fluid properties. The fluid properties can be made explicit using the following equation:

$$V = \left(\frac{w}{\mu} k'\right) \frac{h_L}{L} \qquad (9.4)$$

The permeability k' has the physical dimensions of area L^2. When defined, as in Equation 9.4, the permeability k' is influenced only by the size and shape of the soil particles and pores (soil texture and structure) and is independent of the fluid properties: (specific weight w, and viscosity μ). For most studies of the flow of ground water in irrigation and drainage, the influence of specific weight and viscosity is relatively small; hence, explicit inclusion of w and μ as in Equation 9.4 is not essential for general irrigation studies. The k in Equation 9.3, used hereafter, is equal to (w/μ) k' of Equation 9.4.

Using the value of V from Equation 9.3 in the basic rational equation for quantity of flow, $Q = AV$, it follows that

$$Q = Ak \frac{\iota h}{L} \qquad (9.5)$$

in which A is the gross area at right angles to the flow direction. The pressure intensity at any point is

$$p = wh$$

and the pressure head

$$h = p/w$$

Let the hydraulic head at point 1 be h_1 and at point 2 be h_2; then:

$$h_1 = \frac{p_1}{w} + y_1$$

$$h_2 = \frac{p_2}{w} + y_2$$

Assume that h_1 is greater than h_2 and that the two points are a distance L apart. Then the hydraulic slope becomes:

$$\frac{H_L}{L} = \frac{h_1 - h_2}{L} = \frac{[(p_1/w) + y_1] - [p_2/w + y_2]}{L} \qquad (9.6)$$

In Equation 9.5, w is the specific weight of the water and p_1/w and p_2/w are the pressure heads; y_1 and y_2 are the elevation heads with respect to a selected datum plane.

Applications of Equations 9.3, 9.5, and 9.6 for ground-water flow in saturated soils are illustrated by two examples of field conditions: the first for flow of unconfined or free ground water in sand under a small piezometric slope, Fig. 9.2, and the second for upward flow of water through a 40-foot stratum of clay over an artesian aquifer of gravel in which the water is under pressure or confined, Fig. 9.3.

Figure 9.2 illustrates unconfined ground water flowing through sandy soil overlying a compact clay. The piezometer at A shows a piezometric head, $h_1 = (p_1/w) + y = 50$ ft, and at B, $h_2 = 40$ ft. Therefore, as the flow distance is 100 feet, the hydraulic slope, by Equation 9.2, is 10/100 and the velocity is $(1/10)k$. Selecting an average k of 1200 ft per yr, or 3.8×10^{-5} ft per sec, the approximate velocity of flow through the sand is $V = 120$ ft per yr, and the quantity of flow in a section 1000 feet long and 20 feet deep, by Equation 9.5, is

$$Q = Ak\frac{h_L}{L} = (1000 \times 20) \times \frac{3.8}{100,000} \times \frac{10}{100} = 0.076 \text{ cfs} = 34 \text{ gpm}$$

For a second example, the results of piezometer measurements at different elevations in a clay soil overlying an artesian gravel aquifer are presented in Fig. 9.3. The permeability of the clay soil has been measured and found to average 5 feet per year. Then, to find the

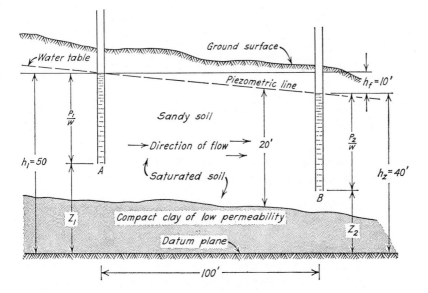

Fig. 9.2 Flow of unconfined ground water in saturated sand overlying a compact clay.

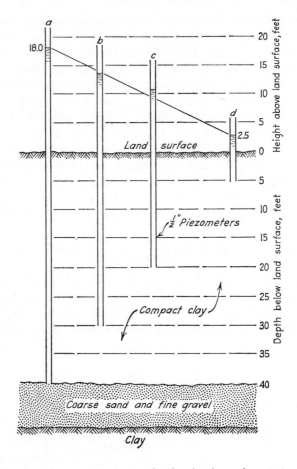

Fig. 9.3 Showing the average pressure head, p/w, in a clay stratum overlying an artesian aquifer.

average velocity of the flow from the 40-foot soil depth up to the 5-foot depth, in feet per year, use Equations 9.2 and 9.5:

$$\frac{h_L}{L} = \frac{h_1 - h_2}{L} = \frac{18.0 - 2.5}{35} = \frac{15.5}{35} = 0.44$$

$$V = k\frac{h_L}{L} = 2.5 \times 0.44 = 2.2\,\text{ft per yr}$$

Note that the velocity of 2.2 feet per year is the gross apparent velocity since the cross-sectional area A was a gross area rather than the net cross-sectional area of the pores.

If the piezometric slope and the soil permeability, as measured at the place represented by the data of Fig. 9.3, were the same for one section of land, 640 acres, then the annual movement of water from the artesian aquifer, due to upward flow through the clay soil and surface consumptive use, would be (640 × 2.2) = 1408 ac-ft. The values of pressure head and elevation head above recorded are presented to show the application of the Bernoulli and Darcy equations.

9.2 PIEZOMETERS TO MEASURE PRESSURE HEADS IN SATURATED SOILS

The pressure heads p/w illustrated in Figs. 9.2 and 9.3 are measured by driving a small-diameter pipe, designated piezometer, to the desired depth. A much used driving hammer and driving head for either ¼ inch or ⅜ inch pipe is shown in Fig. 9.4.

When the piezometers are driven, a plug of soil from 6 to 12 inches in length forms in the lower end of the pipe. This plug is removed by flushing, Fig. 9.5, which consists of pumping water down to the bottom of the piezometer through plastic tubing. This water flows up through the annular space between the tubing and wall of the pipe and carries with it soil in suspension. Flushing is continued until a small cavity has been formed below the end of the pipe and the water

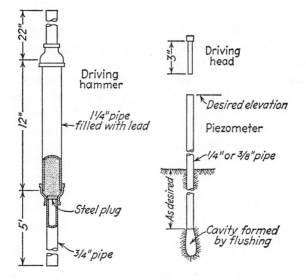

Fig. 9.4 Detail of driving hammer and piezometer.

Fig. 9.5   Flushing the piezometer.   (Courtesy U.S. Regional Salinity Lab.)

becomes clear. The length of the piezometer is marked on the plastic tubing, and care is taken not to push the tubing more than 3 or 4 inches below the end of the piezometer.

In soils of average permeability, water in the piezometer reaches an equilibrium level in a few minutes, but in soils of low permeability several hours may be required for water to reach an equilibrium level.

In highly compacted soils, and in those containing fine gravel that cannot pass upward through the annular space during the flushing operation, a rivet is inserted in the end of the piezometer before it is driven. The pipe is then driven to a point about 3 inches below the desired level and pulled up 3 inches so that the rivet can easily be punched out with a rod before flushing. This eliminates difficulty sometimes encountered in removing the soil plug formed in the end of the pipe. In most soils, however, this plug can be removed by flushing in less time than is required to punch out the rivet.

Piezometers of an inch in diameter are now being used, but the larger the pipe, the larger and more expensive the installation equipment and the slower the reaction time due to storage. Jetting equipment is also used to install piezometers, especially for the larger sizes. A cavity is formed by the jet ahead of the pipe and the pipe is forced into the soil. Such a technique permits installation of piezometers to deeper depths in heavy clay soils, but this method does have the liability that a good seal is not always obtained around the outside of the piezometer. Hence, in stratified soils having considerable vertical flow under sizable pressures, driven piezometers are generally more reliable than are jetted piezometers.

Reading piezometers with an electric sounder is shown in Fig. 9.6. Electric sounders generally consist of a weighted insulated probe, the tip of which is a conductor connected to the pipe. With a battery in the circuit, a microampmeter will register a current when the tip of the probe touches the surface of the water. The depth at which this occurs is noted on the marked conductor leading from the probe to the surface. Piezometers can also be read by blowing on a plastic-tubing air line. The depth of the end of the tube can be adjusted until the end just touches the water surface and the depth recorded.

Fig. 9.6 Reading the piezometer with an electrical sounder gage. (Courtesy U.S. Regional Salinity Lab.)

9.3   SOIL-PERMEABILITY MEASUREMENT

The permeability of saturated soils varies greatly. In irrigation and drainage studies, permeability is the dominant variable, some soil being as much as 100,000 times as permeable as others.

Knowledge of soil permeabilities is essential to progress in studies of water conveyance and water application efficiencies, and in the design of drainage systems for the reclamation of saline and alkali soils.

Permeabilities are influenced by the size and shape of pore spaces through which water flows and by the specific weight and viscosity of the soil water, as shown in Equation 9.4. For ordinary irrigation applications, it is impractical to measure all of the factors that influence permeability, but it is practical and very essential to measure permeability of soils in the laboratory and in the field.

Two of the many types of equipment for measuring permeability are the constant-head and variable-head permeameters.

Constant-Head Permeameter

With a constant head maintained by either continuous inflow or frequent additions of water, steady flow through the soil is obtained. Figure 9.7 illustrates two constant-head permeameters, one for laboratory tests and the other for field studies. Darcy's law for flow of water

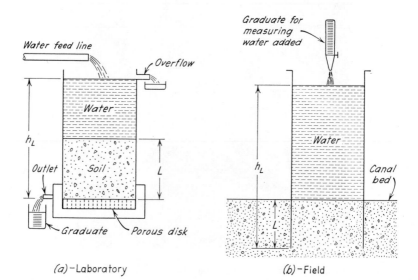

(a)–Laboratory   (b)–Field

Fig. 9.7   Constant-head permeameters.

in soils is applied for computing permeability after measuring volume of flow in unit time t, gross soil cross-sectional area A at right angles to flow, loss of hydraulic head h_L, and flow length L.

In field studies on undisturbed soil, loss of head and flow length sometimes cannot be measured accurately at reasonable cost. If the surface soil consists of a thin layer of low-permeability soil overlying a layer of highly permeable soil, then the loss of hydraulic head may be considered as the distance from water surface to the highly permeable soil and the flow length, as equal to the thickness of the top layer of soil, as indicated in Fig. 9.7. Using symbols defined above and solving the Darcy flow Equation 9.5 for the permeability, it follows that

$$k = \frac{QL}{Ah_L} \tag{9.7}$$

For example, where a permeameter having an area of 1.19 sq ft was used, the flow of water was 0.336 cu ft in 0.4 hr, the loss in hydraulic head flowing through 1 ft of soil was 2.4 ft; therefore, the permeability is

$$k = \frac{QL}{Ah_L} = \frac{0.336 \times 1 \times 12}{0.40 \times 1.19 \times 2.4} = 3.5 \text{ in. per hr, or } 2600 \text{ ft per yr}$$

Irrigation engineers utilize one or more of several different units of volume, area, and time with the term permeability. The units cubic feet per square foot per day, or acre-inches per acre per hour (or simply surface feet per day and inches per hour), are typical. When water flows in saturated soils vertically downward under the force of gravity, the piezometric slope is considered as unity.

Table 9.1, comparing permeabilities expressed in different units, includes permeabilities from 0.005 to 25 in. per hr (column 2) and from 0.01 to 50 cu ft per sq ft of soil in 24 hr (column 1), the relative variation being from 1 to 5000.

One cubic foot per second is equal approximately to 1 ac-in. per hr. On this basis, column 2 of Table 9.1 gives the number of cubic feet per second that would be required to maintain a stream of water percolating vertically downward through 1 acre of soil having different permeabilities. For example, line 24 shows that a stream of 1 cfs per ac would equal 2 cu ft per sq ft per 24 hr.

Variable-Head Permeameter

The variable-head permeameter is adapted to the measurement of permeability of fine-textured, compact soils of low permeability. It consists of a cylinder with a conical top to which is attached a vertical

TABLE 9.1

COMPARISON OF PERMEABILITY OF SOIL TO WATER STATED IN DIFFERENT
UNITS

	1	2	3
	Permeability in		
Line No.	Cu Ft per Sq Ft per 24 Hr	Surface Inches per Hour	Cfs per Acre
1	0.01	0.005	0.005
2	.02	.010	.010
3	.03	.015	.015
4	.04	.020	.020
5	.05	.025	.025
6	.06	.030	.030
7	.07	.03⁵	.035
8	.08	.04	.040
9	.09	.04	.045
10	.1	.05	.05
11	.2	.10	.10
12	.3	.15	.15
13	.4	.20	.20
14	.5	.25	.25
15	.6	.30	.30
16	.7	.35	.35
17	.8	.40	.40
18	.9	.45	.45
19	1.0	.50	.50
20	1.2	.60	.60
21	1.4	.70	.70
22	1.6	.80	.80
23	1.8	.90	.90
24	2.0	1.00	1.00
25	2.2	1.10	1.10
26	2.4	1.20	1.20
27	2.6	1.30	1.30
28	2.8	1.40	1.40
29	3.0	1.50	1.50
30	3.2	1.60	1.60
31	3.4	1.70	1.70
32	3.6	1.80	1.80
33	3.8	1.90	1.90
34	4.0	2.00	2.00
35	4.5	2.25	2.25
36	5.0	2.50	2.50
37	6.0	3.00	3.00
38	7.0	3.50	3.50
39	8.0	4.00	4.00
40	9.0	4.50	4.50
41	10	5.00	5.00
42	15	7.50	7.50
43	20	10.00	10.00
44	25	12.5	12.5
45	30	15.0	15.0
46	35	17.5	17.5
47	40	20.0	20.0
48	45	22.5	22.5
49	50	25.0	25.0

glass tube of small diameter. The cylinder is pressed into the soil to a known depth, and then the whole apparatus is filled with water. As the water percolates through the disk of soil in the cylinder, the water in the glass tube drops. Since the cylinder is usually made with an area 100 or more times that of the glass tube, a small volume of water percolation registers as a large drop in the glass tube. The permeability k is computed from the initial and final reading of the water head in the glass tube $(h_1 - h_2)$, the time interval t, the thickness of the soil in the cylinder or flow length L, and the ratio of the area of the glass tube to that of the cylinder a/A. The formula is:

$$k = \frac{2.3aL}{At} \log_{10} \frac{h_1}{h_2} \tag{9.8}$$

For example, for the measurement of the permeability of a clay lining of a canal, the following values were obtained:

$a = 0.26$ sq in.	$t = 6.2$ hr
$A = 153.20$ sq in.	$h_1 = 66.1$ in.
$L = 5.16$ in.	$h_2 = 59.1$ in.

Then,

$$k = \frac{2.3 \times 0.26 \times 5.16}{153.2 \times 6.2} \times \log_{10} \frac{66.1}{59.0} = 1.6 \times 10^{-4} \text{ in./hr} = 0.13 \text{ ft/yr}$$

The principal difficulty encountered in field tests with this permeameter was the tendency of the cylinder to rise because of the pressure exerted on the inside of the cylinder by the column of water in the glass tube. To overcome this tendency a load was placed on top of the cylinder. Temperature changes causing a change of volume within the cylinder will also produce significant changes in readings.

A variable-head permeameter measures permeability of a canal bed or lining, but not the seepage rate from a canal when in use. In order to determine seepage rate, it is necessary also to know the hydraulic slope causing flow through the canal bed and the effective permeability through the sides of the canal. The depth of water should be approximately the depth to be expected during normal operation. When comparative results from different locations are needed the depth should be essentially the same in all cylinders. Care should be taken not to puddle the soil when water is added.

9.4 WATER FLOW IN UNSATURATED SOILS

Darcy's law for velocity of flow in saturated soils applies also to unsaturated soils. However, as moisture content decreases, cross-sectional area through which flow occurs also decreases causing permeability to decrease. In soils, whether saturated or unsaturated, water flow is caused by a difference in potential.

Thus, flow occurs in unsaturated soil much as it does in saturated soil. That is, flow always occurs in the direction of decreasing energy, Equation 9.3, and the total flow is the product of the velocity of flow and the cross-sectional area, Equation 9.5. Complications in unsaturated flow come from the decreasing cross-sectional area as the moisture content decreases, and from the increasing difficulty of measuring negative pressure.

For saturated flow through soils, a simple opening which will allow water to move into a pipe is sufficient to measure the energy at the end of the pipe. This method has been referred to as the piezometer method. However, for a suction within the soil, it is necessary to provide a barrier to movement of air from the soil to the measuring system. This barrier is usually a porous cup and has been referred to in Chapter 8 as a tensiometer.

A further complication of unsaturated flow rises from the movement of water through the soil as vapor, which is even more difficult to measure. As moisture content decreases, the percentage of water moving through the soil as vapor increases. Movement in the vapor phase occurs in the direction of decreasing vapor pressure. Vapor pressure differences can occur from variations in both moisture content and soil temperature. Near permanent wilting percentages, temperature differences are the principal cause of vapor pressure differences, and hence, of moisture movement.

9.5 INTAKE CHARACTERISTICS OF SOILS

In all irrigation methods, except sub-irrigation, water is applied to the surface of land where it subsequently enters the soil and is stored for later use by plants. The object of irrigation is to get water into soil where it can be stored. Thus, rate of entry of water into soil under field conditions, called intake rate, is of fundamental importance. One of the major problems on irrigated farms is low intake rates of fine-textured soils caused generally by excessive working of soil and by salinity. Unless water enters the soil and is stored therein, crops cannot grow.

Intake rate varies with many factors, including depth of water on the surface, temperature of water and soil, soil structure and texture, and moisture content of the soil. Retarding layers of soil will greatly influence the rate. Configuration of the surface such as furrow shape and size, as well as the method of application, will be influencing factors also. Hence, intake rate varies from place to place on a field and it also varies with time. Sandy soils may have rates in excess of 10 inches per hour, whereas clay soils may have rates approaching zero when soil structure has been practically destroyed by faulty management. A set of typical intake curves and ways of presenting these data are shown in Fig. 9.8. Usually the intake rate plotted against time

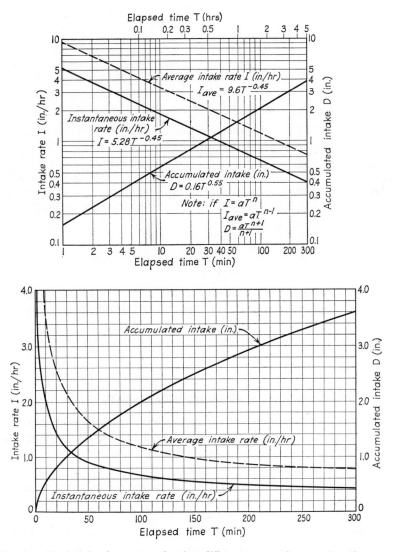

Fig. 9.8 Typical intake curves showing different ways of presenting the same data. The method to be used depends upon the use to be made of the results.

on a semi-logarithmic scale will show as a straight line, and therefore can be represented by the following equation:

$$I = aT^n \qquad (9.9)$$

When the observation of intake extends over long periods, a better representation of the data can usually be obtained by using the equation:

$$I = aT^n + b \qquad (9.10)$$

Since n is negative, I decreases with an increase in T. Therefore, the intake rate I will approach a constant value b as time increases. Generally, the intake does approach a constant rate, referred to as final intake rate. How final intake rate varies with soil texture is shown in Table 7.4.

Caution should be observed in using final intake rates for irrigation design. When rather light, frequent irrigations are applied, the irrigation may be completed before the final intake rate is reached. Since initial rates are considerably in excess of final rates, the amount of water entering the soil can best be represented by the accumulated depth of water that has entered the soil. This quantity is represented by the integral of Equation 9.9 as

$$D = \frac{a}{n+1} T^{n+1} = CT^m \qquad (9.11)$$

or, integrating Equation 9.10 when this equation represents more accurately the intake function, the accumulated depth of water applied becomes

$$D = \frac{a}{n+1} T^{n+1} + bt = CT^m + bT \qquad (9.12)$$

9.6   MEASURING INTAKE RATE

The best method of measuring intake is to obtain direct measurements by recording the water applied less the water flowing from the field. When direct measurements are not feasible, intake cylinders can be used with reasonable success.

The cylinders should be at least 9 inches in diameter, made of smooth tough steel, strong enough to drive, but thin enough to enter the soil with a minimum of disturbance. Cold rolled steel about 0.08 inch thick (14 gage) works well. Cylinders should be about 12 inches long. Since approximately five tests should be made in a given location, it is well to make five cylinders of somewhat different diameters so that all five can be nested together and thereby be made less bulky for handling.

Care should be taken to place the cylinders in an area representative of the field to be evaluated. Cylinders should be carefully driven into the soil to a depth of about 6 inches. The soil profile should be examined, the moisture estimated or measured, the notes made of the soil cover and surface condition.

After water penetrates to the bottom of the cylinder, it will begin to spread radially and the rate of intake will change accordingly. When a principal restricting layer does not lie within the depth of penetration of the cylinder, this radial flow will cause considerable change in the intake rate. Buffer ponds surrounding the cylinder are used to minimize this effect. Buffer ponds can be constructed by forming an earth dike around the cylinder or by driving into the soil a larger diameter cylinder concentric with the intake cylinder. The water levels in both cylinders should be approximately the depth to be expected during the irrigation. When comparative results from different locations are needed, the depth should be essentially the same in all cylinders. Care should be taken not to puddle the soil when water is added to the cylinder or the buffer pond.

The results obtained with cylinders are indicative of the rates to be expected during irrigation. Close correspondence is obtained when the soil surface is covered by the irrigation water. Considerable departure usually occurs when the irrigation water is applied by furrows or sprinklers. Hence, the cylinders are generally used to obtain an index from which design values can be obtained on the basis of local experience.

Other methods simulating ponding, furrows, rainfall, or sprinkling have been developed to measure intake rate. Good results have been obtained using these methods, but their practicality depends upon the accuracy desired, and upon the time and money available.

Equation 7.9, $qt = ad$, can be used to obtain the following equation:

$$I_{ave} = \frac{\text{gpm}}{\text{spacing}} \tag{9.13}$$

where I_{ave} is the average intake in inches per hour that would result if the actual intake occurring in 100 feet of furrow, measured in gallons per minute, was spread over the spacing between furrows measured in feet. Thus, if 3 gpm entered the soil in 100 feet of furrow and the furrows were 2 feet apart, the average intake, considering the entire soil surface, would be 3/2 or 1.5 inches per hour.

9.7 INFILTRATION AND SOIL-WATER MOVEMENT DURING IRRIGATION

Having considered the intake phenomenon and the empirical equations describing its behavior, an analysis of water movements during irrigation will shed further light on the irrigation process.

Experiments point conclusively to the existence of three distinct zones. Since these zones are produced by a dynamic continuum, they are inseparably connected, but nevertheless show distinct characteristics pertaining to the nature of flow through soil. These three zones are the transmission zone, wetting zone, and wetting front.

Within the so-called transmission zone, moisture content is essentially constant regardless of the direction of flow. Degree of saturation (approximately 80 percent) is considerably less than the 100 percent commonly assumed. Since experiments indicate an approximately linear relationship between moisture content and permeability over this range of saturation, it is quite possible that a simple expression for permeability in terms of moisture content can be obtained for any particular homogeneous soil to represent effective transmissibility of the transmission zone.

Associated with the constant moisture content throughout the transmission zone is the experimentally verified principle that the gradient of the driving force is also constant throughout this zone. Furthermore, a constant potential gradient implied by a constant permeability throughout the transmission zone aids materially in the development which follows.

When the surface layer of soil is disturbed, as is the case in most irrigations, the degree of saturation and resistance to flow is not the same as in the soil beneath the surface layer. Usually the resistance to flow increases considerably due to structural breakdown and segregation of particles as well as packing of the surface soil.

Moisture content within the wetting zone reduces rapidly as the wetting front is approached. This in turn causes a corresponding decrease in permeability. In a moist soil the wetting zone is deeper or occupies more soil than in a dry soil. This is so because the hydraulic conductivity is greater; also there is a tendency for the displacement of water ahead of the actual incoming flow.

Analysis indicates that the energy consumed within the wetting zone is approximately constant when the moisture ahead of the wetting zone is constant. Hence, for a given available energy, the gradient of the driving force will be much less in a moist soil than in a dry soil. In other words, dry soil with a low permeability will result in a greater loss of energy per unit length, causing a shallower wetting zone.

Visual observation of the wetting phenomena which take place within the wetting zone indicates that the flow occurs in a very erratic manner. The movement which might be described as a "jumping" movement takes place essentially in increments. This "jumping" movement is more pronounced in drier soils. The advancing or wetting front is in effect a capillary fringe, the foremost part of which has an average moisture content equal to the original soil, while the moisture content of each succeeding cross section behind the wetting front increases until approximately 80 percent of the conducting pores are filled with moisture. The lateral extent of this transition is a measure of the capillary fringe.

When the position of the wetting front is recorded, whether from visual observation or for tensiometer or moisture data, it is usually the forepart of the capillary fringe that is interpreted as the wetting front.

Using the basic principles of flow through soils, it can be shown that the rate of downward movement V_y can be defined by the following equation:

$$V_y = \frac{I_f}{ns} \qquad (9.14)$$

where I_f is the final intake rate, n is the total pore space of the soil, and s is the change in degree of saturation during the irrigation. Hence, for a sandy soil having a final intake rate I_f of 2 in. per hr, a porosity of 38 percent, and a degree of saturation of 22 percent, the rate of movement of the water into the soil will be 24 in. per hr. Likewise, in a loam soil with $I_f = 0.5$ in. per hr, $n = 47$ percent, and $s = 24$ percent, the rate of downward movement is 4.5 in. per hr.

It can also be shown that for downward movement the final intake rate I_f is equal to the permeability of the soil at the degree of saturation (about 80 percent) which occurs during irrigation.

When water enters the soil and encounters a restricting soil layer, causing it to move laterally, the intake rate and the rate of movement throughout the soil will both decrease with the square root of time. Thus, when observations of intake are being made, it is well to observe the trend with time as an indication of the nature of movement below the surface. When the intake rate approaches a constant, it can be assumed that water is moving downward in an unrestricted manner. However, when the rate of intake continues to decrease with time, not approaching a constant value, then it can be assumed that the flow is moving laterally, due to restricting layers beneath the surface.

Observation of the soil profile should readily verify the existing position and nature of restricting layers.

9.8 NON-HOMOGENEOUS ANISOTROPIC SOIL

Soils are seldom homogeneous. Usually they are non-homogeneous and also anisotropic, meaning that the permeability not only varies from point to point, but that the value of the permeability in one direction is different from the value in another direction, even at a given point within the soil. This lack of uniformity becomes most extreme in alluvial soils. Frequently the horizontal permeability of an alluvial soil is ten times greater than the vertical permeability because of the natural horizontal orientation of the major axis of particles deposited under water. The presence of tight clay layers further decreases the relative vertical permeability. Thus, it is very important to recognize the non-homogeneous as well as the anisotropic nature of the soil when permeability measurements are being taken. For example, an undisturbed vertical core of soil may be tested in the laboratory and found to have essentially no vertical permeability. However, in the field this same soil may drain readily due to the presence of interconnected sandy layers of soil. Flow of water toward wells and drains is predominantly horizontal, and the permeability for horizontal flow is more important than for vertical flow.

Also, because of the extensive variability which can, and does, frequently occur from point to point in a soil, the results of intake or permeability measurements which are essentially point observations should be used cautiously. A sample of soil represents only a very, very small portion of the soil to be evaluated.

Because of these problems of measurement caused by variability in soil and permeability, it is generally not feasible to use precise techniques of permeability measurement for irrigation purposes. More measurements made more quickly at less cost are generally more valuable than a few precise determinations on a few samples which may not be, and probably are not, representative of the field to be evaluated.

9.9 LIMITATIONS OF EQUATIONS

Because of variability which exists in soils, judgment and discretion are essential in applying the equations of this chapter to the solution of practical irrigation and drainage problems. The equations are valid, but the values applied to those equations are not precise. A major challenge exists for the researcher to develop more practical methods of measuring soil properties over extensive areas rather than more precise determinations at a point in the soil.

REFERENCES

Bodman, G. B. and E. A. Colman, "Moisture and Energy Conditions during Downward Entry of Water into Soils," *Proc. Soil Sci. Am.*, Vol. 8, pp. 116–122, 1943.

Bondurant, James A., "Developing Furrow Infiltrometer," *Agr. Eng.*, August 1957.

Brooks, R. H. and R. C. Reeve, "Measurement of Air and Water Permeability of Soils," *Trans. Am. Soc. Agr. Eng.*, 1959.

Browning, G. M. and F. M. Milam, "The Lateral Movement of Water in Relation to Pasture Contour Furrows," *Proc. Soil Sci. Soc. Am.*, Vol. 5, pp. 386–389, 1940.

Colman, E. A. and G. B. Bodman, "Moisture and Energy Conditions during Downward Entry of Water into Moist and Layered Soils," *Proc. Soil Sci. Soc. Am.*, Vol. 9, pp. 3–11, 1944.

Donnan, W. W. and J. E. Christiansen, "Ground Water Determinations," *Western Construction News*, Vol. 19, No. 11, pp. 77–79, 1944.

Duley, F. L. and C. E. Domingo, "Effect of Water Temperature on Rate of Infiltration," *Proc. Soil Sci. Soc. Am.*, Vol. 8, pp. 129–134, 1943.

Edlefsen, N. E. and G. B. Bodman, "Field Measurements of Water Movement through a Silt Loam Soil," *J. Am. Soc. Agron.*, Vol. 33, pp. 713–731, 1941.

Fletcher, Joel E., "Some Properties of Water Solutions that Influence Infiltration," *Trans. Am. Geophysical Union*, Vol. 30, No. 4, August 1949.

Hansen, Vaughn E., "Infiltration and Soil Water Movement During Irrigation," *Soil Sci.*, Vol. 78, No. 2, February 1955.

Hendricks, David W., "Conductivity Variation on Anisotropic and Non-homogeneous Conducting Media Defined by Stream Function and Potential Function Relationships Using a Variable Resistance Electrical Analogue," Utah State University (Thesis), 1960.

Kirkham, Don, "Proposed Method for Field Measurement of Permeability of Soil below the Water Table," *Proc. Soil Sci. Soc. Am.*, Vol. 10, pp. 55–68, 1945.

McCalla, T. M., "Factors Affecting the Percolation of Water through a Layer of Loessial Soil," *Proc. Soil Sci. Soc. Am.*, Vol. 9, pp. 12–16, 1944.

Merriam, John L., "Field Method of Approximating Soil Moisture for Irrigation," *Trans. Am. Soc. Agr. Eng.*, Vol. 3, No. 1. Sp. Soil and Water Edition, pp. 31–32, 1960.

Moore, Ross E., "Water Conduction from Shallow Water Tables," *Hilgardia*, Vol. 12, No. 6, pp. 383–426, 1939.

Moore, R. E. and Kenneth R. Goodwin, "Hydraulic Head Measurements in Soils with High Water Tables," *Agr. Eng.*, Vol. 22, No. 7, pp. 263–364, 1941.

Nelson, W. R. and L. D. Baver, "Movement of Water through Soils in Relation to the Nature of the Pores," *Proc. Soil Sci. Soc. Am.*, Vol. 5, pp. 69–76, 1940.

Pillsbury, A. F. and J. E. Christiansen, "Installing Ground Water Piezometers by Jetting for Drainage Investigations in Coachella Valley, California," *Agr. Eng.*, Vol. 28 pp. 409–410, illus. (U.S. Regional Salinity Lab. 76), 1947.

Richards, L. A., "The Usefulness of Capillary Potential to Soil-Moisture and Plant Investigators," *J. Agr. Research*, Vol. 37, No. 12, pp. 719–741, 1940.

Richards, L. A., "Concerning Permeability Units for Soils," *Proc. Soil Sci. Soc. Am.*, Vol. 5, pp. 49–53, 1940.

Taylor, Sterling A., "How Water Moves in Soil," *Farm and Home Sci.*, Vol. 18, No. 4, pp. 84–85, December 1957.

"The Use of Cylinder Infiltrometers to Determine the Intake Characteristics of Irrigated Soils," U.S.D.A.—A.R.S. and S.C.S., May 1956.

CHAPTER 10 SALT PROBLEMS IN SOIL AND WATER

Excess soluble salts and alkali and the occurrence of drought cause sterility and barrenness of arid-region soils. Although saline and alkali lands are characteristic of arid regions, they are neither of general occurrence nor of uniform distribution. Reclamation of saline and alkali soils, and prevention of excess salts in fertile areas now irrigated, are of paramount importance in arid regions.

Saline soils are soils having excess soluble salts that make the soil solution sufficiently concentrated to injure plants and impair soil productivity. The term "alkali soil" is applied to soils which have an excess of exchangeable sodium either with or without excessive total soluble salts.

10.1 CLIMATE AND SALINITY

Arid-region soils contain relatively large amounts of soluble salts. The heavy annual rainfalls of humid regions cause water to percolate through the soil and carry to the streams, rivers, and oceans large amounts of soluble mineral substances. The scanty rains of arid regions do not penetrate the virgin arid soils deeply enough to cause appreciable percolation. The greatest depth of penetration of water from natural precipitation in arid regions, either melted winter snow or wet-season rains, is found to vary from 1 to 4 feet, depending on the depth and time of the precipitation and the nature of the soil. Lack of percolation through arid-region soils, together with excessive evaporation of water in arid regions, gives rise to accumulation on and in the soils of soluble salts that are injurious to plant life. Hence, the basic cause for saline and alkaline soil is insufficient application of water.

10.2   SOURCES AND ACCUMULATION OF SOLUBLE SALTS

Mineral soils are derived largely from the weathering of rocks. There is an intimate relation between salt accumulation and the chemical composition of rocks from which soils are formed. Soils formed directly from salt-bearing rocks usually contain excessive salts.

Some arid-climate soils that were free from excess salts before cultivation have been rendered non-productive by the use of irrigation water containing excessive quantities of salt. Water that percolates through salt-bearing rocks usually contains an appreciable amount of salt.

Salts continue to accumulate in soils of irrigated areas where greater amounts are brought in than are removed. Irrigation waters contain from $\frac{1}{10}$ to 5 tons of salt per acre-foot of water. Some irrigation farmers apply only a depth of 2 feet of water per season; others, where the summers are long and hot, apply up to 5 feet of water or more. Where drainage is not provided, the irrigation water may add as much as 1 to 10 or more tons of salt each year to an acre of land.

The most effective method for the removal of salt from soil is by means of water which passes through the root zone of the soil; but, if the amount carried away is less than the amount brought in by irrigation water, salt will accumulate. To prevent salt accumulation, and consequent decrease in crop yields, irrigators must remove as much salt as is brought in. In some areas, an effort is made to spread limited irrigation water supplies over too many acres, with the result that the soil is not wetted below a depth of a few feet. In other areas, the ground-water table is so close to the surface as to retard or prevent the leaching of salt from the root-zone soils. In soils having shallow water tables, upward flow of saline ground water results in a continuing accumulation of salt in the surface soil.

The barrenness that results from excessive accumulation of soluble salts is illustrated in Figs. 10.1 and 10.2.

10.3   RELATION BETWEEN CONCENTRATION
AND CONDUCTIVITY

The salt content of soil solutions or irrigation waters is usually represented in one of three ways; parts per million (ppm), milliequivalents per liter (MEQ/L), or as electrical conductivity expressed in micromhos per centimeter (EC $\times$ 10^6). Conductivity has the advantage of simplicity since it can be measured readily in the field or laboratory by using a portable electrical conductance meter. The

Fig. 10.1 Surface accumulation of salts on lands near Delta, Utah, prevents the growth of crops without leaching. (*Utah Agr. Exp. Sta. Bul.* **335**.)

relationship between concentration and conductivity is shown in Fig. 10.3.

10.4   SOME BASIC TERMS

In a contribution entitled "Diagnosis and Improvement of Saline and Alkali Soils," the United States Regional Salinity Laboratory presents a glossary of sixty terms together with definitions.   The

Fig. 10.2 Punjab flax fails to grow on saline spots at Meloland Station, Imperial Valley, California. (*U.S.D.A. Circ.* **707**.)

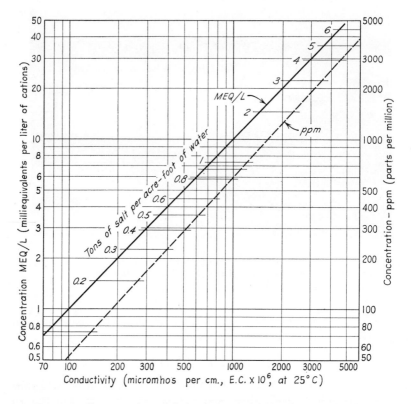

Fig. 10.3 Concentration of irrigation waters related to conductivity.

twelve terms most used in describing soils having excessive soluble salts and exchangeable sodium are presented here:

Alkali Soil. Soil which has an excessive degree of saturation with exchangeable sodium, either with or without appreciable amounts of soluble salts.

Alkaline. A chemical term referring to "basic" soil reaction where the pH is above 7. The pH is the logarithm of the reciprocal of the hydrogen ion concentration, pH 7 is neutral.

Electrical Conductivity. The reciprocal of electrical resistivity. The resistivity is the resistance of ohms of a conductor which is 1 sq cm. Therefore, electrical conductivity is expressed in reciprocal ohms per centimeter or mhos per centimeter. The terms "electrical conductivity" and "specific electrical conductance" have identical meaning.

Equivalent Weight (or combining weight with hydrogen). Atomic or formula weight divided by its valence. For example:

$$Na^+ = \frac{23}{1} = 23 \qquad Ca^{++} = \frac{40}{2} = 20$$

$$Cl^- = \frac{35.5}{1} = 35.5 \qquad SO_4^= = \frac{96}{2} = 48$$

Exchangeable-Sodium Percentage. Degree of saturation of the soil-exchange complex with sodium, defined as follows:

ESP =

$$\frac{\text{Exchangeable sodium (milliequivalents per 100gm soil)}}{\text{Cation-exchange capacity (milliequivalents per 100gm soil)}} \times 100$$

Non-Saline-Alkali Soil.[1] Soil for which the exchangeable-sodium percentage is greater than 15, and the conductivity of the saturation extract is less than 4 millimhos per centimeter (at 25°C). The pH values for these soils generally range between 8.5 and 10.

Saline-Alkali Soil. A soil for which the conductivity of the saturation extract is greater than 4 millimhos per centimeter (at 25°C), and the exchangeable-sodium percentage is greater than 15. The pH of the saturated-soil paste may somewhat exceed 8.5.

Saline Soil. A soil for which the conductivity of the saturation extract is greater than 4 millimhos per centimeter (at 25°C), and the exchangeable-sodium percentage is less than 15. The pH of the saturated-soil paste is usually less than 8.5.

Soil Reaction, pH. The pH scale is a measure of the effective concentration of the hydrogen ion. It has long been used as an index of the acidity or alkalinity of soils. A pH value of 7.5 to 8.0 usually indicates the presence of carbonates of calcium and magnesium, and a pH of 8.5 or above usually indicates appreciable exchangeable sodium.

Reference to these terms and definitions will be helpful in a study of this chapter. A summary of conditions for saline, saline-alkali, and alkali soils is shown in Table 10.1.

Saline Soils. For saline soils the electrical conductivity of the saturation extract is more than 4 millimhos per centimeter, and the exchangeable sodium percentage is less than 15. These soils corre-

---

[1] The term commonly used is simply "alkali" soil. Recently the term "sodic" is being used in lieu of "alkali" for clarity of nomenclature.

TABLE 10.1

CONDITIONS FOR SALINE, SALINE-ALKALI, AND ALKALI SOILS

Salt Condition	Common Term	Salt Index — Conductivity of Saturation Extract in Millimhos per Cm	Sodium Index — Exchangeable Sodium Percentage	Hydrogen Ion Index[1] — pH	Reclamation
Saline	White alkali	>4	<15	≦ 8.5	Leaching
Saline-Alkali		>4	>15	generally about 8.5	Leaching necessary and possible, but as salts are removed the sodium must be replaced to prevent dispersion of soil particles and reduction of permeability.
Alkali (Sodic)	Black alkali[2]	<4	>15	generally between 8.5–10.0	Low permeability due to dispersion of soil by the sodium requires replacing the sodium to improve the permeability so that leaching can proceed.

[1] Since a pH of 7.0 is neutral, pH values less than 7.0 indicate an acid soil which is common in the non-arid regions. Note that the larger the pH, the less the concentration of hydrogen ions, since pH is the logarithm of the reciprocal of the hydrogen ion concentration.

[2] A black crust forms on the surface of an alkali soil only if organic matter is present. Hence, the term may be misleading in soils of low organic matter.

spond to Hilgard's "white alkali" soils. When adequate drainage is established and the soluble salts are removed by leaching, the soils become normal, that is, non-saline.

Saline soils may develop from normal soils through the accumulation of salts from applied irrigation water, or by upward moving ground water, or by a combination of both processes.

The chemical characteristics of saline soils are determined by the kinds and amounts of salts present which largely control the concentration and osmotic pressure of the soil solution. Sodium seldom comprises more than half of the soluble cations, and thus is not absorbed to any appreciable extent. The relative amounts of calcium and magnesium in the soil solution may vary considerably. Soluble and exchangeable potassium are ordinarily only minor constituents but may occasionally occur in excess. The chief anions are chloride, sulfate, and sometimes nitrate. Small amounts of bicarbonate may occur, but because the pH is 8.5 or less, soluble carbonates are usually absent. In addition to the readily soluble salts, saline soils may contain relatively insoluble salts such as calcium sulfate (gypsum), and calcium and magnesium carbonates.

Owing to the presence of excess salts, and the absence of appreciable exchangeable sodium, the colloids in some saline soils are highly flocculated, thus providing a favorable structure and improved permeability to water and air. However, some saline soils have very low permeabilities.

Saline-Alkali Soils. Soils for which the conductivity of the saturation extract is greater than 4 millimhos per centimeter, and the exchangeable-sodium percentage is greater than 15, are designated saline-alkali soils. As long as excess salts are present, the appearance and properties of these soils are generally similar to those of saline soils. Under conditions of excess salt, the pH value is seldom higher than 8.5, and the colloids remain flocculated. If the excess soluble salts are temporarily leached downward, the properties of these soils may change markedly (unless gypsum is present) and become similar to those of non-saline-alkali soils. As the concentration of the salts in the soil solution is lowered, some of the exchangeable sodium hydrolyzes and forms sodium hydroxide. The sodium hydroxide may change to sodium carbonate upon reaction with carbon dioxide. Then the soil usually becomes strongly alkaline (pH above 8.5), the colloids disperse, and the soil develops a structure unfavorable for water infiltration and percolation, and for tillage. Although the return of the soluble salts may lower the pH value and restore the colloids to a flocculated condition, the management of saline-alkali soils continues to be a problem until the excess salts and exchangeable sodium are removed from the root zone.

Alkali Soils. When the exchangeable-sodium percentage of soils is greater than 15, and the conductivity of the saturation extract is less than 4 millimhos per centimeter, the pH values generally range between 8.5 and 10 and the soils are designated alkali soils. These soils correspond to Hilgard's "black alkali" soils. They frequently occur in semi-arid and arid regions in small irregular areas, which are referred to as "slick spots." Except when gypsum or another source of soluble calcium is present, the drainage and leaching of saline-alkali soils develop alkali soils. The removal of excess salts in such soils permits hydrolysis of the exchangeable sodium, and may lead to the formation of small amounts of sodium carbonate. The soil organic matter is highly dispersed and distributed over the soil particles, thereby darkening the color. When the soil contains appreciable organic matter, its surface may be quite dark; hence the term "black alkali."

Clay which is partially sodium-saturated is highly dispersible in the absence of flocculating salts and has a tendency to migrate downward through the soil and accumulate at lower levels. As a result, the surface few inches of soil may be relatively coarse in texture and friable; but, below where the clay accumulates, the soil develops a dense layer of low permeability. The physical and chemical properties of alkali soils are largely determined by the exchangeable sodium present. As the proportion of exchangeable sodium increases, the soil tends to disperse more and the pH value increases, becoming as high as 10. The soil solution of alkali soils, although relatively low in soluble salts, has a composition which differs greatly from that of normal and saline soils. The anions present consist mostly of chloride, sulfate, and bicarbonate, but small amounts of the normal carbonate usually occur. At high pH values, and in the presence of normal carbonate, calcium and magnesium are precipitated so that the soil solutions of alkali soils usually contain only traces of these cations, sodium being the predominant one. Considerable amounts of exchangeable and soluble potassium occur in some alkali soils.

10.5 CHEMISTRY OF SALTY SOILS

The following clarifies the above statements regarding the sodium (Na) and hydrogen (H) ions:

$$\text{Ca} \boxed{\begin{matrix} \text{Na} \\ \text{Clay Particle} \\ \text{Na} \end{matrix}} \text{Na} + \text{H}_2\text{O} \rightleftharpoons \text{NaOH} + \text{H} \boxed{\begin{matrix} \text{Na} \\ \text{Clay} \\ \text{Na} \end{matrix}} \text{Ca} \quad (10.1)$$

A clay particle with sodium and calcium ions attached tends to hydrolyze. When a sodium ion is exchanged for a hydrogen ion, and the sodium ion combines with a molecule of water, sodium hydroxide (NaOH) is formed. When carbon dioxide (CO_2) is present in the soil air, it readily reacts with the water to form hydrogen carbonate (H_2CO_3). However, the sodium hydroxide (NaOH) reacts readily with the hydrogen carbonate (H_2CO_3) to form sodium carbonate (Na_2CO_3).

$$H_2CO_3 + 2NaOH \rightleftharpoons Na_2CO_3 + 2H_2O \quad (10.2)$$

The Na_2CO_3 is gradually removed with extensive leaching, and the soil is left with hydrogen ions having replaced sodium ions. This increase of hydrogen ions is reflected in a lower pH (pH being a reciprocal index of hydrogen ion concentration).

Where soils contain calcium carbonate ($CaCO_3$) or gypsum, calcium is dissolved into the soil solution. This available calcium is exchanged for sodium during the leaching process to obtain a normal soil.

$$\text{Ca} \boxed{\begin{matrix} \text{Na} \\ \text{Clay Particle} \\ \text{Na} \end{matrix}} \begin{matrix} \text{Na} \\ \text{Na} \end{matrix} + \text{CaCO}_3 \rightleftharpoons \text{Na}_2\text{CO}_3 + \text{Ca} \boxed{\begin{matrix} \text{Na} \\ \text{Clay Particle} \\ \text{Na} \end{matrix}} \text{Ca}$$
$$(10.3)$$

When chemicals must be added to the soil to accomplish the exchange of sodium for other ions, the following products can be used:

gypsum ($CaSO_4 \cdot 2H_2O$)
soil sulfur (S)
sulfuric acid (H_2SO_4)
concentrated lime sulfur solution ($CaS_x + H_2O$)
dry lime sulfur (CaS_x)

Availability, cost, and solubility should be considered when determining the most feasible product to use.

10.6 MOVEMENT OF SALTS IN SOILS

If it were possible to maintain a moisture distribution in irrigated soils such that water flow would be continuously downward, there would be relatively little trouble from salinity on irrigated farms. A continuous downward flow of water with adequate drainage would gradually decrease the soluble salts in the upper few feet of soil where plants obtain most of their moisture and food. However, without adequate drainage downward, percolating waters fill the lower soil spaces and cause the water table to rise. During periods between irrigations, a high water table favors the upward capillary flow of water to the land surface, where the water evaporates. The soluble salts carried by the upward-moving water cannot be evaporated, hence they are deposited on or near the soil surface. Salts so deposited may come from salty soil horizons well below the surface. The mere concentration on the surface of salts that are normally distributed through the upper few feet of soil may cause serious salinity.

10.7 INFLUENCE OF WATER TABLE

Irrigation farmers sometimes point out the advantages of keeping the water table within a few feet of the soil surface because of the high crop yields obtained during the early years after it has risen from great depths. The favorable moisture supply from a water table near the soil surface may cause high crop yields; but as a rule, in areas where alkali salts occur, the temporarily favorable condition of the high water table is followed by a serious decrease in yield, if not by complete non-productivity due to salt concentration.

Water tables can be lowered by preventing excessive seepage losses from canals, by careful and efficient application of water on the farms, and by providing artificial drainage on areas for which natural drainage is inadequate. Keeping the water well below the soil surface is very urgent and not likely to be overemphasized.

10.8 MANAGEMENT OF A HIGH WATER TABLE

Despite hazards of a high water table, good management can obtain a permanently good production with a high water table. The major need is for lighter and more frequent irrigations during the growing season. Sprinkler irrigation is especially well adapted to this condition because of the ability of the irrigator to apply a limited amount of water. Special care must be taken not to apply an excessive irrigation during the growing season, which would result in a rise of water table within the root zone of the crop.

During the dormant season, one or more heavy leaching irrigations will be necessary to remove the excess salts from the soil. During dormancy, a temporary rise of the water table generally will not seriously damage even the perennial crops.

Since a crop will naturally obtain a considerable portion of its needed water from a high water table, the amount of water that must be applied by irrigation is reduced, resulting in a reduction in cost of irrigation. Hence, there are distinct economic benefits arising from a reasonably high water table, but good management is essential if a profitable agriculture is to be maintained. Not to be overlooked is the reduced root zone available to the plants which usually causes a reduced yield. Therefore, the economic depth for a water table depends upon management, the crops to be grown, and the cost of irrigation. The hazard to productive farming is definitely greater with a higher water table.

10.9   STEPS ESSENTIAL TO RECLAMATION

Temporary control of salts on irrigated land is sometimes practiced by one or more of the following methods:

(a) Plowing salt-surface crusts deeply into the soil.
(b) Removing surface accumulation from the soil.
(c) Neutralizing the effects of certain salts by use of other salts or acids.

Permanent reclamation of saline and alkali land requires four essential steps and the attainment of four basic conditions, namely:

(a) Adequate lowering of water table.
(b) Satisfactory water infiltration.
(c) Leaching excess salts out of the soil.
(d) Intelligent management of soil.

Adequate Lowering of Water Table

All waterlogged lands, whether or not impregnated with alkali, are improved for ordinary crops by lowering the water table. This means a permanent lowering under the farmer's control so that a rise of water above a given elevation in the soil for any length of time may be wholly prevented. The first step in lowering the water table is to learn the source of the water that caused it to rise. In isolated cases on small tracts, it is sometimes possible for one farmer alone, or a small group of farmers, to find the water source and cut it off by construction of one or more intercepting ditches or drains. Usually, in irrigated regions, small waterlogged areas are caused by surface or underground water flowing to them from higher irrigated lands, or from canals, ponds, or reservoirs. The farmer whose holdings are located within large areas of comparatively level waterlogged land cannot, as a rule, lower the water table by his own efforts. For such areas community action is essential.

Satisfactory Intake Rates

The rate of water infiltration into soils depends on soil texture, structure, degree of dispersion, and also on the depth to the water table. When adequate drainage has been provided, the structure of the soil and other soil properties are dominant in influencing intake rates. The time required for an adequate depth of water to percolate through the soil may well limit the feasibility of reclamation.

Alkali soils disperse during leaching and often become impermeable as the soluble salts are removed. Chemical amendments are then required to bring about improvement of the soil by replacing the exchangeable sodium with calcium. Gypsum, and under certain conditions, sulfur, may be used for this purpose. Considerable gypsum occurs naturally in the soils of some areas, but the amounts and distribution within the soil profile are variable.

Usually the cheapest method of obtaining satisfactory intake rates is to minimize the surface tillage. Leaving the soil surface rough and cloddy will materially increase the rate of entry of water through the soil. Avoiding tillage when the soil is either too wet or too dry is essential. Working a dry powdery soil is especially harmful.

A saline water, not containing appreciable quantities of sodium, will also improve intake rates. For leaching purposes, a saline water may be very useful for increased intake rates; also the saline water often contains the salts that would otherwise be applied as chemical soil amendments. Hence, the quality of water may greatly influence the intake rates.

Adding chemical amendments to reduce dispersion of soil particles is not the only method of increasing the intake rate. Good tillage practice will invariably increase the rate significantly. Also, the value of deep-rooted crops like clover and alfalfa should not be neglected. Getting plant roots to penetrate soil will also materially improve soil permeability to air and water. Not only do these roots leave cavities when they decay, but they also remove water from the soil, causing alternate wetting and drying which are essential for building soil structure.

Information on the rates at which water enters and moves through soil is useful in connection with irrigation methods as well as with improvement of saline and alkali soils. Many factors which are not yet fully understood and correlated influence these rates, but it is not difficult to make measurements that have practical significance. Infiltration and permeability are expressed in terms of the velocity of water flow, and for irrigation use are usually reported either in inches per hour or feet per year. Intake rates are generally measured in the field. The principal method is flooding or impounding water on the soil surface.

Water of the same quality as the irrigation or leaching water should be utilized for infiltration tests in the field; otherwise, the measurements may be misleading. The length of time the test should be conducted, or the depth of water to be applied, will depend on the purpose of the test and the kind of information sought. If it is a matter of appraising an irrigation problem, then the depth corresponding to one irrigation may be sufficient. If getting advance information on infiltration for planning a heavy leaching program is involved, then the full leaching depth is recommended. It often happens that subsurface drainage in the soil is sufficiently slow to reduce infiltration rates considerably. Although small-area tests will give useful information on infiltration during leaching, infiltration values thus obtained will apply to large areas only if underdrainage is adequate.

Permeability as measured in the laboratory is influenced by many factors. Some of these, such as dispersion of the soil, base status, air saturation, and microbial sealing, have been separately studied and partially appraised. The structure and packing of the sample enter directly into the measurement. It appears that air-dried and screened samples may be used if interest is centered primarily on the cultivated layer of soil. Natural structure may be required for significant measurements on the sub-surface layers.

Leaching Excess Salts

It is usually essential that large depths of water be applied to saline and alkali lands and be made to percolate through the soil in order to leach out excess salts. Coarse-textured soils of open structure as a rule have sufficiently high permeability to make leaching of alkali salts an easy task after the water table has been sufficiently lowered. Fine-textured compact soils of low permeability predominate in the low-lying waterlogged areas. Consequently, soil permeability is a factor of paramount importance in the leaching of excess soluble salts from most waterlogged soils. Permeability is influenced not only by the texture and compactness of the soil, but also by flocculation or dispersion of the soil particles. The dispersion, and consequently the permeability, is greatly influenced by certain chemical compounds. A very low permeability sometimes follows the leaching of alkali salts, and this decreases the productivity of the soil because of the difficulty of getting air and water to penetrate it.

Maintaining a favorable salt balance in the soil requires proper and efficient irrigation methods. Irrigation must provide water for growth of crops and at the same time allow enough water to pass through the soil to leach out the excess salts. Excessive leaching, however, may be detrimental in that plant nutrients, especially nitrates, may be removed from the soil. Overuse of water also adds to the drainage problem. Water applications should be sufficient to maintain a favorable salt balance in the soil, but without excessive leaching of the plant nutrients and without materially adding to the drainage problems.

The effectiveness of leaching on the yield of grain in bushels per acre for each of three experimental sites is shown by the curves of Fig. 10.4. At site A, for example, the yield was increased from about 12 bu per acre with 1-foot depth of water up to more than 40 bu with 4 feet. Without leaching, the land was barren.

10.10 LEACHING REQUIREMENTS

The amount of water which must be leached through the root zone to keep a favorable salt balance has been formulated by the staff of the U.S. Salinity Laboratory. The leaching requirement is defined as the fraction of the irrigation water that must be leached through

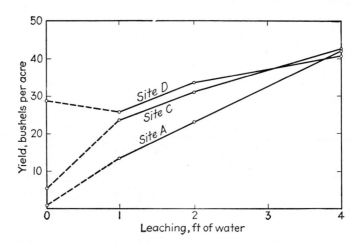

Fig. 10.4 Relation of grain yields to depth of water used for leaching. (*Utah Agr. Exp. Sta. Bul.* 335.)

the root zone to control soil salinity at any specified level. The concept is most useful when applied to steady-state water-flow rates, or to total depths of water used for irrigation and leaching over a long period of time. The amount of water required to leach a saline soil will be several times greater than the leaching requirement determined from the following formula to maintain a healthy soil.

$$LR = \frac{D_d}{D_i} = \frac{EC_i}{EC_d} \tag{10.4}$$

where LR is the leaching requirement, D_d and D_i are the depths of drainage water and irrigation water respectively, and EC_i and EC_d are the electrical conductivities of the irrigation and drainage waters. When rainfall during the period of observation is appreciable, or when water moves into the root zone from the water table, a weighted average of conductivities for all sources of water should be used.

The leaching requirements for different conductivities and concentrations of irrigation water are shown in Table 10.2. Note that the leaching requirement increases at least 10 percent for each 1000 parts per million of dissolved solids in the irrigation water. Hence, irrigation water having 4000 parts per million will necessitate that approximately 50 percent of the applied water passes through the root zone.

The leaching requirement represents, therefore, a necessary water use for a permanent irrigated agriculture. When water needs for a project are being planned, the leaching requirements should be considered.

10.11   RECLAMATION AND MANAGEMENT OF SALINE AND ALKALI SOILS

Sometimes irrigation farmers who own alkali land on which the water table has been adequately lowered, and from which the excess salts have been leached, erroneously conclude that the task of reclamation has been completed. Reclamation is attained only when the lands are made to produce large crop yields. Usually, restoration to full productive capacity, or its establishment in virgin salty lands, requires both chemical and physical improvements in the soils.

Alkali soils usually contain excessive amounts of sodium, held by the finer soil particles, and therefore have a poor physical condition. To be reclaimed, these soils must be changed chemically and improved physically.

The chemical changes consist of exchanging calcium for sodium and then leaching away the sodium salts. Any soluble calcium com-

TABLE 10.2

LEACHING REQUIREMENT RELATED TO QUALITY OF DRAINAGE AND IRRIGATION WATER FOR AN EQUILIBRIUM CONDITION, IGNORING INFLUENCE OF RAINFALL

Quality of Irrigation Water				Percentage Leaching Requirement for Given Qualities of Drainage Water			
Micromhos/cm $EC \times 10^6$	Ppm	MEQ/L	Tons/Acre-Foot	5 millimhos/cm $EC \times 10^3$	10 millimhos/cm $EC \times 10^3$	15 millimhos/cm $EC \times 10^3$	20 millimhos/cm $EC \times 10^3$
100	63	1	.09	2	1	.7	.5
200	125	2	.17	4	2	1	1
400	250	4	.34	8	4	3	2
800	600	8	.82	16	8	5	4
1600	1000	16	1.36	32	16	11	8
3200	2000	32	2.72	64	32	21	16
6400	4000	64	5.44	128	64	43	32

pound will do, and the greater the concentration of calcium in the reclamation water, or in the soil solution bathing the soil particles during reclamation, the more rapidly this process will take place. If the soil does not contain gypsum, it can be added and dissolved by irrigation water. Other soluble calcium compounds such as calcium chloride are also very helpful but are expensive.

Some alkali and saline soils are low in available phosphorus and will give better crop yields if phosphate fertilizers are used. Super-phosphate, treble super-phosphate, and ammonium phosphate are among those generally recommended. Liberally applying barnyard manure, plowing under cover crops, the avoiding of plowing or of other farm operations when the soil is too wet or too dry are all helpful. Applying enough water to assure adequate penetration into heavy soils, keeping drains open and in good repair, preventing excessive evaporation, and other important precautions are urgently essential to the management of salty lands in order to assure permanent relief from waterlogging and excess salts, and to assure perpetuation of soil productivity.

10.12   SALINE WATER FOR IRRIGATION

It is probable that population pressures and public welfare will require the utilization of all available water in the arid regions of the world. Domestic water supplies are of first importance, and next in importance are the uses of water for irrigation and for industrial purposes. Proper quality of water is essential for satisfactory domestic use and irrigation. The intensity of demand for water is greater in some irrigated sections than in others, but in only a very few of the arid regions of the world is the total water supply, when fully stored and controlled, large enough to completely irrigate all the arable land. In some sections all the available water is used, and the so-called saturation point of irrigation expansion has been reached.

As the saturation point in irrigation expansion is approached, the need to use saline waters is increased. However, excessive amounts of salts in irrigation water result in fatality to crops and financial losses to water users. On the other hand, small amounts of salts in water not only are harmless, but also actually stimulate crop growth under some conditions. A study of the safe limits of salts in irrigation water, and of the conditions under which saline water may best be used for irrigation, is essential to complete an intelligent utilization of arid-region resources.

Since water is a vital resource essential to agriculture, saline irrigation water should not be pronounced unfit for irrigation without careful consideration of all factors concerned. Frequently, the salty water is the only water available. Then it becomes a matter of necessity to accept its limitation, use it wisely, and be grateful for a water supply even though it is not entirely satisfactory.

Too often saline waters are pronounced unsuitable without considering the need for water in the area. Permanent irrigated agriculture can be maintained using rather salty waters, with proper management. For example, the greater the salt content of the water, the larger is the percentage of applied water that must be leached through the soil to maintain a favorable salt balance. Hence, the amount and kinds of salt present in the irrigation water should be a guide to necessary management if the water is to be used successfully.

10.13   TYPICAL ANALYSES OF IRRIGATION WATERS

The analyses of selected irrigation waters shown in Table 10.3 brings out the following points:

TABLE 10.3

ANALYSES OF SELECTED IRRIGATION WATERS

Sample	Source	Electrical Conductivity, $EC \times 10^5$	Concentration		Constituents (m.e. per liter)									Sodium, %	Boron, ppm
			Per Acre-Foot, tons	Total Solids, ppm	Cations (bases)				Anions (acids)					
					Ca	Mg	Na	K	HCO_3	SO_4	Cl	NO_3		
1	Rio Grande, Colorado	85.0	0.1	81	0.56	0.27	0.30		0.70	0.32	0.12	(²)	26	
2	Rio Grande, New Mexico	870.0	.8	641	3.76	1.34	4.03		2.97	4.69	1.53		44	
3	Pecos River, Texas	9150.0	8.4	6198	30.62	17.19	53.62			44.52	54.00		52	
4	Snake River, Idaho		.5	380	2.60	1.80	1.74	0.14	3.28	1.69	1.18	0.04	28	
5	Colorado River, Arizona	1135.0	1.1	795	5.08	2.21	4.54		2.56	6.97	2.31	.04	35	0.14
6	Well in Coachella Valley, California	1740.0	1.2	910	2.14	.08	12.67		1.02	1.80	12.04	.14	85	.71

Sample 1, from the Rio Grande at Del Norte, Colorado, represents a mountain water low in total salts, whereas sample 2, taken from the same river in southern New Mexico, shows the marked increase in concentration between Colorado and New Mexico.

Sample 3, from the Pecos River near Barstow, Texas, is an example of one of the most concentrated waters used for irrigation purposes. The water is applied to a soil high in gypsum. Fair to good crops of cotton and alfalfa are obtained.

Sample 4, from the Snake River, Idaho, is representative of irrigation waters in the Northwest. They are, as a whole, low in total salt content.

Sample 5, from the Colorado River, Arizona, was selected to show the type of irrigation water used on fairly heavy soil in the Imperial Valley. The water contains 1.1 tons of salt per acre-foot, which is a little more than average for irrigation water.

Sample 6, from an irrigation well, was selected as a water of high sodium percentage. Used on sandy soils it does some injury to grapes. The boron content is rather high.

10.14 STANDARDS FOR IRRIGATION WATERS

Plants in saline soils are adversely affected by high concentrations of salts in the soil solution and by poor physical condition of the soil. Both soil conditions are affected by the irrigation water. An irrigation water having a high sodium percentage will, after a time, give rise to a soil having a large proportion of replaceable sodium in the colloid, often designated as "black alkali" soil. Even on sandy soils with good drainage, waters of 85 percent sodium[2] or higher are likely to make soils impermeable after prolonged use. With higher total salt content, there is a flocculating action that tends to counterbalance the poor physical condition caused by a high sodium concentration in the water. On a heavy soil already high in replaceable sodium, the poorest water would be that which is low in total salts but which has a high sodium percentage.

The concentration of the soil solution also modifies plant growth, and is usually 2 to 100 times the irrigation water concentration and seldom more diluted. In heavily irrigated sandy soils, the soil solution will tend to approach the same concentration as that of the irrigation water. On heavy soils, where evaporation may greatly exceed drainage, the concentration of the soil solution may be 100 times that of the irrigation water, and may become too great for plant growth.

In most arid-region valleys neither the irrigation companies nor farmers can modify their irrigation water to any substantial degree. However, the standards for waters presented in Table 10.4 can be a guide to interested agencies in appraising and developing irrigation and drainage practices to harmonize with qualities of water.

Class 1 waters are considered by the Salinity Laboratory Staff as

[2] Percent sodium $= \dfrac{100\ \text{Na}}{\text{Ca} + \text{Mg} + \text{Na} + \text{K}}$.

TABLE 10.4

STANDARDS FOR IRRIGATION WATERS

Water Class	Electrical Conductivity $EC \times 10^6$	Salt Content		Sodium, %	Boron, ppm
		Total, ppm	Tons per Acre-Foot		
1	0–1000	0–700	1	60	0.0–0.5
2	1000–3000	700–2000	1–3	60–75	0.5–2.0
3	Over 3000	Over 2000	Over 3	75	Over 2.0

excellent to good, suitable for most plants under most conditions. Class 2 waters are mentioned as good to injurious, probably harmful to the more sensitive crops. Also the laboratory considers class 3 waters injurious to unsatisfactory, probably harmful to most crops. Class 3 waters are considered unsuitable under most conditions.

If the salts present are largely sulfates, the values for salt content in each class can be raised 50 percent. Because soil, crop, climate, drainage, and soil management all influence the suitability of water for irrigation, no simple classification scheme will hold for all cases.

Some writers have indicated that waters of 70 percent sodium are unsuitable under most conditions, and a few have suggested an even lower limit. Yet, on sandy soils in the Coachella Valley, California, waters of more than 80 percent sodium are used and the farmers stay in business. For class 3 waters a low limit sodium percentage of 75 has therefore been selected. Scofield, Wilcox, and Magistad have selected a specific electrical conductivity ($K \times 10^5$) of 300 as being the upper limit for good production on most soils and waters. A marked exception occurs in the Pecos Valley, where a good growth of alfalfa and cotton is found on lands irrigated with water having a conductance of 915. The Pecos Valley may be considered special because of the heavy content of gypsum and lime in the soil.

The boron content of water is of great importance for many crops. Some crops, as beans, are very sensitive to an excess of boron; others, like sugar beets, will tolerate large quantities. Water containing more than 2.0 ppm will, in time, usually cause trouble with many of the crops grown.

10.15 SOURCES OF SALINITY IN WATER

Salinity in irrigation water that is obtained from gravity canals originates in one or more of three sources:

(a) In the natural drainage water yielded by watersheds that contain large amounts of alkali salts in the soils and rocks; or

(b) In the conveyance of rivers or canals through soil or rock formations that are highly impregnated with alkali salts; or

(c) In the diversion of canals from the lower reaches of streams and rivers that receive large quantities of seepage and return flow from irrigated lands.

The amounts of salts in natural drainage water near the stream sources are usually so small as to be of little concern. However, the Malad River, a small stream in southern Idaho and northern Utah, contains so large an amount of soluble salts that the application of its water for irrigation proved early to be seriously harmful to trees and general farm crops. In the Uintah Basin, Utah, irrigation water that was almost entirely free from salts at the diversion works absorbed excessive amounts of salts as it was conveyed through canals constructed in beds of Mancos shale.

The greater and more dangerous source of salts in irrigation water is from seepage and return flow. This fact is illustrated by a study of the salt content of water at different points along rivers that traverse irrigated lands and receive return seepage waters. Table 10.5, re-

TABLE 10.5

INCREASE IN THE SALT CONTENT OF THE WATER IN TYPICAL WESTERN RIVERS DUE TO SEEPAGE AND RETURN FLOW OF WATER FROM IRRIGATED LANDS, AS SUMMARIZED BY HARRIS

River	State	Salt Content, ppm		Increase, ppm	Distance, in miles	Increase per mile, ppm
		Upper	Lower			
	Colorado	110	1178	1068	20	53.4
Jordan	Utah	890	1970	1080	14	77.0
Sevier	Utah	205	831	626	60	10.4
Sevier	Utah	205	1316	1111	150	7.4
Pecos	New Mexico	760	2020	1260	30	42.0
Pecos	New Mexico	760	5000	4240	180	24.1
Arkansas	Colorado	trace	2200	2200	120	18.3

porting some determinations of total alkali salts in five western rivers at points separated by 14 to 180 miles, shows appreciable increases in alkali from the higher to the lower points on the river. The maximum increase per mile, 77.0 ppm, occurred in the Jordan River, Utah, and

the minimum in the Sevier River, also in Utah. The owners of irrigation projects that divert water from the lower reaches of streams
which receive seepage and return flow from upstream alkali lands
should know the salt content of the water, and if necessary, take
special precautions to avoid injury to crops and soils from this
source of alkali.

The salt content of irrigation water in several streams of the West
varies appreciably from one part of the irrigation season to another,
as is shown by the data of Table 10.6.

TABLE 10.6

Seasonal Variation of Total Salt Content of Typical Western
Rivers. Amounts of Salts Expressed as Parts per Million of Water
(ppm)

Salt River, Arizona	ppm	Gila River, Arizona	ppm	Sevier River, Utah	ppm
Aug. 1–Sept. 1, 1899	724	Nov. 28, 1899–Jan. 18, 1900	1168	July 29	958
Sept. 2–Sept. 9, 1899	1100	Feb. 1–Mar. 7, 1900	1136	Aug. 12	1104
Sept. 10–Oct. 9, 1899	1142	Aug. 1–Aug. 14, 1900	541	Aug. 24	1268
Oct. 10–Oct. 17, 1899	952	Aug. 15–Aug. 28, 1900	925	Sept. 18	1190
Oct. 18–Dec. 30, 1899	1026	Sept. 1–Sept. 28, 1900	471	Sept. 21	1426
Feb. 17–May 30, 1900	1069	Sept. 29–Nov. 5, 1900	1085	Oct. 5	1406
June 1–Aug. 4, 1900	1391			Oct. 19	1436
				Nov. 9	1376

10.16 TOLERANCE OF CROPS TO SALINITY

Some plants can survive in a waterlogged soil for short periods while
others cannot survive under the same conditions. For soils having a
high water table as well as considerable salinity, plants should be
selected which can tolerate the waterlogged soils as well as excess salts.
Strawberry clover, Bermuda grass, and sweet clover owe part of their
popularity to this characteristic.

Selection of salt-tolerant crops depends on the intended use of the
crop, moisture conditions of the soil, climate, farm-management practices, and other local factors. Three general tolerance groups are
listed by the Salinity Laboratory. Group I includes those plants which
have good tolerance; group II, moderate tolerance; and group III,
poor or slight salt tolerance.

The plants listed in Table 10.7 have been classified into three broad
divisions; namely, (1) common fruit and vine crops, (2) field and truck
crops, and (3) forage crops such as grasses, legumes, and cereals,
which are used primarily for pasture or hay production. In each group
the plants first named are considered to be more tolerant and the

TABLE 10.7

TOLERANCE OF THREE TYPES OF CROPS FOR SALINITY AS DETERMINED
BY THE UNITED STATES SALINITY LABORATORY

Under each of the three types of crops the most tolerant
crops are listed first and the least tolerant last.

Type of Crop	Salt Tolerance			
	Good (Group I)	Moderate (Group II)		Poor (Group III)
Fruit	Date palm	Pomegranate Fig Grape Olive		Grapefruit Pear Almond Apricot Peach Plum Apple Orange Lemon
Field and truck	Sugar beet Garden beet Milo Rape Kale Cotton	Alfalfa Flax Tomato Asparagus Foxtail millet Sorghum (grain) Barley (grain) Rye (grain) Oats (grain) Rice	Cantaloupe Lettuce Sunflower Carrot Spinach Squash Onion Pepper Wheat (grain)	Vetch Peas Celery Cabbage Artichoke Egg plant Sweet potato Potato Green beans
Forage	Alkali sacaton Salt grasses Nuttall alkali Bermuda Rhodes Rescue Canada wild rye Beardless wild rye Western wheat grass	White sweet clover Yellow sweet clover Perennial rye grass Mountain brome Barley (hay) Birdsfoot trefoil Strawberry clover Dallis grass Sudan grass Hubam clover Alfalfa (California common) Tall fescue Rye (hay)	Wheat (hay) Oats (hay) Orchard grass Blue grama Meadow fescue Reed canary Big trefoil Smooth brome Tall meadow oat grass Cicer milk vetch Sour clover Sickle milk vetch	White Dutch clover Meadow foxtail Alsike clover Red clover Ladino clover Burnet

last named more sensitive to salinity. The opinion emphasized by the
Laboratory is that further research in crop tolerance to waterlogging
and salinity is essential to making the classification as presented in
Table 10.7 more complete and more nearly final.

REFERENCES

Hayward, H. E., "The Control of Salinity," *U.S.D.A. Yearbook of Agriculture,* pp. 547–553, 1947.

Hayward, H. E. and O. C. Magistad, "The Salt Problem in Irrigation Agriculture," U.S.D.A. Misc. Pub. 607, 1946.

Hill, Raymond A., "Leaching Requirements in Irrigation" (publication pending *Am. Soc. Civil Eng.*).

Hill, Raymond A., "Salts in Irrigation Water," *Trans. Am. Soc. Civil Eng.,* p. 1478, 1942.

Krimgold, D. B., "Kostiakov on Prevention of Waterlogging and Salinity of Irrigated Land," *Agr. Eng.,* Vol. 26, pp. 327–328, 1945.

Larson, C. A., "Reclamation of Saline (Alkali) Soil in the Yakima Valley, Washington," *Wash. Agr. Exp. Sta. Bul.* 376, 39 pp., illus., 1942.

Magistad, O. C. and J. E. Christiansen, "Saline Soils, Their Nature and Management," *U.S.D.A. Circ.* 707, 1944.

Reeve, R. C., L. E. Allison, and D. F. Peterson, Jr., "Reclamation of Saline-Alkali Soils by Leaching," *Utah Agr. Exp. Sta. and U.S. Regional Salinity and Rubidoux Lab. Bul.* 335, December 1948.

Reitemeier, R. F. and J. E. Christiansen, "The Effect of Organic Matter, Gypsum, and Drying on the Infiltration Rate and Permeability of a Soil Irrigated with a High Sodium Water," *Trans. Am. Geophys. Union,* Vol. 27, No. 2, pp. 181–186, 1946. (U.S. Regional Salinity Lab. 61.)

Richards, L. A., "Availability of Water to Crops on Saline Soils," *U.S.D.A. Res. Ser. Bul.* 210, November 1959.

United States Regional Salinity Laboratory, "Diagnosis and Improvement of Saline and Alkali Soils," by Laboratory Staff, L. A. Richards, Editor, *Agr. Handbook* 60, U.S.D.A., February 1954.

United States Salinity Laboratory Staff, "Salt Problems in Irrigated Soils," *U.S.D.A. Bul.* 190, August 1958.

Wilcox, L. V., "Explanation and Interpretation of Analyses of Irrigation Waters," *U.S.D.A. Circ.* 784, May 1948.

CHAPTER 11 CONSUMPTIVE USE OF WATER[1]

Consumptive use of water involves problems of water supply, both surface and underground, as well as problems of management and economics of irrigation projects. Consumptive use has become a highly important factor in the arbitration of controversies over major stream systems where the public welfare of valleys, states, and nations is involved. Before available water resources of a drainage basin in arid and semi-arid regions can be satisfactorily ascertained, careful consideration must be given to consumptive-use requirements for water in various sub-basins.

11.1 DEFINITION OF CONSUMPTIVE USE

Consumptive use, or evapo-transpiration, is the sum of two terms; (1) transpiration, which is water entering plant roots and used to build plant tissue or being passed through leaves of the plant into the atmosphere, (2) evaporation, which is water evaporating from adjacent soil, water surfaces, or from the surfaces of leaves of the plant. Water deposited by dew, rainfall, or sprinkler irrigation and subsequently evaporating without entering the plant system is part of consumptive use.

Consumptive use can apply to water requirements of a crop, a field, a farm, a project, or a valley. When the consumptive use of the crop is known, the water use of larger units can be calculated. Hence, the term "consumptive use" and the following discussion generally refer to the crop.

[1] In the second edition, this chapter on consumptive use of water was written by Harry F. Blaney and O. W. Israelsen. Mr. Blaney is Senior Irrigation Engineer, Agricultural Research Service, U.S.D.A.

231

11.2 CONDITIONS AFFECTING CONSUMPTIVE
USE OF WATER

Evapo-transpiration is influenced by temperature, irrigation practices, length of growing season, precipitation, and other factors. The volume of water transpired by plants depends in part on the water at their disposal, and also on temperature and humidity of the air, wind movement, intensity and duration of sunlight, stage of development of the plant, type of foliage, and nature of the leaves.

Consumptive use by native vegetation of non-cropped land is fairly constant from year to year where the water supply is adequate, but surface evaporation increases in wet years because of the expansion of water surface and moist soil areas. In some irrigated valleys irrigation is lavish when the water supply is abundant, and increased waste water and return flow are consequently available for non-productive consumption. Also, evaporation is increased from the cropped land soils. Plant diseases and pests reduce consumptive use by inhibiting plant growth. Spreading of noxious weeds may reduce the area irrigated if crops cannot be grown in infested areas. Indirect results of infestation may appear as widespread changes in crops.

Size of a farm may affect water usage appreciably since borders or furrows, size of stream, and crop rotations may be adjusted differently to acreages relatively large or small.

Evaporation from Soils

Where the water table is near the ground, surface evaporation from the soil is almost equal to evaporation from a free-water surface; whereas, with water levels at greater depths below the surface, evaporation from soil decreases until it becomes negligible when moisture no longer reaches the surface by capillary action.

A comparison between evaporation from a water surface and evaporation from different soil and river-bed materials was made by R. B. Sleight (1917). Depths to water table were varied in his tank experiments. Table 11.1 gives results obtained from fine sandy loam and also from river-bed sand.

Where irrigation water is applied by flooding methods, large amounts of water are lost by direct evaporation from soil surfaces without having passed through the roots, stems, and leaves of plants. After light showers during the growing season, considerable rainfall, if not all, is retained on the leaves where it subsequently evaporates without entering the roots of the plants. Showers of 0.1 inch seldom enter the soil to a sufficient depth to be used by the plants. Therefore, water

TABLE 11.1

EVAPORATION FROM WATER SURFACE AND FROM DIFFERENT SOILS
WITH WATER TABLE AT VARIOUS DEPTHS BELOW THE
SURFACE IN TANKS 2 FT IN DIAMETER

Period Ending	Evaporation from Water Surface, Inches	Evaporation from Soil, Inches					

Fine Sandy Loam

		Water Table Depths Below the Soil Surface					
		4 in.	16 in.	28 in.	38 in.	43 in.	50 in.
Aug. 30*	3.84	2.98	2.69	2.19	1.23	0.24	0.19
Sept. 15	4.77	4.37	4.09	3.12	1.98	0.37	0.32
25	2.10	1.81	1.67	1.44	0.30	0.16	0.16
29	0.91	1.04	0.84	0.69	0.43	0.11	0.16
Oct. 4	1.23	1.14	0.97	0.58	0.30	0.10	0.10
Total	12.85	11.34	10.26	8.02	4.24	0.98	0.93
Percentage	100.0	88.2	79.8	62.4	33.0	7.63	7.24

River-Bed Sand

		Water Table Depths Below Soil Surface			
		3 in.	6 in.	10½ in.	24 in.
Aug. 4†	1.01	0.62	0.67	0.50	0.19
9	1.12	1.07	0.80	0.74	0.15
12	0.85	0.90	0.80	0.69	0.18
15	0.69	0.35	0.32	0.28	0.04
17	0.54	0.34	0.29	0.21	0.04
29	3.54	2.54	2.42	2.22	0.80
Sept. 12	4.44	2.71	2.62	2.47	0.34
25	2.83	2.16	2.06	1.82	0.25
29	0.91	0.42	0.40	0.36	0.12
Oct. 4	1.23	0.69	0.67	0.63	0.00
10	0.99	0.54	0.50	0.50	0.00
16	0.88	0.62	0.62	0.50	0.16
Total	18.93	13.06	12.21	10.92	2.27
Percentage	100.0	69.0	64.5	57.7	11.3

* The period began Aug. 17.
† The period began July 31.

received from light rain showers or from light sprinkler irrigations may not be used by the plants in the transpiration process. Subsequent discussion based upon the energy available for evaporation shows, however, that under normal conditions, water that evaporates directly from the soil or leaves is beneficially used, reducing by a like amount water that would otherwise have been transpired by the plants. This is true provided a normal mature crop exists over the soil surface. Early or late in the growing period, when the crop is not transpiring much water, evaporation from a wet soil surface may exceed normal consumptive use.

There are still decided differences in opinion on the effect of cultivation on direct evaporation of water from soils. Recent studies seem to throw considerable doubt on advantages of soil mulches for conserving water through reduction of evaporation losses from soils that are not excessively wet or in contact with a water table at a shallow depth. Therefore, broad general conclusions concerning influence of cultivation on direct evaporation losses from soils may be misleading. The large number of variable factors involved—notably the differences in distances to free-water sources, in original moisture content of unsaturated soils, in texture, structure, and water conductivity—make it hazardous to generalize.

Transpiration

The process by which water vapor escapes from living plants, principally the leaves, and enters the atmosphere is known as transpiration. During the growing period of a crop, there is a continuous movement of water from the soil into roots, up stems, and out of leaves of the plants. Velocity of water flowing through plants varies widely from 1 to 6 feet per hour; but, under conditions of unusually high temperature, dry atmosphere, and wind, velocity of the stream may be greatly increased. A very small proportion of water absorbed by the roots is retained in the plants. If the rate of evaporation at the leaves exceeds the rate of absorption by the roots, wilting occurs and growth of the plant is impeded. On the other hand the available water is not used efficiently, if conditions are such as to stimulate excessive transpiration.

When water changes from liquid to vapor and passes from the plants into the atmosphere, 540 calories of heat are required for every cubic centimeter being converted from liquid to vapor. Hence, if heat is not available, transpiration ceases; and conversely, as available heat increases, transpiration increases. Heat can come from the soil, the plant, or the air. The majority of heat used for transpiration, even over short periods of time, comes directly from the sun as radiant energy.

Thus, a given increment of heat will vaporize a given quantity of water. If the available heat is used for vaporizing water from the ground surface, it is not available for vaporizing water within the cavities of the leaf. Therefore, a light shower of rain remaining on the surface of leaves or on the ground can evaporate, utilizing the bulk of available energy. When this occurs, the transpiration of water through the plant will be reduced accordingly.

For example, suppose a rainfall of only 0.1 inch occurred during the dry season and all the moisture was intercepted by leaves of the growing crop. Assume also that daily consumptive use was 0.25 inch per day. Experiments show that the bulk of consumptive use occurs during the day. Therefore, if consumptive use of 0.25 inch per day occurs over say 12 hours, the average use per hour would be 0.02 inch. During midday, the rate is likely to nearly double the average or 0.04 inch per hour. Hence, in 2.5 hours the rainfall of 0.1 inch would be evaporated and would have been fully utilized. Often overlooked is the fact that rainfall of 0.1 inch may represent only $\frac{2}{5}$ of a normal daily consumptive use. Observations of the recovery of a wilted crop following a light shower will show that the rainfall is not without effect.

Rainfalls of less than 0.5 inch have frequently been neglected in project planning on the erroneous assumption that rain must get into the root zone to be fully effective. When the crop covers the ground surface, a light rain or a light irrigation will reduce by a like amount moisture removed from the soil by the roots, provided of course that an excessive wind does not exist.

11.3 DIRECT MEASUREMENTS OF CONSUMPTIVE USE

Various methods have been used to determine the amount of water consumed by agricultural crops and natural vegetation. Regardless of the method, the problems encountered are numerous. The source of water used by plant life whether from precipitation alone, irrigation plus rainfall, or ground water plus precipitation is a factor in selecting a method. Principal methods are: tank and lysimeter experiments, field experimental plots, soil moisture studies, integration, and inflow-outflow for large areas. These five methods of measuring consumptive use of water are described in the following articles.

Tank and Lysimeter Experiments

Reliability of consumptive-use determinations by means of tanks or lysimeters is dependent on nearness of reproduction of natural conditions. Artificial conditions are caused by limitations of soil, size of tank, regulation of water supply, and sometimes environment.

Tanks should be placed in surroundings of natural growth of the same species, that is, in their natural environment, so that consumptive use of water will presumably be the same as for similar growth outside the tank. It has been found that all tank vegetation must be protected from the elements by surrounding growth of the same species.

Weighing is the precise means of determining consumptive use from tanks. However, conditions and facilities will not always permit the weighing of tanks. Soil tanks equipped with Mariotte water supply tanks have proved successful in consumptive-use measurements from water tables at various depths. Double-type soil tanks, with an annular space between the inner and outer shells, are considered best.

The Mariotte supply system furnishes water as needed to maintain a fixed water level in the annular space in the soil tank. The amount of water withdrawn is determined by differences in daily or weekly readings of the glass gage attached to the supply tank. The value of a Mariotte-equipped tank lies in the ease with which periodic measurements of water used may be made, since the system operates automatically.

Field Experimental Plots

Measurements by soil moisture studies in field plots are usually more dependable than measurements with tanks or lysimeters. Tank and lysimeter experiments for individual crops do not always represent the natural conditions of the soil, as there are many ways of preparing and arranging soil materials.

Dr. John A. Widtsoe pioneered measurement of consumptive use in field plots at Utah State University, beginning in 1902. His work was done on land having a water table about 75 feet below the surface; hence, it is reasonably safe to conclude that crops obtained no ground water, and that crop-season rainfall, the draft on stored capillary soil moisture, and irrigation water furnished all water to which crops had access. There was no runoff from experimental plots used by Widtsoe, and the deep percolation losses were not measured and were assumed to be negligible. Widtsoe measured the water used by 14 crops during a 10-year period, 1902–1911 inclusive. Yields obtained by Widtsoe were plotted against the total water used, and, as a basis for arriving at the consumptive use, those yields were selected which appeared to be most profitable. With nearly every crop, yield increased rapidly with an increase of water used to a certain point, and then decreased with further increase in water. At his "break in the curve," the amount of water used was considered as the consumptive use.

Widtsoe's work indicates the importance of yield in determining consumptive use. It is probable that Widtsoe's values are too high rather than too low because of the deep percolation losses which undoubtedly occurred.

Soil Moisture Studies

Consumptive use of water for various crops has been determined by intensive soil moisture studies. This method is usually suitable for areas where soil is fairly uniform and the depth to ground water is such that it will not influence soil moisture fluctuation within the root zone. Soil moisture is determined before and after each irrigation with some measurements between irrigations in the major root zone. Usually a great number of measurements must be taken to obtain the desired accuracy.

Acre-inches of water extracted per day from the soil are computed for each period. When rate of use is plotted against time, a curve can be drawn from which the seasonal use can be obtained.

Integration Method

The integration method is the summation of the products of unit consumptive use for each crop times its area, plus the unit consumptive use of native vegetation times its area, plus water surface evaporation times water surface area, plus evaporation from bare land times its area.

Before this method can successfully be applied, it is necessary to know unit consumptive use of water and the areas of various classes of agricultural crops, native vegetation, bare land, and water surfaces. By means of aerial maps and field surveys areas of various types of native vegetative cover, bare land, and water surfaces can be determined.

Results of determinations of consumptive use by this method in the Mesilla Valley, New Mexico, and in Texas are presented in Table 11.2.

Inflow-Outflow for Large Areas

Applying this method, valley consumptive use U is equal to the water that flows into the valley during a 12-month year I, plus yearly precipitation on the valley floor P, plus water in ground storage at the beginning of the year G_s, minus water in ground storage at the end of the year G_e, minus yearly outflow R—all volumes measured in acre-feet.

TABLE 11.2

AREAS OF DIFFERENT CROPS AND CONSUMPTIVE USE OF WATER IN MESILLA
VALLEY AREA, NEW MEXICO, AND TEXAS, AS ESTIMATED BY INTEGRATION
METHOD, USING DIFFERENT UNITS, 1936 (ISRAELSEN AND BLANEY)

Land Classification	1936 Area, Acres (a)	Consumptive Use*	
		Unit, Feet (c)	Annual, Acre-Feet (ca)
Irrigated crops:			
Alfalfa and clover	17,077	4.0	68,308
Cotton	54,513	2.5	136,282
Native hay and irrigated pasture	216	2.3	497
Miscellaneous crops	11,117	2.0	22,234
Entire irrigated area	82,923	2.74	227,321
Natural vegetation:			
Grass	2,733	2.3	6,286
Brush	6,933	2.5	17,332
Trees-Bosque	3,532	5.0	17,660
Entire area	13,198	3.13	41,278
Miscellaneous:			
Temporarily out of cropping	5,569	1.5	8,354
Towns	1,523	2.0	3,046
Water surfaces, pooled, river, and canals	4,081	4.5	18,364
Bare lands, roads, etc.	3,124	0.7	2,187
Total (entire area)	110,418	2.72	300,550

* *ca* = the product of unit consumptive use in feet (*c*) times area in acres (*a*).

Thus
$$U = (I + P) + (G_s - G_e) - R \qquad (11.1)$$

The difference between storage of capillary water at the beginning
of the year and at the end of the year is usually considered to be
negligible. It is assumed that stream measurements are made on
bedrock controls, and that the sub-surface inflow is about the same as
sub-surface outflow. The quantity $(G_s - G_e)$ is considered as a unit
so that absolute evaluation of either G_s or G_e is unnecessary, only
the difference being needed. This is the product of the difference in
the average depth of water table during the year, measured in feet,

and multiplied by the specific yield[2] of the soil and area of the valley floor. The quantity P is obtained by multiplying the average annual precipitation in feet by the area of the valley floor in acres. Unit consumptive use of the entire valley in acre-feet per acre is obtained by dividing the total consumptive use by the area of the valley floor.

Results of typical inflow-outflow measurements in several areas are given in Table 11.3.

TABLE 11.3

VALLEY CONSUMPTIVE USE OF WATER AS DETERMINED BY THE
INFLOW-OUTFLOW METHOD IN SEVERAL AREAS OF THE WEST

Location	Year	Area, Acres	Annual Consumptive Use		Authority
			Total, Acre-Feet	Unit, Feet	
San Luis Valley, Colo.	1925–1935	400,000	664,900	1.66	Blaney-Rohwer
San Luis Valley, Colo.	1936	400,000	685,423	1.71	Blaney-Rohwer
San Luis Valley, Colo.	1930–1932	17,300	26,215	1.52	Tipton-Hart
Isleta-Belen, N. Mex.	1936	17,500	38,700	2.28	Blaney-Morin
Mesilla Valley, N. Mex.	1919–1935	109,000	297,756	2.73	Israelsen-Blaney
Mesilla Valley, N. Mex.	1936	110,418	303,683	2.75	Israelsen-Blaney
Carlsbad, N. Mex.	1921–1939	51,700	129,752	2.51	Blaney-Morin
Carlsbad, N. Mex.	1940	51,700	119,898	2.33	Blaney-Morin
New Fork, Wyo.	1939–1940	25,000	......	1.50	Lowry-Johnson
Michigan-Illinois, Colo.	1938–1940	43,000	......	1.50	Lowry-Johnson
Uncompahgre, Colo.	1938–1940	137,700	......	2.28	Lowry-Johnson

II.4 CONSUMPTIVE USE BY NATURAL VEGETATION

Water consumed by natural vegetation cannot be made available for irrigation of crops. Therefore, in considering the water supply of a region, water consumed by natural vegetation (such as salt grass, willows, cottonwoods, tamarisk, and tules) growing in irrigated valleys, moist areas, and along streams becomes of increasing importance as greater land areas are irrigated, especially during periods of drought. The value of data on consumptive use by these non-crop plants is recognized by administrators and engineers in regions where water rights are in dispute, or where interstate water supply and water use

[2] Specific yield is defined as the total pore space of the soil less the moisture content at field capacity, both expressed as volume percentages of the total soil volume.

are not in balance. In planning new irrigation projects, consideration must often be given to differences in consumptive uses of water by irrigated crops and by natural vegetation replaced by the crops.

The relation of plant communities to moisture supply is one of the outstanding characteristics of the growth of natural vegetation. Whereas individual species are largely restricted to physical environments, the principal condition that governs distribution of vegetative groups in irrigated areas is the availability of water. Each species responds to water conditions for its most favorable growth and its widest distribution. Temperatures, moisture, and chemical and physical properties of soils are contributing factors in the distribution of natural vegetation. Quantity of water available for plant use and the effect of plant growth on supply are of great interest.

Measurements of consumptive use indicate that water-loving natural vegetation uses from 50 to 100 percent more water than most crop plants. Tules and cattails growing in or near irrigation canals and drainage ditches are exposed in narrow strips to sun and wind so that their consumption of water is high. Under such circumstances the natural vegetation along a mile of canal or ditch may consume enough water to irrigate 8 or 10 acres of alfalfa or a greater acreage of other field crops or of fruit. Consumptive use of water by water-loving (phreatophytic) plants usually equals, or exceeds, the evaporation from a free-water surface.

11.5 CLIMATIC OBSERVATIONS AS AN INDEX TO CONSUMPTIVE USE

How temperature, humidity, wind velocity, vapor pressure, and solar radiation influence consumptive use has been studied by several researchers. Penman, in England, has made the most complete analysis using several climatic variables, whereas temperature has been used as the principal variable to obtain an index to consumptive use by Thornthwaite in the humid eastern United States, by Lowry and Johnson, and by Blaney and Criddle in the arid western United States. These methods and their significance are outlined in following sections. An excellent digest of these methods has been prepared by Criddle, in "Methods of Computing Use of Water," Proceedings Paper 1507, American Society of Civil Engineers, January 1958. Formulas, charts, and examples to follow were taken from this publication.

Penman Method

Penman has made the most complete theoretical approach, showing that consumptive use is inseparably connected to incoming solar energy.

His formula representing the potential evapo-transpiration (consumptive use) is as follows:

$$E_T = \frac{\Delta H + 0.27E_a}{\Delta - 0.27} \tag{11.2}$$

with values of H and E_a given by

$$H = R_A(1 - r)(0.18 + 0.55n/N)$$
$$- \sigma T_a^4(0.56 - 0.092\sqrt{e_d})(0.10 + 0.90n/N) \tag{11.3}$$

$$E_a = 0.35(e_a - e_d)(1 + 0.0098u_2) \tag{11.4}$$

where H = daily heat budget at surface in mm H_2O/day
   R_A = mean monthly extra terrestrial radiation in mm H_2O/day
   r = reflection coefficient of surface
   n = actual duration of bright sunshine
   N = maximum possible duration of bright sunshine
   σ = Boltzmann constant
   σT_a^4 = mm H_2O/day (see Table 11.6)
   e_d = saturation vapor pressure at mean dew point (i.e., actual vapor pressure in the air) mm Hg
   E_a = evaporation in (mm) H_2O/day
   e_a = saturation vapor pressure at mean air temperature in mm Hg
   u_2 = mean windspeed at 2 meters above the ground (miles/day)[3]
   E_T = evapo-transpiration in mm H_2O/day
   u_1 = measured windspeed in miles per day at height h in feet
   Δ = slope of saturated vapor pressure curve of air at absolute temperature T_a in °F(mm Hg/°F)

In order that various factors might be systematically evaluated in the complex formula, Table 11.4 was developed by Criddle and contains an example of the method.   Tables 11.5 and 11.6 and Figs. 11.1 and 11.2 will be found to be helpful in solving Penman's equation.

Penman found that coefficients were necessary to reduce the potential rate of consumptive use to the actual water use of pasture growing in England.   Very good correlation has been obtained between computed and measured consumptive-use values.

The principal limitation of the Penman approach is the lack of sufficient weather measurements in most localities.   Only a few

[3] Wind measurements taken at other heights can be corrected to the 2-meter elevation by use of the formula. $u_2 = u_1 \left(\dfrac{\log 6.6}{\log h}\right)$.

TABLE 11.4

COMPUTATION SHEET FOR PENMAN METHOD OF COMPUTING EVAPO-
TRANSPIRATION (AFTER CRIDDLE)

Location: Boise, Idaho   Latitude: 43°34′   Crop: Alfalfa
Frost-free period: 4/23–10/17

	July
A. Data:	
1. Air Temp. —	72.5
2. Relative humidity — % (Est.)	40
3. Sunshine n/N — % (Est.)	70
4. Windspeed, u_2 — Mi/day at 2 m (Est.)	135
5. Radiation rate, R_A — mm H_2O/day (see Table 11.5)	16.2
6. Reflection coefficient — % (Est.)	25
B. Solving expression: $R_A(1 - r)(0.18 + 0.55n/N)$	
7. $(1 - r)$	0.75
8. $(0.18 + 0.55n/N)$	0.565
9. Item 6 × item 7 × item 8	6.86
C. Solving expression: $\sigma T_a^4(0.56 - 0.092 \sqrt{e_d})(0.10 + 0.90\, n/N)$	
10. Vapor pressure	
(a) Saturated, e_a (see Fig. 11.1)	21.0
(b) Actual $e_d = (R.H. \times e_a)$	8.4
(c) $\sqrt{e_d}$	2.9
11. σT_a^4 (see Table 11.6)	15.37
12. $(0.56 - 0.092\sqrt{e_d})$	0.29
13. $(0.10 + 0.90\, n/N)$	0.73
14. Item 11 × item 12 × item 13	3.25
D. Solving for H	
15. Item 9 minus item 14	3.61
E. Solving for $E_a = 0.35\,(e_a - e_d)(1 + 0.0098u_2)$	
16. $= 0.35(e_a - e_d)$	4.41
17. $= (1 + 0.0098u_2)$	2.32
18. Item 16 × item 17	10.2
F. Solving for $E_T = \dfrac{\Delta H + 0.27E_a}{\Delta + 0.27}$	
19. Δ (see Fig. 11.2)	0.65
20. ΔH	2.23
21. $0.27E_a$	2.75
22. $\Delta + 0.27$	0.92
23. $E_T =$ (mm of water per day)	6.67
(in. of water per day)	0.26
(in. of water per month)	7.18

TABLE 11.5

MID-MONTHLY INTENSITY OF SOLAR RADIATION (R_A) ON A HORIZONTAL SURFACE IN MM OF WATER EVAPORATED PER DAY.[1]
(AFTER CRIDDLE)

	Northern Hemisphere										Southern Hemisphere									
	90°	80°	70°	60°	50°	40°	30°	20°	10°	0°	0°	10°	20°	30°	40°	50°	60°	70°	80°	90°
Jan.	..	..	..	1.3	3.6	6.0	8.5	10.8	12.8	14.5	14.5	15.8	16.8	17.3	17.3	17.1	16.6	16.5	17.3	17.6
Feb.	..	..	1.1	3.5	5.9	8.3	10.5	12.3	13.9	15.0	15.0	15.7	16.0	15.8	15.2	14.1	12.7	11.2	10.5	10.7
Mar.	..	1.8	4.3	6.8	9.1	11.0	12.7	13.9	14.8	15.2	15.2	15.1	14.6	13.6	12.2	10.5	8.4	6.1	3.6	1.9
Apr.	7.9	7.8	9.1	11.1	12.7	13.9	14.8	15.2	15.2	14.7	14.7	13.8	12.5	10.8	8.8	6.6	4.3	1.9	..	..
May	14.9	14.6	13.6	14.6	15.4	15.9	16.0	15.7	15.0	13.9	13.9	12.4	10.7	8.7	6.4	4.1	1.9	0.1	..	..
June	18.1	17.8	17.0	16.5	16.7	16.7	16.5	15.8	14.8	13.4	13.4	11.6	9.6	7.4	5.1	2.8	0.8	..	..	..
July	16.8	16.5	15.8	15.7	16.1	16.3	16.2	15.7	14.8	13.5	13.5	11.9	10.0	7.8	5.6	3.3	1.2	..	..	..
Aug.	11.2	10.6	11.4	12.7	13.9	14.8	15.3	15.3	15.0	14.2	14.2	13.0	11.5	9.6	7.5	5.2	2.9	0.8	..	..
Sept.	2.6	4.0	6.8	8.5	10.5	12.2	13.5	14.4	14.9	14.9	14.9	14.4	13.5	12.1	10.5	8.5	6.2	3.8	1.3	..
Oct.	..	0.2	2.4	4.7	7.1	9.3	11.3	12.9	14.1	15.0	15.0	15.3	15.3	14.8	13.8	12.5	10.7	8.8	7.1	7.0
Nov.	..	..	0.1	1.9	4.3	6.7	9.1	11.2	13.1	14.6	14.6	15.7	16.4	16.7	16.5	16.0	15.2	14.5	15.0	15.3
Dec.	..	..	..	0.9	3.0	5.5	7.9	10.3	12.4	14.3	14.3	15.8	16.9	17.6	17.8	17.8	17.5	18.1	18.9	19.3

[1] Computed from "Manual of Meteorology" by Napier Shaw, Vol. II, Comparative Meteorology, 2nd Edition, Cambridge University Press, 1936, pp. 4 and 5.

Note: Values from the table by Shaw multiplied by 0.86 and divided by 59 give the radiation in mm of water per day.

TABLE 11.6

VALUES OF σT_a^4 FOR VARIOUS TEMPERATURES WHEN COMPUTING EVAPO-
TRANSPIRATION BY THE PENMAN METHOD (AFTER CRIDDLE)

Temperature	σT_a^4	Temperature	σT_a^4
°Abs	mm H$_2$O/day	°F	mm H$_2$O/day
270	10.73	35	11.48
275	11.51	40	11.96
280	12.40	45	12.45
285	13.20	50	12.94
290	14.26	55	13.45
295	15.30	60	13.96
300	16.34	65	14.52
305	17.46	70	15.10
310	18.60	75	15.65
315	19.85	80	16.25
320	21.15	85	16.85
325	22.50	90	17.46
		95	18.10
		100	18.80

Note: Heat of vaporization was assumed to be constant at 590 gal/gm of H$_2$O.

climatological stations record the needed data. Hence, the formula, even though quite reliable, does have serious practical limitations.

Another fact to be remembered in use of the Penman equation is that coefficients were determined for a rather humid area not far

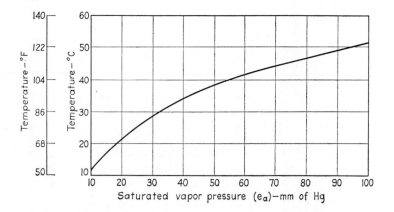

Fig. 11.1 Temperature vs. saturated vapor pressure. (After Criddle.)

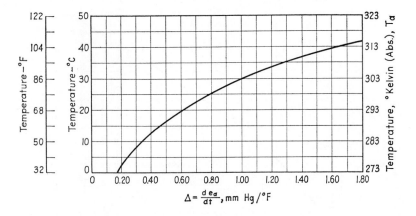

Fig. 11.2 Temperature vs. Δ

$$\left(\frac{d \text{ Saturation Vapor Pressure, mm Hg.}}{d \text{ Temperature, }^\circ\text{F}}\right) \text{ (After Criddle.)}$$

from the ocean and essentially covered with growing vegetation. Experience indicates that the Penman formula applies better under these conditions than in arid, low-humidity areas where temperature and radiant energy may not be as nearly balanced as in England. Oftentimes irrigated agriculture in arid regions is surrounded by vast desert areas with little vegetative cover and no free-water surfaces. This condition results in a greater quantity of energy being available for consumptive use than would be indicated by the incoming solar energy.

Thornthwaite Method

Thornthwaite, realizing the need for a simpler expression that would utilize readily available climatological data, developed an empirical formula based on temperature. He theorized that temperature was a good index to energy in a zone of essential equilibrium. Latitude is also used in his computational procedure, which is illustrated in the following example by Criddle:

Monthly values of heat index are related to monthly temperatures in Fig. 11.3. This relationship is used to determine the seasonal heat index value shown in Table 11.7. The value of 43.3 was plotted on Fig. 11.4 for Boise, Idaho. A straight line drawn from the "index point" through this "heat-index point" gives the relationship between temperature and evapo-transpiration.

With a July temperature of 72.5°F (22.5°C), the uncorrected

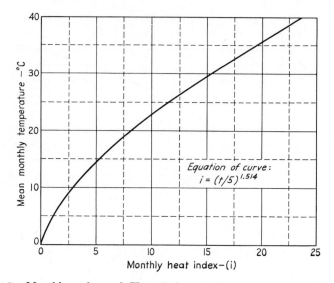

Fig. 11.3 Monthly values of Heat Index, *I*, for use in computing evapo-transpiration by Thornthwaite Method. (After Criddle.)

TABLE 11.7

COMPUTATION OF "HEAT INDEX" FOR BOISE, IDAHO (AFTER CRIDDLE)

Month	Temperature		Heat Index
	°F	°C	i*
January	27.9	−2.2	0
February	33.6	1.0	0.5
March	41.4	5.2	1.0
April	49.1	9.5	2.5
May	56.1	13.3	4.0
June	64.5	18.0	7.0
July	72.5	22.5	10.0
August	71.0	21.6	8.5
September	61.2	16.3	6.0
October	50.1	10.0	2.8
November	39.7	4.3	1.0
December	30.4	−1.0	. .
Annual			43.3

* Obtained from Fig. 11.3.

potential evapo-transpiration is about 11.5 cm. This potential use is corrected for sunlight and days of the month from Table 11.8 using a

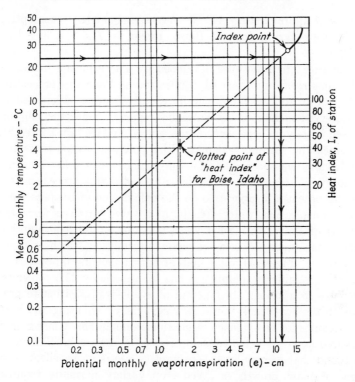

Fig. 11.4 Nomograph for computing monthly evapo-transpiration by Thornthwaite method. (After Criddle.)

TABLE 11.8

MEAN POSSIBLE DURATION OF SUNLIGHT FOR THE THORNTHWAITE METHOD
IN THE NORTHERN AND SOUTHERN HEMISPHERES EXPRESSED IN UNITS
OF 30 DAYS OF 12 HOURS EACH (AFTER CRIDDLE)

N. Lat.	J	F	M	A	M	J	J	A	S	O	N	D
0	1.04	0.94	1.04	1.01	1.04	1.01	1.04	1.04	1.01	1.04	1.01	1.04
10	1.00	0.91	1.03	1.03	1.08	1.06	1.08	1.07	1.02	1.02	0.98	0.99
20	0.95	0.90	1.03	1.05	1.13	1.11	1.14	1.11	1.02	1.00	0.93	0.94
30	0.90	0.87	1.03	1.08	1.18	1.17	1.20	1.14	1.03	0.98	0.89	0.88
35	0.87	0.85	1.03	1.09	1.21	1.21	1.23	1.16	1.03	0.97	0.86	0.85
40	0.84	0.83	1.03	1.11	1.24	1.25	1.27	1.18	1.04	0.96	0.83	0.81
45	0.80	0.81	1.02	1.13	1.28	1.29	1.31	1.21	1.04	0.94	0.79	0.75
50	0.74	0.78	1.02	1.15	1.33	1.36	1.37	1.25	1.06	0.92	0.76	0.70

latitude of approximately 44° N. Average computed consumptive use of all crops during July at Boise, Idaho, is $11.5 \times 1.30 = 15.0$ cm or 5.9 in.

In essence, the procedure developed by Thornthwaite has the same limitations regarding areas of application as Penman's procedure. It applies quite well to humid, well-vegetated areas. Increased errors are observed in humid, low-humidity regions. Since the relationship is based largely on experience in the central and eastern United States, and since the formula calculates the potential evapo-transpiration, no allowance has been made for different crops or other land use.

Thornthwaite and associates have extended this procedure for calculating potential evaporation to the calculation of average potential evapo-transpiration, water deficits, and water surpluses for extensive portions of the earth. Such an application of the consumptive-use concept has produced very valuable information which is useful in many ways. Figure 11.5 is a map of Africa showing this information.

Another significant adaptation has been made by Thornthwaite and associates in applying their potential evapo-transpiration to growing crops on the Sea-Brook Farms in New Jersey. Planting and harvesting schedules for an extensive vegetable cropping and canning industry are based upon predicted consumptive use and growth. Thornthwaite has found that accumulated consumptive use is an excellent index to stages of plant growth—a given crop variety normally requiring a given amount of water to mature. He has shown that accelerated consumptive use results in accelerated maturity.

Lowry-Johnson Method

Lowry and Johnson developed a procedure for estimating water requirements for irrigation projects developed by the United States Bureau of Reclamation. This method applies to a valley, not to an individual farm, and has been widely used by the Bureau of Reclamation in the arid western portion of the United States with good results. It is essentially an empirical procedure based upon data collected from the area of application.

A linear relationship is assumed between "effective heat" and consumptive use. Effective heat is defined as the accumulation, in day-degrees, of maximum daily growing season temperatures above 32°F.

The approximate relationship

$$U = 0.8 + 0.156F \qquad (11.5)$$

is used in estimating the valley consumptive use by the Lowry-Johnson method, where

U = consumptive use in acre-feet per acre
F = effective heat in thousands of day-degrees.

Although the Lowry-Johnson method was not developed to estimate monthly use, or individual crop uses, Criddle has applied it at Boise for July by using a simple proportion of monthly heat units to annual heat units.

The effective heat-day-degrees F at Boise as used by Lowry and Johnson in their original paper was 8940. Consumptive use was 26 in. Taking July only, the mean maximum temperature in degrees Fahren-

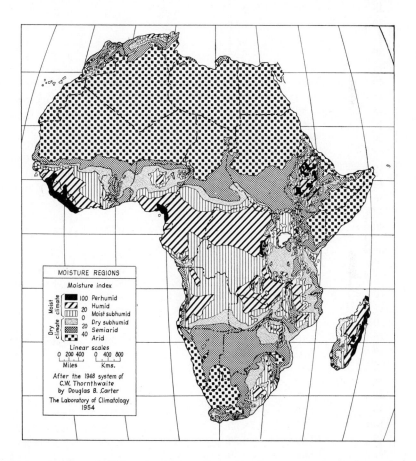

Fig. 11.5 Map of Africa showing average potential evapo-transpiration, water deficiency, and water surplus after the system of Thornthwaite.

Latitude 0° North	January	February	March	April	May	June
0	8.50	7.66	8.49	8.21	8.50	8.22
5	8.32	7.57	8.47	8.29	8.65	8.41
10	8.13	7.47	8.45	8.37	8.81	8.60
15	7.94	7.36	8.43	8.44	8.98	8.80
20	7.74	7.25	8.41	8.52	9.15	9.00
25	7.53	7.14	8.39	8.61	9.33	9.23
30	7.30	7.03	8.38	8.72	9.53	9.49
32	7.20	6.97	8.37	8.76	9.62	9.59
34	7.10	6.91	8.36	8.80	9.72	9.70
36	6.99	6.85	8.35	8.85	9.82	9.82
38	6.87	6.79	8.34	8.90	9.92	9.95
40	6.76	6.72	8.33	8.95	10.02	10.08
42	6.63	6.65	8.31	9.00	10.14	10.22
44	6.49	6.58	8.30	9.06	10.26	10.38
46	6.34	6.50	8.29	9.12	10.39	10.54
48	6.17	6.41	8.27	9.18	10.53	10.71
50	5.98	6.30	8.24	9.24	10.68	10.91
52	5.77	6.19	8.21	9.29	10.85	11.13
54	5.55	6.08	8.18	9.36	11.03	11.38
56	5.30	5.95	8.15	9.45	11.22	11.67
58	5.01	5.81	8.12	9.55	11.46	12.00
60	4.67	5.65	8.08	9.65	11.74	12.39
South						
0	8.50	7.66	8.49	8.21	8.50	8.22
5	8.68	7.76	8.51	8.15	8.34	8.05
10	8.86	7.87	8.53	8.09	8.18	7.86
15	9.05	7.98	8.55	8.02	8.02	7.65
20	9.24	8.09	8.57	7.94	7.85	7.43
25	9.46	8.21	8.60	7.84	7.66	7.20
30	9.70	8.33	8.62	7.73	7.45	6.96
32	9.81	8.39	8.63	7.69	7.36	6.85
34	9.92	8.45	8.64	7.64	7.27	6.74
36	10.03	8.51	8.65	7.59	7.18	6.62
38	10.15	8.57	8.66	7.54	7.08	6.50
40	10.27	8.63	8.67	7.49	6.97	6.37
42	10.40	8.70	8.68	7.44	6.85	6.23
44	10.54	8.78	8.69	7.38	6.73	6.08
46	10.69	8.86	8.70	7.32	6.61	5.92

11.9

LIGHT HOURS
60° NORTH LATITUDE

July	August	September	October	November	December
8.50	8.49	8.21	8.50	8.22	8.50
8.67	8.60	8.23	8.42	8.07	8.30
8.86	8.71	8.25	8.34	7.91	8.10
9.05	8.83	8.28	8.26	7.75	7.88
9.25	8.96	8.30	8.18	7.58	7.66
9.45	9.09	8.32	8.09	7.40	7.42
9.67	9.22	8.33	7.99	7.19	7.15
9.77	9.27	8.34	7.95	7.11	7.05
9.88	9.33	8.36	7.90	7.02	6.92
9.99	9.40	8.37	7.85	6.92	6.79
10.10	9.47	8.38	7.80	6.82	6.66
10.22	9.54	8.39	7.75	6.72	7.52
10.35	9.62	8.40	7.69	6.62	6.37
10.49	9.70	8.41	7.63	6.49	6.21
10.64	9.79	8.42	7.57	6.36	6.04
10.80	9.89	8.44	7.51	6.23	5.86
10.99	10.00	8.46	7.45	6.10	5.65
11.20	10.12	8.49	7.39	5.93	5.43
11.43	10.26	8.51	7.30	5.74	5.18
11.69	10.40	8.53	7.21	5.54	4.89
11.98	10.55	8.55	7.10	4.31	4.56
12.31	10.70	8.57	6.98	5.04	4.22
8.50	8.49	8.21	8.50	8.22	8.50
8.33	8.38	8.19	8.56	8.37	8.68
8.14	8.27	8.17	8.62	8.53	8.88
7.95	8.15	8.15	8.68	8.70	9.10
7.76	8.03	8.13	8.76	8.87	9.33
7.54	7.90	8.11	8.86	9.04	9.58
7.31	7.76	8.07	8.97	9.24	9.85
7.21	7.70	8.06	9.01	9.33	9.96
7.10	7.63	8.05	9.06	9.42	10.08
6.99	7.56	8.04	9.11	9.51	10.21
6.87	7.49	8.03	9.16	9.61	10.34
6.76	7.41	8.02	9.21	9.71	10.49
6.64	7.33	8.01	9.26	9.82	10.64
6.51	7.25	7.99	9.31	9.94	10.80
6.37	7.16	7.96	9.37	10.07	10.97

heit is about 88. Thus, effective heat would be $(88 - 32)\ 31 = 1736°F$. By proportion, average consumptive use for the entire valley during July would be $(1736/8940) \times 26 = 5.1$ in. depth. Perhaps the use of water by alfalfa during July might be 1.5 times the average use in the valley or 7.7 in. depth.

Blaney-Criddle Method

Blaney and Criddle developed a simplified formula using temperature and day-time hours for the arid western portion of the United States. Their formula has been used extensively by the Soil Conservation Service of the United States Department of Agriculture, wherein considerable data has been collected to determine the values of the coefficients to be used for various crops.

By multiplying the mean monthly temperature t by the monthly percentage of day-time hours of the year p, there is obtained a consumptive use factor F. Expressed mathematically, it is

$$U = K \sum pt = KF \tag{11.6}$$

where the following quantities must be determined for the same period:

$U =$ consumptive use of crop; inches for a given time period
$F =$ sum of the consumptive use factors for the period (sum of the products of mean temperature and percent of annual day-time hours) $(t \times p)/100$
$K =$ empirical coefficient (annual, irrigation season, or growing season)
$t =$ mean temperature in degrees Fahrenheit
$p =$ percentage of day-time hours of the year, occurring during the period. (See Table 11.9.)

For monthly calculations lower case letters are frequently used for clarity as follows:

$f =$ monthly consumptive-use factors, $(t \times p)/100$
$k =$ monthly coefficient, u/f
$u = kf =$ monthly consumptive use, inches

The consumptive use of water by a particular crop in some areas being known, an estimate of the use by the same crop in some other area may be made by application of the formula $U = KF$.

Table 11.10 illustrates use of the method to compute monthly consumptive use by alfalfa in Salinas Valley, California. Monthly coefficients k were developed from measured consumptive use u and temperature t in San Fernando Valley, California.

TABLE 11.10

COMPUTED NORMAL UNIT CONSUMPTIVE USE OF WATER BY ALFALFA,
UPPER SALINAS VALLEY, CALIFORNIA

Month	Mean Monthly Temperature, °F	Daytime Hours, %	Consumptive-Use Factor	Coefficient	Consumptive Use, Inches
	(t)	(p)	(f)	(k)	(u)
April	57.9	8.85	5.12	0.60	3.07
May	62.5	9.82	6.14	.70	4.30
June	65.7	9.84	6.46	.80	5.17
July	68.4	10.00	6.84	.85	5.81
August	67.8	9.41	6.38	.85	5.42
September	66.6	8.36	5.57	.85	4.73
October	62.2	7.84	4.88	.70	3.42

Total consumptive use for irrigation season   31.92

$f = \dfrac{t \times p}{100}$ = monthly consumptive-use factor.

k = monthly coefficients developed from observed data on alfalfa in San Fernando Valley.

$u = kf$ = monthly consumptive use.

Values of the normal seasonal consumptive-use coefficients for several crops are shown in Table 11.11. The last column also shows maximum monthly values that can be expected. Usually, irrigation systems must be designed to meet the maximum monthly water needs rather than the seasonal average.

The Blaney-Criddle formula has limitations similar to those discussed previously. It is an empirical formula developed to fit arid conditions and will give good estimates of seasonal water needs under these conditions.

11.6 EVAPORATION AS AN INDEX TO CONSUMPTIVE USE

Evaporation from a United States Weather Bureau pan has been found to be a good index to peak consumptive use. Some observations in low-humidity areas indicate that peak use may be as much as 10 to 20 percent higher than the evaporation from a Weather Bureau pan for a short period of time. Exposure of the pan and the crop vary the rate considerably.

Piche evaporometers, used extensively in many countries, also give good indices to maximum rates of consumptive use. The Piche unit is essentially a test tube filled with water, inverted with a blotter over the end, and installed in a conventional shelter with other weather

TABLE 11.11

NORMAL SEASONAL CONSUMPTIVE-USE COEFFICIENTS FOR THE MORE
IMPORTANT IRRIGATED CROPS OF THE WEST (AFTER CRIDDLE)

Item	Length of Growing Season or Period	Consumptive-Use Coefficients Seasonal (K)	Maximum Monthly[1] (k)
Alfalfa	frost-free	0.85	0.95–1.25
Beans	3 months	0.65	0.75–0.85
Corn	4 months	0.75	0.80–1.20
Cotton	7 months	0.70	0.75–1.10
Citrus orchard	7 months	0.60	0.65–0.75
Deciduous orchard	frost-free	0.65	0.70–0.95
Pasture, grass, hay annuals	frost-free	0.75	0.85–1.15
Potatoes	3 months	0.70	0.85–1.00
Rice	3 to 4 months	1.00	1.10–1.30
Small grains	3 months	0.75	0.85–1.00
Sorghum	5 months	0.70	0.85–1.10
Sugar beets	5½ months	0.70	0.85–1.00

[1] Dependent upon mean monthly temperature and stage of growth of crop.

instruments. Because of the small size of Piche units, rates of evaporation are in excess of the rates of water use by crops. Piche values are also larger than those obtained from a weather bureau pan. Multiplying Piche readings by 0.7 gives average comparable values, although the coefficient does change with climate, season, and exposure.

Atmometers, which are porous ceramic bulbs, have also been used successfully as indices to consumptive use. Differences between readings obtained from black and white bulb atmometers are indicative of maximum consumptive use rates that can be expected at the time.

Principal limitations of evaporation pans are that they are expensive and require considerable water. Piche evaporometers and atmometers have one drawback in common. The evaporating surface of each is subject to contamination by dust, oil, and other foreign material. Blotters for Piche units are inexpensive and can be changed readily. However, since atmometers are relatively expensive porous ceramic bulbs, they are not subject to ready replacement when dirty. Fingerprints, which leave oil on the surface, are especially harmful. Atmometers must be recalibrated when cleaned. All of these evaporative devices are independent, of course, of the physiology of the plant.

They are indices to potential consumptive use. Adjustment must be made for the kind of crop and stage of growth. The importance of these physiological plant factors is discussed next.

11.7 CLIMATE AND PLANT PHYSIOLOGY RELATED TO CONSUMPTIVE USE

Kind of crop and stage of growth certainly have an influence upon transpiration and hence, upon consumptive use. Typical consumptive use and evaporation rates during the growing season are shown in Fig. 11.6. Note that the rate increases to a peak rate of consumptive use and then diminishes. The peak rate for earlier maturing crops is not as high as for later maturing crops. The maximum rate of use occurs in August in the northern hemisphere when days are hottest. The shape of the consumptive-use curves and trend for other crops will be similar.

Two principal variables are not considered in Fig. 11.6. These variables are causing variation in rate from day to day and change during the season. The first is climate and the second is the physiology of plants.

Variations in consumptive-use rate occur from day to day because of changes in weather. A hot, dry, windy day will increase the rate; whereas a cool day will decrease the rate of consumptive use. Characterizing this weather influence is not easy. Several methods were

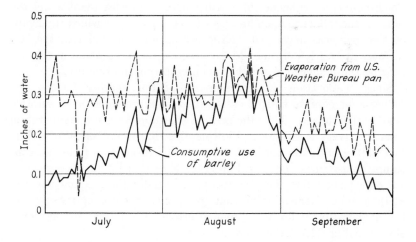

Fig. 11.6 Typical consumptive use and evaporation rates during the growing season.

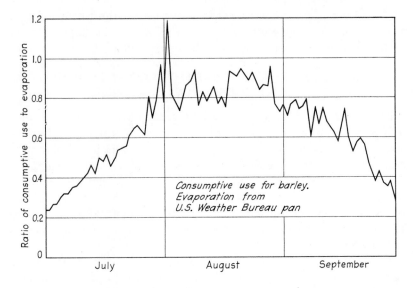

Fig. 11.7 Typical variation of ratio of consumptive use to evaporation during the growing season.

outlined in the previous sections. Penman has used several variables; whereas Thornthwaite, Lowry-Johnson, and Blaney-Criddle have relied upon temperature. Evaporation devices have been used as well to characterize climatic changes. All are helpful, but none are adequate. For example, the influence of weather upon a growing plant is not the same as the influence upon an open-water surface, for which the evaporation pan was developed. Exposure, height, and color of the leaf surfaces are all important. Much progress has been made in characterizing the influence of weather upon crop consumptive use.

Since evaporation integrates many of the weather factors, the influence of climate will be assumed to be represented by a suitable evaporation device. Then the ratio of consumptive use to evaporation should be very significant. Plotting this dimensionless ratio in Fig. 11.7 as the ordinate, much of the variation disappears from the consumptive-use curve shown in Fig. 11.6. Note that the ratio approaches 1.0 as a peak value; i.e., the use is essentially the evaporation rate at that time. Remembering that the measured consumptive use and the evaporation index both involve errors, fluctuations in the curve shown in Fig. 11.7 are not unexpected.

Abscissa of both figures still has the dimension of time, and the water-use curve for each crop is unique. By changing the horizontal scale to a scale of maturity or stage of growth, much of this seasonal variation is eliminated. The curves are essentially superimposed.

Analyses of consumptive use and weather data for many crops grown in many countries of the world give strong support for the empirical curve shown in Fig. 11.8. Most crops which have rapidly expanding root systems or an expanding vegetative system follow this basic curve. The zero point on the scale of maturity is obtained by extrapolating the curves to zero.

The physiology of the growing plant can be characterized by flowering, fruiting, and other distinctive characteristics like tasseling that occurs in corn. Peak use comes at the beginning of flowering and at the end of the vegetative stage of growth. How irrigation and other management practices should be modified according to the state of growth is the subject of the next chapter.

Referring to methods of computing consumptive use, the evaporation index used in the ratio of consumptive use to evaporation in Fig. 11.8 can be Penman's or Thornthwaite's potential evapotranspiration; Blaney and Criddle's product of sunshine and temperature; or the values of potential consumptive use obtained from evaporation pans, Piche evaporometers, or atmometers. Any suitable measure of weather can be used as denominator of the consumptive use evaporation ratio.

For an example of the use of Fig. 11.8, suppose that at 40 percent

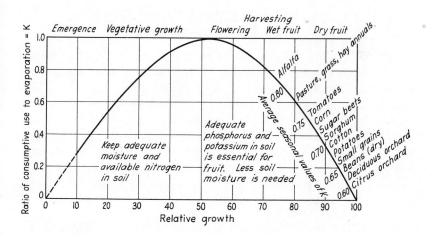

Fig. 11.8 Generalized curve comparing consumptive use-evaporation ratio to relative growth of crop.

of the growing period, perhaps June 25, evaporation was 0.26 inch per day. Since the ratio of consumptive use to evaporation at 40 percent growing season is 0.7 as shown in Fig. 11.8, the consumptive use will be $0.7 \times 0.26 = 0.18$ in. per day.

Variation of Consumptive-Use Coefficient During the Growing Season

The change of crop coefficient during the growing season can best be illustrated using the Blaney-Criddle formula. Note that their formula $U = KF$ can be written as $K = U/F$, which is the ratio of consumptive use to the weather index spoken of previously. Hence, the Blaney-Criddle K is the consumptive-use evaporation ratio shown in Fig. 11.8. Observe the seasonal values of K in Table 11.11 and the curve of Fig. 11.8. Dry beans, for example, are generally allowed to grow during most of the growing season (the average value of the ordinate K for beans is given in Table 11.11 as 0.65), whereas alfalfa is cut shortly after peak use is reached, resulting in an average ordinate for alfalfa of 0.85, which is considerably higher than for dry beans. Hence, the value of K is indicative of the point on the descending portion of the curve at which harvest occurs. Figure 11.8 shows these values on the descending portion of the curve for a series of typical crops. Values for other crops can be estimated from their similarity to the crops shown.

Naturally, crops like paddy rice and varieties of cacti do not have consumptive-use rates corresponding to those shown here, but the principles are very similar.

11.8 RELATIONSHIP BETWEEN FROST-FREE DAYS AND JULY TEMPERATURES

Guy O. Woodward has obtained an interesting and useful relationship between the average number of frost-free days in the growing season and the average temperatures during July, Fig. 11.9. This concept can be extended further to improve accuracy and can be applied to other areas. Such a relationship established for areas of new development where limited data is available would allow estimates to be made as a basis for subsequent agricultural development. Furthermore, July temperatures in the Northern Hemisphere or January temperatures in the Southern Hemisphere can also be used to obtain estimates of peak consumptive use.

11.9 CONSUMPTIVE-USE REQUIREMENTS OF TYPICAL CROPS

Seasonal consumptive-use requirements of common irrigated crops in relation to requirements for alfalfa are shown in Table 11.12 for

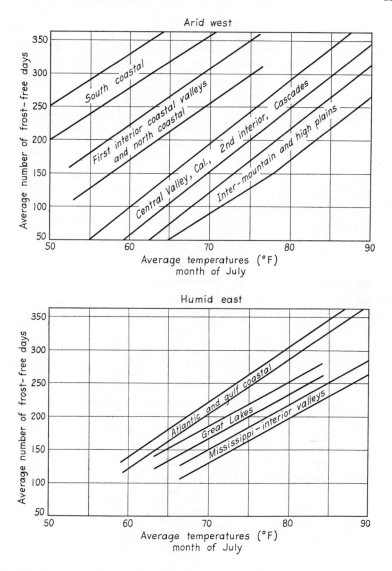

Fig. 11.9 Relation between frost-free days and July temperatures for arid western and humid eastern United States. (After Woodward.)

arid regions in the United States. Seasonal use and peak daily use values have also been prepared as Table 11.13 for seven regions shown in Fig. 11.9. These tables developed for the United States can be used in other areas by determining the length of the growing season and

TABLE 11.12

SEASONAL CONSUMPTIVE-USE REQUIREMENTS OF COMMON IRRIGATED CROPS
IN RELATION TO THE REQUIREMENTS FOR ALFALFA—WESTERN STATES
(AFTER WOODWARD)

Crops	Relation of Seasonal Consumptive Use of Crops listed to the Seasonal Consumptive Use of Alfalfa— (Optimum Yield)
Alfalfa	1.00
*Almonds	0.42
*Apricots	0.50
Artichokes	0.50
*Avocados	0.52
Beans	0.42
Berries	0.40
Clover	0.90
Carrots	0.40
Corn	0.65
*Citrus	0.70
Cotton	0.65
Grain	0.50 (variable)
Grain-Sorghum	0.35 to 0.40
Grapes	(variable)
Hops	0.50
Lettuce	0.15 (Plant use only)**
Melons	0.40
Pasture	0.90 (variable)
*Peaches	0.65
Potatoes—Winter	0.50
Potatoes—Spring or Seed	0.30
Seed Peas	0.30
Sugar Beets	0.82
Strawberries	0.60
Tomatoes	0.45 to 0.50
*Walnuts	0.60

*Add 20–25% for permanent grass—legume cover.
**Water may be applied in addition to that required for growth for the
purpose of quality improvement.

average July temperatures for that area. Thus, if a given arid region
had 150 frost-free days and an average July temperature of 75°F, it
would be comparable to the Intermountain and High Plains, or area
No. 4. Data for Table 11.13 for the Intermountain and High Plains

TABLE 11.13. TOTAL CONSUMPTIVE USE AND PEAK DAILY USE, WESTERN UNITED STATES (AFTER WOODWARD)

Crops	Southern Coastal* (1)				South Pacific Coastal Interior and North Coastal (2)							
	300 Days Plus		250–300 Days		250–300 Days		210–250 Days		180–210 Days		150–180 Days	
	Season Use in.	Daily Use in./d	Season Use in.	Daily Use in./d	Season Use in.	Daily Use in./d	Season Use in.	Daily Use in./d	Season Use in.	Daily Use in./d	Season Use in.	Daily Use in./d
Alfalfa	36.0	0.20	30.0	0.17	37.0	0.27	32.0	0.22	26.0	0.20	22.0	0.18
Pasture	33.5	0.20	28.0	0.17	33.0	0.27	30.0	0.22	24.0	0.20	20.0	0.18
Grain—small	16.0	0.18	14.0	0.16	17.0	0.22	14.5	0.20	12.0	0.20	10.0	0.18
Beets—sugar	29.0	0.20	25.0	0.18	30.0	0.27	26.0	0.22	...	...	...	...
Beans—field	12.0	0.18	10.0	0.16	13.0	0.22	11.0	0.18	...	...	...	...
Corn—field	...	...	...	...	...	0.20	19.0	0.25	16.0	0.22	14.0	0.20
Potatoes	...	...	...	...	24.0	0.20	20.0	0.18	...	...	...	...
Peas—green	10.0	0.18	8.0	0.16	11.0	0.20	9.0	0.18	8.0	0.16	7.0	0.16
Legume seed	25.0	0.20	22.0	0.18	26.0	0.25	22.0	0.22	20.0	0.20	18.0	0.18
Tomatoes	18.0	0.16	15.0	0.16	19.0	0.20	16.0	0.18	13.0	0.16	11.0	0.16
Vegetable seed	16.0	0.16	14.0	0.16	18.0	0.18	16.0	0.18	14.0	0.16	12.0	0.16
Beans—pole	...	...	...	...	18.0	0.20	16.0	0.20	14.0	0.18	12.0	0.16
Corn—sweet	16.0	0.18	14.0	0.16	16.0	0.20	14.0	0.18	12.0	0.18	11.0	0.16
Apples	...	...	...	...	24.0	0.20	22.0	0.20	20.0	0.18	18.0	0.18
Cherries	...	...	...	...	24.0	0.20	22.0	0.20	...	...	...	...
Peaches	...	...	...	...	24.0	0.20	22.0	0.20	...	...	...	...
Prunes	...	...	...	...	22.0	0.20	20.0	0.20	...	...	...	...
Apricots	...	...	...	...	22.0	0.20	20.0	0.20	...	...	...	...
Oranges	20.0	0.16	18.0	0.16	22.0	0.18	...	...	...	...	...	...
Avocados	18.0	0.16	...	...	...	...	...	...	...	...	...	...
Walnuts	22.0	0.20	18.0	0.18	24.0	0.25	22.0	0.22	20.0	0.20	...	...
Strawberries	22.0	0.20	18.0	0.18	23.0	0.25	18.0	0.22	16.0	0.20	14.0	0.18
Lettuce	4.0	0.16	4.0	0.16	6.0	0.18	5.0	0.18	...	...	...	...
Mint	...	...	...	...	23.0	0.24	21.0	0.22	19.0	0.20	17.0	0.18
Hops	...	...	...	...	20.0	0.22	20.0	0.20	...	...	...	...

* Fog Belt.

TABLE 11.13. TOTAL CONSUMPTIVE USE AND PEAK DAILY USE, WESTERN UNITED STATES (Continued)

Central Valley—California and Valleys East Side of Cascade Mountains (3)

Crops	250–300 Days		210–250 Days		180–210 Days		150–180 Days		120–150 Days		90–120 Days	
	Season Use in.	Daily Use in./d	Season Use in.	Daily Use in./d	Season Use in.	Daily Use in./d	Season Use in.	Daily Use in./d	Season Use in.	Daily Use in./d	Season Use in.	Daily Use in./d
Alfalfa	40.0	0.30	34.0	0.28	30.0	0.25	26.0	0.22	20.0	0.20	14.0	0.18
Pasture	36.0	0.30	30.0	0.28	28.0	0.25	24.0	0.22	18.0	0.20	13.0	0.18
Grain—small	18.0	0.22	16.0	0.22	15.0	0.20	14.0	0.18	13.0	0.18	12.0	0.16
Beets—sugar	33.0	0.30	28.0	0.25	24.0	0.22	20.0	0.20	18.0	0.18	…	…
Beans—field	17.0	0.22	13.0	0.20	13.0	0.20	12.0	0.18	12.0	0.18	…	…
Corn—field	26.0	0.35	22.0	0.32	22.0	0.30	20.0	0.25	18.0	0.22	17.0	0.20
Potatoes—summer	12.0	0.16	…	…	…	…	…	…	…	…	…	…
Potatoes—fall	…	…	19.0	0.25	18.0	0.22	18.0	0.20	17.0	0.18	16.0	0.16
Peas—green	…	…	8.0	0.18	7.0	0.18	7.0	0.18	7.0	0.16	6.0	0.15
Peas—field	…	…	10.0	0.18	9.0	0.18	9.0	0.18	8.0	0.16	8.0	0.15
Tomatoes	20.0	0.20	18.0	0.18	18.0	0.18	17.0	0.17	16.0	0.16	…	…
Cotton	26.0	0.30	22.0	0.28	…	…	…	…	…	…	8.0	0.15
Grain—sorghum	15.0	0.20	13.0	0.18	12.0	0.17	10.0	0.16	9.0	0.15	…	…
Apples	…	…	26.0	0.20	23.0	0.20	21.0	0.18	…	…	…	…
Cherries	…	…	24.0	0.20	21.0	0.20	19.0	0.18	…	…	…	…
Peaches	22.0	0.22	22.0	0.20	20.0	0.20	18.0	0.18	…	…	…	…
Apricots	20.0	0.22	17.0	0.20	15.0	0.20	…	…	…	…	…	…
Oranges	28.0	0.18	…	…	…	…	…	…	…	…	…	…
Strawberries	24.0	0.20	20.0	0.20	18.0	0.18	…	…	…	…	…	…
Lettuce—winter	4.0	0.20	…	…	…	…	…	…	…	…	…	…
Mint	…	…	20.0	0.22	18.0	0.20	…	…	…	…	…	…
Hops	…	…	18.0	0.20	16.0	0.18	…	…	…	…	…	…
Grapes	30.0	0.25	25.0	0.22	22.0	0.20	…	…	…	…	…	…
Walnuts	24.0	0.22	20.0	0.20	…	…	…	…	…	…	…	…
Almonds	22.0	0.25	20.0	0.22	…	…	…	…	…	…	…	…

TABLE 11.13. TOTAL CONSUMPTIVE USE AND PEAK DAILY USE, WESTERN UNITED STATES (*Concluded*)

Intermountain, Desert, and Western High Plains (4)

Crops	250–300 Days Season Use in.	250–300 Days Daily Use in./d.	210–250 Days Season Use in.	210–250 Days Daily Use in./d.	180–210 Days Season Use in.	180–210 Days Daily Use in./d	150–180 Days Season Use in.	150–180 Days Daily Use in./d	120–150 Days Season Use in.	120–150 Days Daily Use in./d	90–120 Days Season Use in.	90–120 Days Daily Use in./d
Alfalfa	52.0	0.40	44.0	0.32	36.0	0.29	30.0	0.26	24.0	0.22	19.0	0.20
Pasture	48.0	0.40	40.0	0.30	33.0	0.28	28.0	0.25	22.0	0.22	17.0	0.20
Grain—small	21.0	0.25	18.0	0.22	16.0	0.20	16.0	0.20	16.0	0.20	14.0	0.18
Beets—sugar	37.0	0.30	32.0	0.30	30.0	0.28	26.0	0.25	24.0	0.22	18.0	0.20
Beans—field	22.0	0.25	17.0	0.20	14.0	0.20	14.0	0.18	14.0	0.17	12.0	0.15
Corn—field	..	..	30.0	0.35	26.0	0.30	24.0	0.28	22.0	0.24	..	0.20
Potatoes—fall	..	..	23.0	0.30	21.0	0.28	20.0	0.25	19.0	0.22	17.0	0.20
Peas—field	..	..	..	..	10.0	0.19	10.0	0.18	10.0	0.17	9.0	0.15
Tomatoes	..	..	20.0	0.22	18.0	0.20	17.0	0.18	16.0	0.17	..	..
Cotton	32.0	0.30	30.0	0.28	..	..	..	..	..	..	..	..
Grain—sorghum	19.0	0.25	18.0	0.20	16.0	0.20	14.0	0.18	12.0	0.17	..	..
Apples	..	..	..	..	28.0	0.22	24.0	0.20	20.0	0.18	..	..
Cherries	..	..	..	..	26.0	0.22	..	..	..	..	..	..
Peaches	..	0.25	29.0	0.25	27.0	0.22	..	..	..	..	..	..
Apricots	26.0	0.25	24.0	0.25	25.0	0.20	..	..	..	..	..	..
Almonds	22.0	0.25	20.0	0.25	..	..	..	..	..	..	..	..
Vineyards	40.0	0.27	32.0	0.25	26.0	0.22	..	..	..	..	..	..
Legume seed	..	..	..	..	..	..	..	..	16.0	0.18	14.0	0.16
Grass seed	..	..	..	..	..	..	..	..	14.0	0.14	12.0	0.14
Potatoes—seed	..	..	..	..	..	..	..	..	16.0	0.16	14.0	0.15
Grapefruit	45.0	0.20	..	..	..	..	..	..	..	..	..	..
Oranges	36.0	0.18	..	..	..	..	..	..	..	..	..	..
Lettuce—winter	6.0	0.18	..	..	..	..	..	..	..	..	..	..
Melons	22.0	0.25	20.0	0.22	18.0	0.20	16.0	0.18	..	..	..	..
Palm dates	60.0	0.30	..	..	..	..	..	..	..	..	..	..
Truck crops	20.0	0.25	18.0	0.22	14.0	0.20	12.0	0.18	12.0	0.16	10.0	0.15

should apply, therefore, to the new area in question with reasonable accuracy.

REFERENCES

Blaney, Harry F., "Monthly Consumptive Use Requirements for Irrigated Crops," *Am. Soc. Civil Eng. Proc.*, Paper 1963, March 1959. Discussion by Constantine Christopoulos, *Proc. Am. Soc. Civil Eng.*, Part I, December 1959.

Blaney, Harry F., "Climate as an Index of Irrigation Needs," Water—U.S.D.A. Yearbook of Agriculture, 1955.

Blaney, Harry F. and Wayne D. Criddle, "Determining Water Requirements in Irrigated Areas from Climatological and Irrigation Data," U.S.D.A.—S.C.S. T. P.–96, 1950.

Blaney, Harry F. and Wayne D. Criddle, "A Method of Estimating Water Requirements in Irrigated Areas from Climatological Data," U.S.D.A.—S.C.S. (mimeo.), December 1947.

"Consumptive Use of Water"—A Symposium—*Am. Soc. Civil Eng. Trans.*, Paper No. 2524.

   Blaney, Harry F., "Definition, Methods, and Research Data."

   Criddle, Wayne D., "Municipal and Industrial Areas."

   Gleason, George B., "Municipal and Industrial Areas."

   Lowry, Robert L., "Special Case in Rio Grande Basin."

   Rich, L. R., "Forest and Range Vegetation."

Criddle, Wayne D., "Methods of Computing Consumptive Use of Water," *Am. Soc. Civil Eng. Proc.*, Paper 1507, January 1958.

Croft, A. R., "Water Loss by Stream Surface Evaporation and Transpiration by Riparian Vegetation," *Trans. Am. Geophys. Union,* Vol. 29, No. 2, pp. 235–239, April 1948.

Halkias, N. A., F. J. Veihmeyer, and A. H. Hendrickson, "Determining Water Needs for Crops from Climatic Data," *Hilgardia,* Vol. 24, No. 9, December 1955.

Hendricks, David W. and Vaughn E. Hansen, "Mechanics of Evapotranspiration," *Trans. Am. Soc. Civil Eng.,* Vol. 127, 1962.

Lowry, Robert L. Jr. and Arthur F. Johnson, "Consumptive Use of Water for Agriculture," *Trans. Am. Soc. Civil Eng.,* Vol. 107, pp. 1243–1302, 1942.

Penman, H. L., "Natural Evaporation from Open Water, Bare Soil and Grass," *Proc. of the Royal Society,* Series A, 1948.

Sleight, R. B., "Evaporation from the Surfaces of Water and River-Bed Materials," *J. Agr. Research,* Vol. 10, 1917.

Thornthwaite, C. W., "An Approach toward a Rational Classification of Climate," *The Geographical Rev.,* Vol. 38, No. 1, 1948.

Thornthwaite, C. W., "Climate in Relation to Planting and Irrigation of Vegetable Crops," *Drexel Institute of Technology, Publ. in Climatology,* Vol. 5, No. 5, 1952.

Thornthwaite, C. W. and J. R. Mather, "The Water Budget and its Use in Irrigation," Water—U.S.D.A. Yearbook of Agriculture, 1955.

Thornthwaite, C. W. and F. Kenneth Hare, "Climate Classification in Forestry," reprint from *UNASYLVA,* published by the FAO of the United Nations, Rome, Italy, Vol. 9, No. 2, June 1955.

van Hylkama, T. E. A., "The Water Balance of the Earth," *Drexel Institute of Technology, Publ. in Climatology,* Vol. 9, No. 2, 1956.

Woodward, Guy O., "Sprinkler Irrigation," Sprinkler Irrigation Association, Washington, D.C., second edition 1959.

Young, A. A. and Harry F. Blaney, "Use of Water by Native Vegetation," *California State Div. of Water Resources Bul. 50, 1942.*

CHAPTER 12 WHEN TO IRRIGATE—HOW MUCH WATER TO APPLY

Three major considerations influence the time of irrigation and how much water should be applied, namely, (a) water needs of the crop, (b) availability of water with which to irrigate, and (c) capacity of the root-zone soil to store water. Water needs of the crop are of paramount importance in determining the time of irrigation during the crop-growing season on irrigation projects which obtain their water supplies from storage reservoirs or from other dependable sources of water. Some irrigated areas have a limited water supply during the irrigation season, but an abundance of water during late autumn or winter and early spring. Irrigation farmers cannot always apply water when the crop is most in need; sometimes to save water they must apply it even though the crop does not need it, provided the soil has the capacity to store additional water. Therefore, crop needs, available water supply, and storage capacity of the soil must be considered in a discussion of the proper time to irrigate.

Growing crops use water continuously, but the rate of use varies with the kind of crop grown, age of the crop, temperature, and atmospheric conditions—all variable factors. At each irrigation a volume of water sufficient to supply the needs of the crop for a period varying from a few days to several weeks is stored in the unsaturated soil in the form of available soil water. How frequently water should be applied to soils of different properties in order to best supply crop needs is a question of real and practical significance. A factor of major importance in arriving at the desirable frequency and time of irrigation is the water need of the crop.

12.1 LIMITING SOIL MOISTURE CONDITIONS

Growth of most crops produced under irrigation farming is stimulated by moderate quantities of soil moisture and retarded by either

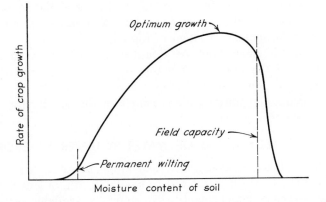

Fig. 12.1 Rate of growth to moisture content of the soil; optimum growth varying somewhat with aeration, water-holding capacity of soil, and crop grown.

excessive or deficient amounts. Air in the soil is essential to satisfactory crop growth; hence, excessive flooding and filling the soil pore spaces with water, thus driving out air, inhibits proper functioning of the plants even though supplying an abundance of available water. On the other hand, soils having deficient amounts of water hold it so tenaciously that plants must expend extra energy to obtain it. If the rate of intake by the plant is not high enough to maintain turgidity of the leaves, permanent wilting follows. At some soil moisture content between these two extreme moisture conditions, designated optimum moisture percentage, plants grow most rapidly (Fig. 12.1).

The objective of considerable research has been to extend knowledge of these two limiting soil moisture conditions; i.e., permanent wilting percentage and optimum plant growth percentage. Because of wide variations in physical properties of different soils, it is easy to understand that the moisture percentage in a clay soil at permanent wilting of a plant may be several times the moisture percentage in a sandy soil when the same plant wilts permanently.

12.2 APPEARANCE OF CROP

A light green color in alfalfa is generally indicative of adequate moisture and satisfactory growth, whereas a dark green color indicates lack of moisture. Among the root crops, sugar beets readily indicate need for water by temporary wilting, particularly during the warmest part of the day. Grain crops also indicate need for water by temporary wilting. In production of fruit crops, it is impractical to detect the need for water by the appearance of the leaves of the trees. It is, therefore, more essential to base the time of irrigation of orchards on observations of moisture content of the soil.

Crop growth should not be retarded by lack of available soil moisture; and the practice of withholding irrigation until the crop definitely shows a need for water is likely to retard growth. It is essential to maintain readily available water in the soil if crops are to make satisfactory growth.

12.3   SEASONAL USE OF WATER BY DIFFERENT CROPS

Irrigation farmers must select their crops, to some extent, on the basis of time at which water will be available. In valleys having no storage reservoirs, larger quantities of water are available early in the season. During the forepart of the crop-growing season, streams in mountain areas fed by melting snow banks and drifts are much larger than during the summer. Under such conditions, alfalfa, wheat, and oats may well be produced as each of these crops requires large amounts of early-season water. Canning peas may be matured before a water shortage begins. Alfalfa continues to grow throughout the late summer months, provided water is available. Sugar beets, potatoes, and corn require less water early in the season, but during the late summer months these crops need an abundance of water. Unless late-season water is assured, it is inadvisable to attempt to grow sugar beets and potatoes.

12.4   AVAILABLE WATER SUPPLY

Irrigation during the dormant or non-growing season, in many localities, is an economical means of storing water in the soil for future use. Immediately after heavy rains which increase the flow in rivers and creeks, large quantities of water for irrigation may be available for short periods of time. It is desirable to build surface reservoirs—both large and small—in which to store water that becomes available from sudden torrential rains or that is available only during the fall, winter, or spring. As a general rule, water that is used in irrigation at times when not really needed by crops is used less efficiently than it would be if it were possible to apply the water to the soil when most needed by crops.

In some localities storage of water in surface reservoirs is impracticable because of high costs and lack of suitable natural facilities. In such cases, it is advantageous to use the soil as a storage reservoir and to apply water whenever available as a means of storing it for future use. Deep percolation from shallow soils underlain by coarse-textured sands and gravels may be excessive unless water is applied carefully. Some areas can be benefited by applying during the dormant season depths of water sufficient to saturate the subsoil gravels and cause the water table to rise to an elevation near the land surface. Where it is not feasible to drain ground water during the irrigation season, excessive depths applied during the dormant season may cause appreciable damage by waterlogging the soil and thus reducing the depth of the root zone.

12.5 FALL IRRIGATION

In many places fall and winter precipitation is insufficient to moisten the soil fully. Fall irrigation under such conditions is a means of saving water and of getting the land in favorable condition for germination of seeds and early growth of crops during the following season.

Alfalfa grown on well-drained soils may be irrigated in the fall; and fall irrigation of meadow forage crops and of pasture lands is usually desirable if the soils become very dry without irrigation. In the practice of irrigation during fall, winter, or early spring, it is important to guard against use of excessive depths of water. As shown in Chapter 7, there are definite limitations to the capacity of any soil to retain water. Permitting water to flow over the lands for many days, and sometimes weeks, is injurious both to land irrigated and to lower-lying areas to which much of the excess water seeps.

Fall irrigation can also be beneficial in land preparation for the subsequent spring planting. By plowing early in the fall and then irrigating, the water causes clods to crumble; also the freezing and thawing of wet soils reduces the need for harrowing, etc., to prepare a good seed bed. The cost of land preparation is reduced thereby, and usually better soil structure is obtained. Thus, fall irrigation may be very helpful in minimizing tillage and improving or maintaining good soil structure.

12.6 WINTER IRRIGATION

In mild climates where the soil does not freeze, winter irrigation may be practiced advantageously as a means of saving water that would otherwise be lost.

At higher elevations and in colder parts of irrigated regions, winter irrigation is of little, if any, practical importance. Frozen soils absorb water slowly, if at all, and it is difficult to spread water over the snow-covered fields effectively. Furthermore, some crops are injured by winter irrigation in cold climates.

12.7 EARLY SPRING IRRIGATION

Some arid-region lands need irrigation during the early spring months in order to supply the moisture essential to satisfactory germination and early growth of annual crops. Arid-region streams usually have ample water to meet the needs for early spring irrigation. Even where the discharge of streams at high mountain elevations is held in storage reservoirs, enough water is frequently available from rains and melting snows on lower elevations to supply the needs for early spring irrigation. The value of early spring irrigation as a means of storing available water in root-zone soil is not yet fully realized. Irrigators are frequently misled when spring rains moisten the soil to a depth of 9 to 15 inches. Some consider the soil "wet enough" when, in fact, there may be 3 to 5 feet of dry soil below the moist surface soil.

Use of a soil auger or of a soil tube will enable irrigators to decide intelligently needs of soils for early spring irrigation. Although it is highly desirable to save water by applying it to inadequately moistened soils, it is undesirable early in the spring to irrigate soils that are already moistened to field capacity.

Fully moistened soils in which the water table is at a shallow depth may be injured rather than benefited by early spring irrigation. Under such conditions it is better to waste water than to apply it to the soil. Also, early spring irrigations may retard growth if irrigation cools the ground considerably. Damage to crops may result from an early irrigation in soils of low soluble nitrogen content, particularly if roots are shallow at the time of irrigation, or on soils susceptible to limited aeration. When nitrogen deficiency and limited aeration occur, the crop may show a yellow, unhealthy appearance. The nitrogen problem can be avoided by maintaining a higher nitrogen level in the soil, or by adding soluble nitrogen to water during spring irrigation.

12.8 SOIL MOISTURE REMOVAL BY PLANT ROOTS

When to irrigate and how much to apply are affected considerably by where and when water is removed from the soil by the plant roots. Shallow-rooted crops will require more frequent irrigation than deep-rooted crops. Soil conditions which restrict the growth of roots will likewise influence irrigation practice. Figure 12.2, showing shallow alfalfa roots resulting from a water table at 3 feet, is in sharp contrast to Fig. 12.3 which shows that less water was extracted from the surface foot than from the second and third foot of soil. However, in Fig. 12.4 more water was extracted by cotton from the first foot than from any

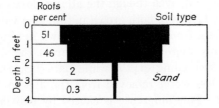

Fig. 12.2 Distribution of alfalfa roots with a water table 3 feet below ground surface.

Depth, ft	Water used	
	Inches per year	% of total
0-1	6.8	⑬
1-2	7.2	⑭
2-3	7.3	⑭
3-4	6.7	⑬
4-5	5.6	⑪
5-6	4.7	⑨
6-7	4.0	⑧
7-8	3.5	⑦
8-9	2.8	⑥
9-10	2.3	⑤

Fig. 12.3 Use of water by alfalfa in Arizona from each foot of the root-zone soil.

Depth, ft	Water used	
	Inches per year	% of total
0-1	8.6	㉚
1-2	7.3	㉖
2-3	5.2	⑱
3-4	3.7	⑬
4-5	2.2	⑧
5-6	1.3	⑤

Fig. 12.4 Use of water from each foot of soil growing cotton in Arizona.

succeeding foot of soil, even though the alfalfa shown in Fig. 12.3 and the cotton in Fig. 12.4 were both grown in the same valley with similar soils. Rooting data from irrigated crops grown in semi-humid regions shows considerably more water removed from the first foot of soil than from any succeeding depth, while data from hot arid regions generally shows less water removed from the first foot than from the next foot of soil. This difference is caused by two factors: first, the depth to which the applied water penetrates, and second, the moisture content of the soil during the growing period. For example, in humid and semi-humid areas several rainfalls occur during the growing season. However, seldom do these rains penetrate beyond the first foot of soil. In the very arid regions, essentially no rainfall occurs during the growing season. Likewise, the cotton shown in Fig. 12.4 is a more shallow-rooted crop than the alfalfa shown in Fig. 12.3. For this reason the cotton is irrigated more frequently, especially during the early growing period. Thus, more water is extracted from the first foot of soil whenever the first foot is kept more moist either by frequent rains or by frequent irrigations.

12.9 EFFECT OF MOISTURE CONTENT ON RATE OF REMOVAL OF SOIL MOISTURE

Other conditions being equal, roots of a plant in moist soil will extract more water than the roots of the same plant growing in dryer soil. Figure 12.5 illustrates the fact that when the soil is wet, most of the crop's moisture is withdrawn from soil near the surface. This is because more roots are normally growing near the surface. However, as the moisture content of soil near the surface decreases, more moisture is extracted from lower depths. Finally, as the moisture content of the soil near the surface approaches permanent wilting percentage, essentially all moisture removed by the roots comes from deeper soil. Since fewer roots exist in the lower portion of a soil profile, more energy must be expended in extracting required moisture, and frequently insufficient moisture is available to the plant to prevent wilting. Moisture may be available in the soil, but the plant roots may not be able to extract this moisture at a sufficient rate to meet transpiration requirements of the plants.

When the upper portion of the root zone is kept moist, most water used consumptively by the plant will be removed from the soil near the surface. However, when infrequent irrigations are applied, and essentially no rainfall occurs, less water may be used from the surface foot than from succeeding depths. Since the soil is wet following a good irrigation and normally dry on and near the soil surface before

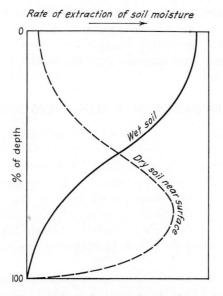

Fig. 12.5 Variation of extraction of soil moisture with moisture content and depth.

another irrigation is applied, the rate of removal will be a composite between the two qualitative curves of Fig. 12.5. Normal arid-region irrigation practice generally results in an average extraction pattern shown in Fig. 12.6 which is essentially triangular in shape. The dis-

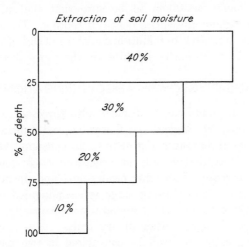

Fig. 12.6 Average extraction of soil moisture by plant roots between irrigations.

tribution of active roots in a normal soil is also approximately triangular in shape, the greatest concentration being near the soil surface. Note that a crop extracting moisture to a depth of 4 feet would remove four times more moisture from the surface foot than from the fourth foot of soil.

12.10 INFLUENCE OF A RESTRICTING SOIL LAYER

Any restricting layer will cause a concentration of roots above that layer. Frequently, roots are restricted by a compacted layer resulting from faulty tillage practices. A 10 percent reduction in soil voids will generally prevent root growth. This much reduction in voids can easily occur using rubber-tired vehicles, discs, and vibrating tillage implements. Restricted roots have less soil available to extract needed soil moisture and plant nutrients. Plants with restricted root systems show symptoms of low fertility and drought, and have deformed tap roots.

If the restricting layer results in a temporarily-perched water table, the saturated soil will cause an aeration problem. Therefore, a restricting layer is particularly harmful. Good farming practices will eliminate and prevent compacted layers, but silty clay layers and other restricting layers are not easily avoided or eliminated. Whenever they occur, heavy irrigations should be given in the dormant season to store water in the deeper soils, and lighter, more frequent irrigations during the active growing period will be beneficial.

Remember also that roots cannot penetrate a dry soil. Special precautions should be taken to be sure that the soil moisture is favorable throughout the entire depth of rooting. For a young crop, it is especially necessary to maintain moisture ahead of the expanding roots so that they will not be inhibited in their growth and expansion.

12.11 STAGE OF GROWTH AFFECTS IRRIGATION PRACTICE

Growth of all plants can be divided into three stages with regard to irrigation practice: vegetative, flowering, and fruiting. The relationship of these three stages of growth to the consumptive use is shown in Fig. 11.8. Note that during the vegetative stage consumptive use continues to increase. Flowering occurs near and during the peak of consumptive use. The fruiting stage is accompanied by a decrease in consumptive use until the transpiration essentially ceases during the latter part of the formation of dry fruit.

The fruiting stage can well be considered in two parts: the wet-fruit stage which follows flowering, and the dry-fruit stage following wet-fruit growth. Dry-fruiting is accompanied by a decrease in consumptive use until transpiration ceases and the plant is dead.

The amount of water applied and the frequency of irrigation must be adjusted to the actual consumptive use of the crop, water-holding capacity of the soil, and depth of rooting. Naturally, a shallow sandy soil will require quite different scheduling of irrigation than will be required for a deep clay loam soil.

Examples of crops harvested in different stages of growth are shown in Table 12.1.

TABLE 12.1

TYPICAL CROPS HARVESTED DURING THE DIFFERENT STAGES OF GROWTH

| Vegetative | Flowering | Fruiting | |
		Wet	Dry
Lettuce	Flowers	Tomatoes	Dry peas
Alfalfa	Cauliflower	Green peas	Dry beans
Grasses	Broccoli	Green beans	Potatoes
Cabbage		Sweet corn	Mandioca
Asparagus		Deciduous fruits	Seeds and grains
Mint		Sugar cane	Field corn
		Artichokes	Cotton
		Sugar beets	Rice
		Berries	Onions
		Bananas	Nuts
		Cantaloupes	
		Citrus	
		Watermelons	

Irrigation During the Vegetative Stage of Growth

A good moisture supply should be available to plants at all stages of vegetative growth. When vegetation is being produced, nitrogen is particularly essential. Light, frequent irrigations are generally desirable because of the need to keep ample moisture in the soil for the relatively shallow root system. For perennial crops like alfalfa with a deep root system, less frequent but heavier irrigations are necessary. Whenever excessive temperatures exist, frequent irrigations may be desirable for cooling the soil and the plants. When a vegetative crop like lettuce is grown in areas where peak consumptive use may be as high as 0.40 inch per day, benefit will result from frequent irrigations for cooling as well as for maintaining ample moisture in the soil.

Irrigation Practice During the Flowering Stage of Growth

Since consumptive use is at or near a maximum during the flowering stage, care must be exercised to insure adequate moisture in the root zone. However, the increased consumptive use is offset by the normally increased depth of roots. Deeper roots have a greater root-zone depth and hence a larger water supply available. In fact, the ratio of consumptive use to depth of extraction for a normally expanding root system remains remarkably constant. This ratio for green peas, potatoes, sugar beets, and grain grown in Logan, Utah, was about 0.06 inch depth of water each day for each foot of root extraction depth. Thus, when grain roots were 3 feet deep, the consumptive use was 0.06×3 or 0.18 inch per day. Hence, when the increasing depth of rooting is considered, the frequency of irrigation for most crops remains remarkably constant during the vegetative and flowering periods.

Actually, best production is obtained if the crop is kept adequately irrigated during both vegetative and flowering stages. However, harvesting procedures and early frosts do modify irrigation practice. Recall that a crop consists of thousands of individual plants, with some more mature than others. The best yield can be obtained when the plant has reached a certain stage of maturity. However, it is not practical to harvest each plant at its optimum stage of growth. Economics generally requires harvesting all plants at a given time; this is true in the harvesting of grain, hay, and sugar cane. Even though tomatoes are harvested by handpicking several times during the season, it is still an advantage to minimize the number of pickings. Also, frost may occur before some of the fruit has matured. Therefore, harvesting methods and frost may justify withholding irrigation for a short period during flowering to force the less mature plants out of the vegetative stage and into the flowering and fruiting stages. Reducing the amount of water available to the plant at this stage of growth will reduce the yield of more mature plants, but this decrease may be offset by an increased harvest from less mature plants. Normally, ample water should be available during fruiting unless maturing is not uniform or unless damaging frosts occur before harvesting is completed.

Special precautions should be taken to prevent the soil from becoming too dry, causing the flowers to fail to mature into seed.

Irrigation Practice During the Fruiting Stage of Growth

The root system is essentially extended to its maximum depth by the time fruiting occurs and the consumptive use has begun to decrease, reducing the water requirements of the crop and the frequency of irrigation. During production of dry fruit, irrigation has essentially ceased—the slight water requirements of the crop are met usually from stored water in the soil. The last heavy irrigation should normally be given during the wet-fruiting stage so that deep roots will have water available during final development of the fruit.

Tubers like potatoes and peanuts require adequate moisture during the entire growth period. Allowing the soil to become too dry between irrigations causes uneven, nobby tubers. Ample aeration is also essential. For this reason, good potatoes are generally produced on sandy loam soils which can be irrigated frequently without creating a problem. Heavier soils offer increased resistance to the growing tubers and thereby restrict growth and result in non-uniform tubers. For this reason, in addition to aeration problems, tuberous crops do better on sandy soils than on clay soils.

It is well to remember that sufficient moisture must be available to fully develop wet fruit. For example, soft fleshy fruits, green peas, and grains will not be firm and fully formed unless ample moisture is available.

Excessive irrigation during the fruiting period will stimulate vegetative growth for some crops and result in a reduction of fruiting. For example, cotton, which is normally planted yearly, is a perennial which will start new vegetative growth during the time the cotton is ready to pick if excess moisture is available in the soil.

Phosphorus and potassium are especially essential during the fruiting stages of growth, and excessive nitrogen may prolong vegetative growth.

12.12 DEPTH OF ROOT ZONE

Assuming a favorable unrestricted root zone, the depth of rooting increases during the entire growing period. The hotter the climate or the longer the growing period, the deeper the roots will penetrate. Crops requiring only 2 months to mature generally do not penetrate more than 2 to 3 feet. Crops requiring 3 to 4 months to mature penetrate 3 to 5 feet. Crops requiring 6 months to mature may penetrate 6 to 10 feet or more. As a rough guide, depth of rooting varies from 1 to 1½ feet depth per month of active growth, depending upon the crop and climate. There exists a tendency to underestimate the depth of rooting.

12.13 FREQUENCY OF IRRIGATION

How often to irrigate depends upon several factors. A number of management factors have been discussed in previous sections of this chapter. The importance of optimum moisture content for crops at a given stage of growth also has been discussed. The available water supply has been shown to be important, making it necessary to irrigate at times to store surplus water in the soil. Also, the need to leach out excess salts and the need to cool the soil may govern the frequency of irrigation.

During the active growing period of the crop, however, the most important factor governing frequency of irrigation is the need to keep adequate moisture in the soil for the crop. Two approaches can be made based upon the need for water by the root-zone soil. Both depend upon the capacity of the soil within the root zone to store water and the amount of water to be depleted. The available water per foot of soil multiplied by the depth to which moisture will be depleted will give the total water-holding capacity. However, maximum production can be obtained on most crops if not more than 50 percent of the available water is removed during the vegetative, flowering, and wet-fruit stages of growth. For some crops, removing not more than 25 percent of the available water will produce maximum yields. Yet, the cost of irrigating frequently usually makes the 50 percent depletion more economical than irrigating when only 25 percent has been depleted. At least 75 percent of the available moisture can generally be removed during the dry-fruit stage without detrimental results.

Frequency of irrigation can be determined by dividing the amount of moisture to be depleted from the soil by the consumptive use per day. Care should be taken to be sure the amount of moisture and the consumptive use are measured in similar units. Frequency of irrigation can also be determined by noting when the amount of moisture consumptively used is equal to the amount to be depleted from the soil.

12.14 DEPTH OF WATER TO BE APPLIED
DURING IRRIGATION

Determining how much water to apply can be calculated either by using the consumptive-use rate or the amount of soil moisture to be

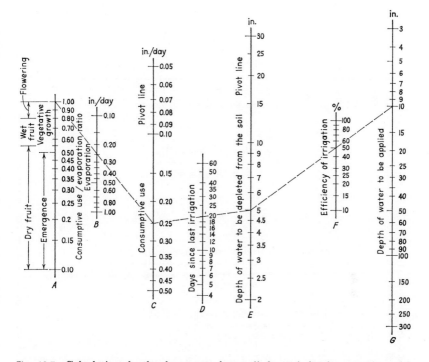

Fig. 12.7 Calculating depth of water to be applied per irrigation, starting with the stage of growth of the crop. (Directions for use are same as for Fig. 12.8.)

depleted from the soil. When the consumptive-use rate is to be used, it is necessary to know how many days since the field was last irrigated and the efficiency of the irrigation. Efficiency of irrigation is influenced by management, by method of irrigation, and by leaching requirements. Irrigation efficiencies are the subject of the following chapter.

Starting with the consumptive use—evaporation ratio discussed in the previous chapter, the depth to be applied can be determined using the nomograph of Fig. 12.7. When the amount of water available per foot of soil is known, the nomograph of Fig. 12.8 can be used to calculate the depth to be applied considering depth of the root zone, percentage of moisture to be depleted, and efficiency of irrigation. Converting the depth to be applied into stream size or the hours required to apply the necessary water can be accomplished using the nomograph as shown in Fig. 12.9 based on Equation 7.9, $qt = da$.

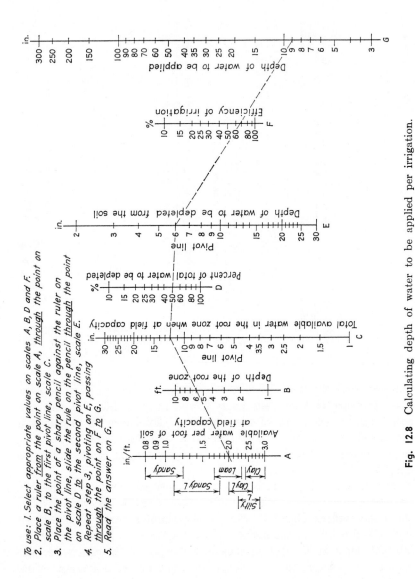

Fig. 12.8 Calculating depth of water to be applied per irrigation.

To use: *1. Select appropriate values on scales A, B, and D*
2. Lay a ruler from the point on scale A through the point on scale B to the pivot line.
3. Place the point of a sharp pencil against the ruler on the pivot line.
4. Slide the ruler on the pencil to the point on scale D
5. The answer appears where the ruler intersects scale E.

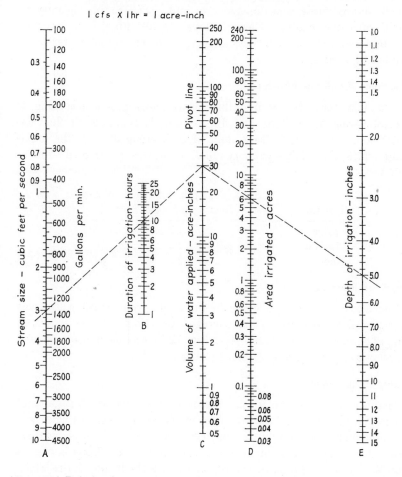

I cfs X I hr = I acre-inch

Fig. 12.9 Relation between stream size, time of irrigation, area to be covered, and depth of water to be applied, $qt = ad$. (After Garton.)

12.15 EXAMPLE OF WHEN TO IRRIGATE AND HOW MUCH TO APPLY

Following is a digest of a detailed report[1] which shows the pattern that can be followed in arriving at a guide for management to estimate when to irrigate and how much to apply. Facts observed in this summary are: water-holding capacities of soil, depth of rooting, water use by crops, peak consumptive use, and variation in consumptive use during the season. These factors are all utilized to ascertain the actual water use and the frequency and amount of water to be applied by irrigation.

Maximum Available Moisture

The soils were divided into three convenient groups according to texture as coarse, medium, and fine. After determining the field capacity and permanent wilting point of each group of soils, the maximum depth of available water per foot of soil was determined to be as follows:

Texture	Maximum Depth of Available Water Inches per Foot of Soil
Coarse	1.2
Medium	1.7
Fine	2.2

Depth of Rooting

Consideration was then given to the depth from which water was extracted from the soil. The average extraction pattern, shown in Fig. 12.6, was considered and then Fig. 12.10 was developed showing the depth of rooting during the growing period.

Water Requirements

Analysis of water use in the basins of the Cuanza, Lucala, and Bengo Rivers indicated that average evaporation from a Piche-type evaporometer at Dondo during the dry season was about 0.14 inch per day. Observations both in Angola and in the United States indicated that the peak consumptive use of water was about 0.8 of the Piche evaporation, or 0.11 inch per day. A balanced crop rotation will per-

[1] Taken from "Design of Irrigation System for Plantacoes do Mucoso, Lda. with an Evaluation of Natural Resources Governing the Feasibility of Irrigated Agricultural Development and Recommended Management Practices," by Vaughn E. Hansen and Jay M. Bagley, Agricultural Development and Engineering Services, Inc., Logan, Utah, Oct. 1955.

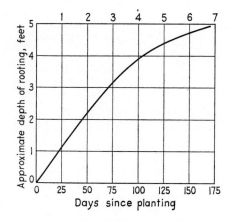

Fig. 12.10 Depth of rooting during the growing period.

mit a reduction in design capacity of approximately 0.75 which resulted in a peak design use of 0.8 inch per day.

Since not all water which is put into the irrigation system is
efficiently used by the plants because of irregular placement of the
water, etc., it is necessary to apply more water than will be needed
by the plants. Thus, it was assumed that the average efficiency for
application of water would be 70 percent for sprinkling and 50 percent
for surface irrigation. Also, approximately 10 percent of the water
was assumed to be lost in conveyance to the field when surface irrigation was used.

It was also necessary to consider that the system probably would
not be operated more than 12 hours per day on the average, considering Sundays, etc. Training and lack of dependability of laborers
prevent extensive irrigating at night. When all these factors were
considered, the design requirements which determined the size of the
pipe lines and pumps were:

	gpm/acre	acres/cfs
Sprinkler irrigation	4.5	100
Surface irrigation	7.0	64

After detailed consideration, surface irrigation was deleted in favor
of steel pipe conveyance lines and sprinkler irrigation. One of the
factors considered in the decision to use sprinkler irrigation was that
surface irrigation may require that 50 percent more water be pumped
from the adjacent river than would be necessary if sprinklers were

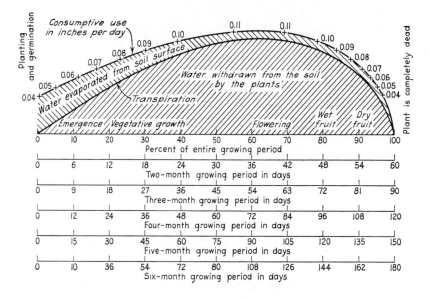

Fig. 12.11 Consumptive use of water for various periods of growth.

used. Hence, the 4.5 gpm per acre was used to estimate water needs.

The concepts embodied in the generalized consumptive-use curve were used to obtain Fig. 12.11 showing the amount of water used by the crops during the growing period.

Table 12.2 was then prepared from Fig. 12.11 to show the accumulated use of water during the growing period. Growing periods from 2 months to 6 months were considered. For example, cotton, which has a 6-month growing period, requires a total of almost 15 inches of water. If cotton is planted about March 15 after the rains have stored 3 inches of water in the soil, it will require an average of about (15 − 3) or 12 inches of irrigation water.

Frequency and Depth of Water to be Applied by Irrigation

The depth of water to be stored in the soil during irrigation affects the efficiency with which it can be applied. Figure 12.12 was prepared considering the irrigation efficiency to show how much water had to be applied by irrigation to store different depths of water in the soil. Note that the efficiency is less when smaller amounts of water are applied. If 4 inches are to be stored, 5.3 inches must be applied because the water is not uniformly distributed over the field.

Combining the growing period, rate of consumptive use, depth of

TABLE 12.2

ACCUMULATED USE OF WATER FOR CROPS WITH VARIOUS PERIODS FROM PLANTING TO COMPLETE MATURITY

Percentage of Growing Period*	Accumulated Consumptive Use of Water in Percentage of Total Use	Total Period of Growth									
		2 Mo.		3 Mo.		4 Mo.		5 Mo.		6 Mo.	
		Days Since Planting	Accum. Water Use Inches	Days Since Planting	Accum. Water Use Inches	Days Since Planting	Accum. Water Use Inches	Days Since Planting	Accum. Water Use Inches	Days Since Planting	Accum. Water Use Inches
10	6.0	6	0.3	9	0.4	12	0.6	15	0.8	18	0.9
20	13.8	12	0.7	18	1.0	24	1.4	30	1.7	36	2.0
30	23.5	18	1.1	27	1.7	36	2.3	45	2.9	54	3.5
40	34.5	24	1.7	36	2.6	48	3.4	60	4.3	72	5.1
50	46.5	30	2.3	45	3.4	60	4.6	75	5.7	90	6.8
60	59.4	36	2.9	54	4.4	72	5.8	90	7.3	108	8.7
70	72.3	42	3.5	63	5.3	84	7.1	105	8.9	126	10.6
80	84.3	48	4.1	72	6.2	96	8.3	120	10.3	144	12.4
90	94.5	54	4.7	81	7.0	108	9.3	135	11.6	162	13.9
100	100.0	60	4.9	90	7.4	120	9.8	150	12.3	180	14.7

* Growing Period is used herein to refer to the entire time from planting to the time the plant dies, which is usually longer than the period from planting to harvesting. Flowering will occur at about 50 to 60 percent of the growing period and fruiting after 60 percent.

TABLE 12.3

IRRIGATION FREQUENCY AND AMOUNT OF WATER TO BE APPLIED BY SPRINKLING IRRIGATION WHEN VARYING AMOUNTS OF AVAILABLE WATER REMAIN ON THE SOIL

Length of Growing Period = 3 months   Soil Texture = Coarse

Time Since Planting in Days	Consumptive Use of Water, Inches per Day	Average Depth of Rooting, Inches	Maximum Depth of Available Water in Root Zone, Inches	Interval in Days Between Irrigations and Depth in Inches to be Applied When Different Amounts of Available Water Remain in the Soil							
				75%		50%		25%		0%	
				Days	Depth Inches	Days	Depth Inches	Days	Depth Inches	Days	Depth Inches
0–9	0.05	2	0.2	1	0.3	2	0.4	3	0.5	4	0.6
9–18	0.07	6	0.6	2	0.5	5	0.7	8	1.0	10	1.3
18–27	0.08	11	1.1	3	0.6	7	1.1	10	1.5	13	1.9
27–36	0.09	16	1.6	4	0.8	8	1.4	13	2.0	17	2.5
36–45	0.10	21	2.1	5	1.0	10	1.7	15	2.5	20	3.1
45–54	0.11	26	2.6	6	1.1	12	2.1	17	2.9	24	3.7
54–63	0.11	31	3.1	7	1.3	14	2.4	21	3.3	28	4.2
63–72	0.10	35	3.5	8	1.5	17	2.7	25	3.7	34	4.7
72–81	0.09	38	3.8	11	1.6	21	2.9	33	4.0	44	5.1
81–90	0.05	41	4.2	20	1.7	41	3.1	61	4.2	82	5.5

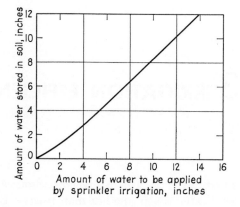

Fig. 12.12 Amount of water that must be applied by sprinkler irrigation in order to store the required amount of water in the soil.

rooting, efficiency of irrigation, and the water-holding capacity of the soils, Table 12.3 was prepared. Similar tables for medium- and fine-textured soils and also for 4-, 5-, and 6-month growing periods were prepared but are not included herein. These tables, together with the preceding material, contain data which are necessary to carefully regulate irrigation practice for different crops grown on either coarse-, medium-, or fine-textured soils. The report also contained recommended management practices similar to the recommendations presented in preceding portions of this book.

REFERENCES

Givan, C. V., "Irrigation Water Supply System Capacities," *Agr. Eng.,* Vol. 23, pp. 281, 1942.

Hansen, Vaughn E. and Jay M. Bagley, "Design of Irrigation System for Plantacoes do Mucoso, Lda. with an Evaluation of Natural Resources Governing the Feasibility of Irrigated Agricultural Development and Recommended Management Practices." Report submitted to Plantacoes do Mucoso, Lda. by Agricultural Development and Engineering Services, Inc. Logan, Utah, October 1955.

Shockley, Dale R., "Capacity of Soil to Hold Moisture," *Agr. Eng.,* Vol. 36, No. 2, pp. 109–112, February 1955.

CHAPTER 13 IRRIGATION EFFICIENCIES

In a world where water is such a precious resource, no man has the right to waste water which another man needs. Efficient use of irrigation water is an obligation of each user. However, efficiency of use will vary from locality to locality. In areas where water is scarce and costly, available water is generally used carefully. Whereas, in areas of abundant water, the value is less and the tendency is to waste water. Also efficiency is influenced by cost and quality of labor, ease of handling water, crops being irrigated, and soil characteristics. For these reasons irrigation efficiency is a broad general term which can be applied to irrigation practices in a qualitative manner. To describe segments of the over-all efficiency picture quantitative evaluations can be made. The objective of these efficiency concepts is to show where improvements can be made which will result in more efficient irrigation. Adequate control and management of irrigation water requires that methods be available to evaluate irrigation practices from the time water leaves the point of diversion until it is utilized by the plants.

13.1 WATER-CONVEYANCE EFFICIENCY

The earliest irrigation efficiency concept for evaluating water losses was water-conveyance efficiency. Most irrigation water then came from diversions from streams or reservoirs. Losses which occurred while conveying water were often excessive. Water-conveyance efficiency formulated to evaluate this loss can be stated as follows:

$$E_c = 100 \frac{W_f}{W_r} \qquad (13.1)$$

where E_c = water-conveyance efficiency
W_f = water delivered to the farm
W_r = water diverted from the river or reservoir

13.2 WATER-APPLICATION EFFICIENCY

Having conveyed the available water to the farm through costly diversions and conveyance structures, the need was apparent to apply the water efficiently. Often considerably more water was applied to the soil than it could possibly hold. The following concept of water-application efficiency was developed to measure and focus attention upon the efficiency with which water delivered was being stored within the root zone of the soil, where it could be used by plants.

$$E_a = 100 \frac{W_s}{W_f} \qquad (13.2)$$

where E_a = water-application efficiency

W_s = water stored in the soil root zone during the irrigation

W_f = water delivered to the farm

The concept of water-application efficiency can be applied to a project, a farm, or a field to evaluate irrigation practice. Irrigation efficiencies can vary from extremely low values to values approaching 100 percent. However, in normal irrigation practice, surface irrigation efficiencies of application are in the range of 60 percent, whereas well-designed sprinkler irrigation systems are generally considered to be approximately 75 percent efficient.

Common sources of loss of irrigation water from the farm during water application are represented thus:

R_f = surface runoff from the farm

D_f = deep percolation below the farm root-zone soil

Neglecting evaporation losses during the time water is being applied and immediately after, it follows that

$$W_f = W_s + R_f + D_f \qquad (13.3)$$

therefore
$$E_a = 100 \frac{W_f - (R_f + D_f)}{W_f} \qquad (13.4)$$

At each irrigation the farmer applies to his land a given volume of water. His irrigation problem is to store this water in the root zone of his soil. He cannot store all the water as soil moisture, for some loss of water is unpreventable. He should store in his root-zone soil the maximum percentage of the water that he applies consistent with good irrigation practice and economy. The most common losses of irrigation water are represented by runoff and deep percolation. Irregular land surfaces, shallow soils underlain by gravels of high permeability, small irrigation streams, non-attendance of water during irrigation, long irrigation runs, excessive single applications—all these factors contribute to large losses and low efficiency. Also excessively large irrigation streams, improper preparation of land, compact impervious soils, steep slopes of land surfaces, and non-attendance contribute to inefficiency.

The depth of water applied in each irrigation is a dominant factor influencing efficiency of application. Even if water were spread uniformly over the land surface, excessive depths of application would result in low efficiencies. Many variable factors such as land uniformity, irrigation method, size of irrigation stream, length of run, soil texture, permeability, and depth influence the time the irrigator keeps water running on his farm and hence the depth he applies. The fact that applying excessive water depth in each irrigation causes low efficiency is shown in Fig. 13.1, based on 133 Utah tests represented by fourteen plotted points.

The curve shows that, when the depth of water exceeded 10 inches, the highest efficiency was only 30 percent, the lowest 12 percent, and in five of the nine averages plotted it was less than 20 percent.

13.3 WATER-USE EFFICIENCY

Having conveyed water to the point of use and having applied the water, the next efficiency concept of concern is the efficiency of water use. What proportion of the water delivered was beneficially used on the project, farm, or field can be calculated using the following formula:

$$E_u = 100 \frac{W_u}{W_d} \tag{13.5}$$

where E_u = water-use efficiency
W_u = water beneficially used
W_d = water delivered

13.4 WATER-STORAGE EFFICIENCY

The increasing scarcity of water and the growing realization of the value to be gained from irrigation has resulted in high-priced water, discouraging excessive use of water. Low financial returns from irrigation often occur not because of excessive water application but because of insufficient application. In many cases only a small fraction of the needed water is being applied. The water application efficiencies under such practices are essentially 100 percent, and yet the irrigation practice is poor. To assist in the evaluation of this problem,

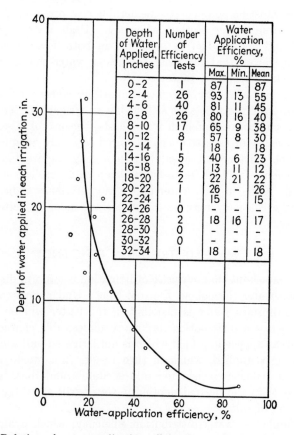

Fig. 13.1 Relation of water-application efficiencies to depths of irrigation water applied in each irrigation. (*Utah Agr. Exp. Sta. Bul.* 311.)

the concept of water-storage efficiency is useful. This concept directs attention to how completely the needed water has been stored in the root zone during the irrigation.

$$E_s = 100 \frac{W_s}{W_n} \tag{13.6}$$

where E_s = water-storage efficiency
W_s = water stored in the root zone during the irrigation
W_n = water needed in the root zone prior to the irrigation

Research has shown that improved storage efficiency will triple production in the Lower Rio Grande Valley of Texas where insufficient water is normally available. Similar research in western Kansas where water must be pumped at considerable cost produced double the production from better storage efficiency. High water application efficiencies and low production measured in both Texas and Kansas gave positive proof that for these conditions water-storage efficiency is the important irrigation concept.

Water-storage efficiency becomes important whenever insufficient water is stored in the root-zone soil during an irrigation. This condition may occur because of high water costs, because of a scarcity of water, or because of excessive time required to secure adequate penetration. The existence of a salt problem may require that water-storage efficiency be high in order to keep salts washed out of the surface soil. Often a high water-application efficiency indicates that water-storage efficiency may be the important index to better irrigation practice.

13.5 WATER-DISTRIBUTION EFFICIENCY

Another important characteristic of irrigation is uniform distribution of irrigation water throughout the root zone. Under most conditions, the more uniformly water is distributed, the better will be the crop response. Uneven distribution has many undesirable characteristics. Drought areas appear in a field which is not irrigated uniformly unless excess water is applied, which in turn results in a waste of water. Whenever a tendency exists for salt accumulation, those areas receiving less than the desired depths of water will show the greatest accumulation of salt.

The formula for water-distribution efficiency, which evaluates the extent to which water is uniformly distributed, is shown below.

$$E_d = 100 \left[1 - \left(\frac{y}{d} \right) \right] \qquad (13.7)$$

where E_d = water-distribution efficiency
 y = average numerical deviation in depth of water stored from average depth stored during the irrigation
 d = average depth of water stored during the irrigation

Note that the following formula for the uniformity coefficient C_u developed by Christiansen in 1942 is identical to Equation 13.5.

$$C_u = 100 \left[1.0 - \left(\Sigma \frac{x}{Mn} \right) \right] \qquad (13.8)$$

where x = deviation of individual observations from the mean value M
 n = number of observations

To illustrate this concept used as an efficiency, suppose two sprinkler systems were each delivering average water applications of 0.3 iph. However, system A had a water-distribution efficiency of 90 percent and system B had a distribution efficiency of 70 percent. Assume the owner of system B cannot afford to improve his system and hence will continue to use it as it is. Would you recommend identical operation of both systems to grow similar crops on.similar soils? Or would you find it advisable for system B to apply more water in order to obtain results approximately comparable to system A? If you consider that system B must apply more water per irrigation than system A to produce satisfactory results, how much more water would you recommend? (Obviously system B is not as efficient as system A.)

When the variation in distribution is considered in terms of water distribution efficiency, the answer to the question of how much to increase the application from system B is obvious. The depth to be applied is divided by the water-distribution efficiency, thus

$$\text{Depth of application} = \frac{\text{Rated depth}}{\text{Water-distribution efficiency}}$$

Therefore, system A would be applying 11 percent more water and system B 43 percent more water to produce comparable results. Obviously the less efficient system will cost much more to operate.

On the average 75 percent of each area irrigated in this example will have received an amount of water equal to or in excess of the rated amount, and 25 percent will still be deficient. It is undoubtedly impractical and uneconomical to apply sufficient excess water to wet all of the area to the desired depth.

Water-distribution efficiency will provide the measure for comparison between systems, whether it be sprinkler compared to sprinkler, surface compared to surface, or sprinkler compared to surface irrigation.

13.6 CONSUMPTIVE-USE EFFICIENCY

Water stored in the soil during an irrigation may not remain in the soil for use by the crops. A wide furrow spacing and considerable exposed ground surface may result in excessive surface evaporation and continual significant downward movement of moisture beyond the root zone. Loss of water by deep penetration and by excessive surface evaporation following an irrigation can be evaluated by the following concept of consumptive-use efficiency:

$$E_{cu} = 100 \frac{W_{cu}}{W_d} \qquad (13.9)$$

where E_{cu} = consumptive-use efficiency
    W_{cu} = normal consumptive use of water
    W_d = net amount of water depleted from root-zone soil

This concept may be extremely important, particularly where high moisture, high permeability, wide spacing, and ridges are combined. When irrigating potatoes under moist conditions in ridges of permeable soil with widely spaced rows, consumptive-use efficiency may be in the order of 50 percent. Efficiency can be materially increased if ridges are not used. Efficiency can also be increased if row spacing is decreased, or if good sprinkler irrigation is used to more uniformly distribute water. Hence, if better irrigation practice is followed, consumptive-use efficiency is increased. Consumptive-use efficiency is another measure of irrigation efficiency which may be very useful under certain conditions.

This concept also involves the efficiency with which roots are able to utilize moisture stored in the soil during an irrigation. Consumptive-use efficiency is affected by soil texture, amount and distribution of vegetation, profile of the soil surface, distribution of roots within the soil, and variation of soil moisture within the root zone. These factors affect the water lost by excessive evaporation from the soil surface and the water lost from percolation downward beyond the root zone. The water lost by excessive evaporation is increased as the amount of bare ground increases, and decreases as the foliage increases.

Consumptive-use efficiency is useful in explaining the differences in crop response from different methods of irrigation. The irrigation research literature is full of field tests comparing yields obtained from surface irrigation. Sometimes differences are significant and other times on the same crops, differences are not significant. Frequently variability lies in the differences in boundary conditions which are reflected in consumptive-use efficiency. When considering which irrigation method to use, the important aspect is which method will place water where it can be most efficiently utilized by plant roots. This concept may not be important for alfalfa, with its well-distributed root system, but may be very important for potatoes, corn, or melons.

Since losses from evaporation and deep percolation following an irrigation are often small, consumptive-use efficiency can best be evaluated on a seasonal basis, so that the effects of these factors are summated for the entire growing season.

13.7 IRRIGATION-EFFICIENCY EXAMPLES

Figure 13.2 contains nine illustrative profiles of common irrigation practices. Examples (a), (b), and (c) are cases of heavier water

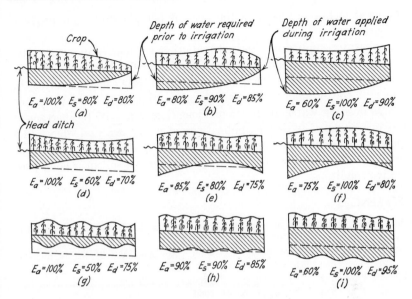

Fig. 13.2 Application, E_a, storage, E_s, and distribution, E_d, efficiencies and the effect upon corn production illustrated by two-dimensional profiles of the soil. Note: Water-application efficiency estimates are made assuming no runoff. (After Hansen, 1960.)

application near the upper end of the field and illustrate how the three efficiencies vary as the amount of water applied increases. Examples (d), (e), and (f) illustrate a common case where heavier applications of water are applied near each end of the field. Examples (g), (h), and (i) illustrate the type of moisture distribution which frequently occurs under sprinkler irrigation. Example (g) is common sprinkler-irrigation practice when insufficient water is applied. The low-water storage efficiency E_s, and the low-water distribution efficiency E_d, show vividly why this is poor irrigation practice. Example (i) is also an undesirable irrigation practice and is well marked by the low-water application efficiency E_a.

Note, however, that efficiencies of 100 percent are not always desirable. Maximum net profit can often be obtained by filling the root zone only every second, third, or fourth irrigation, depending upon the soil and crop. This is primarily because of the capacity of plants to extract more water from the upper portion of the root zone than from the lower portions. Alternating the depth of application also promotes better bacterial growth in fine-textured soils. Thus, a water-storage efficiency of less than 100 percent periodically might produce the best results.

Also, water-application and -distribution efficiencies of nearly 100 percent are not always desirable or practical. The expense necessary to secure high application and distribution efficiencies is often in excess of the economic returns. Another important reason for not always desiring water-application efficiencies of nearly 100 percent is that one or more excessive applications of water per year, depending upon salt content of water and soil, is usually desirable to flush out excess salts in the root zone, thereby keeping the soil in a good productive condition.

REFERENCES

Christiansen, J. E., "Hydraulics of Sprinkling Systems for Irrigation," *Trans. Am. Soc. Civil Eng.*, Vol. 107, pp. 221–239, 1942.

Hansen, Vaughn E., "Water Storage Efficiency," *Agr. Eng.*, Vol. 34, No. 12, pp. 835–836, December 1953.

Hansen, Vaughn E., "New Concepts in Irrigation Efficiency," *Trans. Am. Soc. Agr. Eng.*, Vol. 3, No. 1, pp. 55–61, 1960.

Willardson, Lyman S., "What is Irrigation Efficiency?" *Irrigation Eng. and Maint.*, Vol. 10, No. 4, pp. 13–14, April 1960.

CHAPTER 14 SURFACE AND SUB-SURFACE IRRIGATION

Irrigation water is applied to land by three general methods; namely, (a) surface, by flooding; (b) sub-surface or with furrows, in which the surface is wetted little if any; and (c) sprinkling, in which the soil surface is wetted much as it is by rainfall. Sprinkler irrigation is the subject of Chapter 16. These surface and sub-surface methods are further subdivided as indicated below and as considered in the following articles.

1. Surface:
   a. Uncontrolled or "wild" flooding.
   b. Flooding controlled with corrugations, borders, checks, or basins.
   c. Furrows.
2. Sub-surface:
   a. Controlled by lateral supply ditches.
   b. Uncontrolled, from excess application of water to adjacent or higher lands.

Irrigation methods vary in different parts of the world and on different farms within a community because of differences in soil, topography, water supply, crops, and customs. Forage crops such as alfalfa, clover, hay, and pastures in some areas are irrigated by use of corrugations. Flooding methods of irrigation as well as border strips and basins are suitable for forage crops. Row crops are irrigated by furrows. Any one or a combination of several methods may be best suited to one farm, as illustrated in Fig. 14.1.

14.1 UNCONTROLLED OR "WILD" FLOODING

In the early irrigation of centuries past, throughout Asia and southern Europe, water was applied by flooding extensive areas of

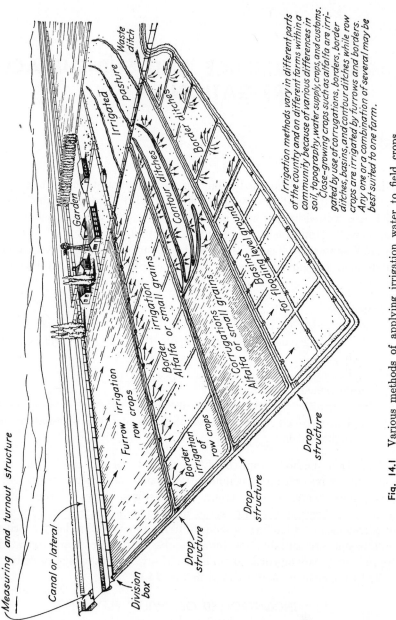

Measuring and turnout structure

Canal or lateral

Division box

Drop structure

Drop structure

Drop structure

Furrow irrigation row crops

Border irrigation of row crops

Garden

Border irrigation Alfalfa or small grains

Corrugations or small grains Alfalfa

Basins level ground for flooding

Irrigated pasture

Waste ditch

Contour ditches

Border ditches

Irrigation methods vary in different parts of the country and on different farms within a community because of various differences in soil, topography, water supply, crops, and customs. Close-growing crops such as alfalfa are irrigated by use of corrugations, borders, border ditches, basins, and contour ditches while row crops are irrigated by furrows and borders. Any one or a combination of several may be best suited to one farm.

Fig. 14.1 Various methods of applying irrigation water to field crops.

rather smooth, flat land. In Egypt, especially, the flooding method was of general adoption, the water being forced to spread over vast tracts, during the season of high stream flow.

In modern irrigation, several improved flooding methods have been developed. Brief descriptions of these methods follow.

Where water is applied from field ditches without any levees to guide its flow, or otherwise restrict its movement, the method is designated uncontrolled or "wild" flooding. It is practiced largely where irrigation water is abundant and inexpensive.

If the water is made to flow over the surface too quickly, an insufficient amount will percolate into the soil. On the other hand, if water is kept on the soil surface too long, waste will result from percolation beyond the root zone. It is an important and difficult problem to apply water efficiently by flooding methods. The size of stream used, the depth of water as it flows over the soil surface, and the rate of intake of water into the soil all influence application efficiency. In uncontrolled flooding, the smoothness of the land surface and the attention and skill of the irrigator are also important.

Water is brought to the field in permanent supply ditches and distributed from ditches built across the field. The spacing of the ditches is determined by the grade of the land, the texture and depth of the soil, the size of stream, and the nature of the crop. The distances between the diversions from ditches down the steepest slope are similarly determined.

Flooding from field ditches is well adapted to some lands that have such irregular surfaces that the other flooding methods are impractical. However, even on lands that may advantageously be irrigated by the other flooding methods, irrigators often continue to use uncontrolled flooding because of the low initial cost of preparation of land for this method. The extra labor cost in the application of water and the greater losses of water by surface runoff and deep percolation usually offset the apparent advantages of low initial cost of preparation of land.

Where land, water, or labor is expensive, where soil is deep and not likely to crust badly, and where the land is not too rough or steep, it is generally advisable to prepare for controlled flooding in border strips or level or contour basins.

14.2 BORDER-STRIP FLOODING

Dividing the farm into a number of strips, preferably not over 30 to 60 feet wide and 330 to 1320 feet long, separated by low levees or borders, is designated the border-strip method. Water is turned from the supply ditch into these strips along which it flows slowly toward the lower end, wetting the soil as it advances.

Figure 14.2 shows farm lot 74 of the California State Durham Colony before it was prepared for irrigation. The contours indicate that the highest land is on the north border in the west half of the tract. In preparing the tract for irrigation by the border-strip method the agricultural engineers divided it into three fields as shown in Fig. 14.3. The border strips in each field are 40 feet wide. In the west fields they are 470 feet long; and in the east field, because of the irregular

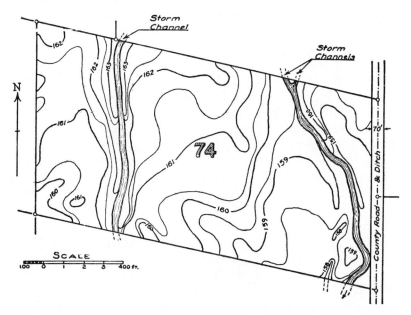

Fig. 14.2 Farm lot enlarged, showing contours. (*U.S.A.D. Farmers' Bul.* 1243.)

position of the storm channels, the border strips vary in length from a minimum of approximately 150 feet to a maximum of nearly 500 feet. While irrigating the two west fields, the water flows from north to south, and in the east field it flows in two directions, as indicated by the position of the borders and the pointing of the small arrows.

The surface is essentially level between levees, so that the advancing sheet of water covers the entire width of land strip; but lengthwise of the levee the surface slopes somewhat according to the natural slope of the land. It is desirable, although not urgently essential, that the slope be uniform within each levee. If practical, it is best to make the border slope from 2 to 4 feet per 1000 feet, but slopes as low as 1 foot per 1000 and as high as 75 feet per 1000 may be used where it is impractical to obtain the more appropriate slopes. Special care is essential to prevent erosion of soil on the steeper slopes.

The size of stream turned into a single border strip varies from ½ to 10 cfs, depending on the kind of soil, the size of border, and the nature of the crop.

Because of the relatively high initial cost of preparing land for the border method, it is desirable to plan the location of the levees and strips so that different forage and grain crops may be irrigated

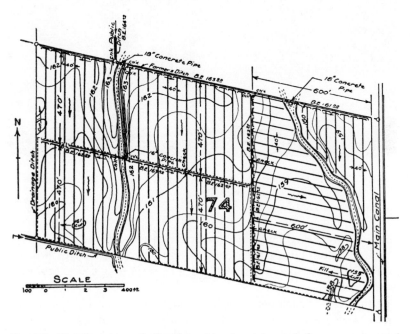

Fig. 14.3 Showing size and direction of border strips and the necessary supply ditches and draws.

with the same borders. Crops which are to be furrow-irrigated, such as sugar beets, potatoes, and corn, may be grown on land on which the forage crops have been irrigated by the border method. Provided the soil conditions are favorable to lateral water movement underneath the low, broad border levees, it is practical to plant and mature crops on the levees. It is difficult to furrow the levees satisfactorily and to keep irrigation water in furrows on the levees.

The border method is suitable to soils of wide variation in texture. It is important, however, to study the physical soil properties in ad-

Fig. 14.4 Borders being strengthened and prepared for planting and irrigation, (Courtesy Soil Conservation Service.)

vance of preparing land for border irrigation. Rather impervious subsoils overlain by compact loams permit long border strips, whereas open soils having highly permeable, gravelly subsoils necessitate short narrow strips.

At the head of each border strip a gate is placed in the supply ditch for convenience in turning water into and out of the strip. Borders being strengthened and prepared for planting and irrigation are shown in Fig. 14.4.

14.3 CHECK FLOODING

The check-flooding method consists of running comparatively large streams into relatively level plots surrounded by levees. This method is well suited to very permeable soils which must be quickly covered with water in order to prevent excessive losses near the supply ditches through deep percolation. It is also suited to heavy soils into which water percolates so slowly that they are not sufficiently moistened during the time a sheet of water flows over them, making it necessary to hold the water on the surface to assure adequate penetration.

Checks are sometimes prepared by constructing levees along contours having vertical intervals of 0.2 to 0.4 foot and connecting them with cross levees at convenient places. These are called contour checks and are formed by building longitudinal levees approximately parallel to the contours, and connecting them at desirable places with levees at right angles.

The check method of irrigation for grain and forage crops is advantageous in localities where large irrigation streams are available and also on projects which depend on direct flow from widely fluctuating streams. In some areas, torrential summer rains suddenly make swollen streams which must be quickly applied to the land to prevent loss of the water. On land of very small slopes the area of each check may be several acres.

The levees should be from 6 to 8 feet wide at the base and not over 10 to 12 inches high because it is essential to avoid obstruction to farm machinery, and also to assure satisfactory growth of crops on the levees.

14.4   BASIN FLOODING

The basin method of flooding is essentially the check method especially adapted to irrigation of orchards. On some farms a basin is made for each tree, but under favorable conditions of soil and surface slope from two to five or more trees are included in one basin. From the supply ditch the water is conveyed to the basin, either by flowing through one basin and into another, or preferably by small ditches constructed so that the water may be turned directly from a ditch into each basin.

14.5   THE FURROW METHOD

In the irrigation methods thus far described, almost the entire land surface is wetted in each irrigation. Using furrows for irrigation, as shown in Fig. 14.5, necessitates the wetting of only a part of the surface (from one-half to one-fifth) thus reducing evaporation losses, lessening the puddling of heavy soils, and making it possible to cultivate the soil sooner after irrigation. Nearly all row crops are irrigated by the furrow method, rather than by flooding.

Grain and alfalfa crops are irrigated by means of small furrows designated corrugations. These corrugations are advantageous when the available irrigation streams are small, and also for land of uneven topography. Furrow irrigation is adaptable to a great variation in slope. It is customary, although on steep slopes inadvisable, to run

Fig. 14.5 Four corrugations supplied from single outlet. (*U.S.D.A. Farmers' Bul.* 1348.)

the furrows down the steepest slope to avoid overflowing the banks of the furrows.

Length of Furrows

On some soils, furrows having slopes of 100 to 150 feet per 1000 feet are successfully used by allowing only very small streams to enter the furrows and by careful inspection to control erosion. Slopes of 5 to 30 feet per 1000 are preferable, but many different classes of soil are satisfactorily irrigated with furrow slope from 30 to 60 feet per 1000 feet.

The length of furrows varies from 100 feet or less for gardens to as much as 1500 feet for field crops. Furrow lengths of 300 to 660 feet are common. Excessive deep percolation losses and soil erosion near the upper end of the field result from use of very long furrows. However, reduction of waste land for ditches and turning of machinery favor longer rows.

Spacing and Depths of Furrows

Spacing of furrows for irrigation of corn, potatoes, sugar beets, and other row crops is determined by the proper spacing of the plant rows, one irrigation furrow being provided for each row. In orchard irrigation, furrows may be spaced from 3 to 6 feet apart. Soils having unusually favorable capillary properties, or impervious subsoils, may permit orchard furrows 10 to 12 feet apart. With the greater spacing it is essential to check on the moisture distribution after each irrigation by making borings with a soil auger or tube to find whether or not the lateral moisture movement from furrows is adequate.

Furrows from 8 to 12 inches deep facilitate control and penetration of water into soils of low permeability. They are well suited to orchards and to some furrow crops. Other furrow crops, such as sugar beets, are best irrigated with furrows from 3 to 5 inches deep. It is highly desirable in irrigating sugar beets and similar root crops to have the furrows deep enough, and the stream in each small enough, so that the water cannot come in contact with the plant.

Water Distribution to Furrows

Water is distributed to the furrows from earth-supply ditches or from wood or concrete flumes or concrete pipe placed underground. The earth-supply ditch is very common. Small openings are made through the bank, and the water flows into one or more furrows. Figure 14.5 shows four small ditches or corrugations supplied from a single outlet. This method necessitates careful supervision to avoid erosion of the supply-ditch openings, and consequent excess flow in some and inadequate flow in others. On the other hand, it provides flexibility, permitting a large stream in each furrow when the water is first turned in, thus wetting the furrow through its entire length quickly, and then decreasing it so that just enough water enters the furrow to keep it wet, thereby reducing to a minimum the runoff from the lower end of the furrow.

The use of small-diameter 48-inch length curved pipe, made of lightweight plastic, aluminum, galvanized iron, or rubber, enables the irrigator to siphon water from the ditch to the furrows as shown in Fig. 14.6 and keep ditch banks solid. This method permits easy and frequent change of water from furrow to furrow.

Irrigators can increase the uniformity of application of water to their furrow-irrigated crops by frequent regulation of the size of stream flowing into the furrow. For this purpose, gated pipe shown in Fig. 14.7 is especially helpful. Small, easily adjusted gates in the pipe facilitate control of the size of stream delivered to the furrow. Streams as small as 1 gpm or as large as 10 gpm or more can be delivered. The lightweight aluminum or galvanized gated pipe is easily placed, easily connected, and easily moved after irrigation.

14.6 SUB-IRRIGATION

In a few localities, natural soil and topographic conditions are favorable to the application of water to soils directly under the surface, a practice known as sub-irrigation. An impervious subsoil at a depth

Fig. 14.6 Plastic siphon tubes used in furrow irrigation. (Courtesy Soil Conservation Service.)

of 6 feet or more, a highly permeable loam or sandy loam surface soil, uniform topographic conditions, and moderate slopes favor sub-irrigation. Under such conditions, proper water control to prevent alkali accumulation or excess waterlogging usually results in economical use of water, high crop yields, and low labor cost in irrigation.

California has several large tracts of low-lying lands in the Sacramento-San Joaquin Delta that are successfully sub-irrigated. Before being reclaimed some of these tracts were flooded every year by the overflow waters of the Sacramento and San Joaquin rivers. Reclamation was made possible by building large levees around tracts of several thousands of acres, followed by installing drainage systems and by pumping the water discharged from the drains over the levees into the river channels. The soils are composed largely of decayed organic matter and are known as peat, tule, or muck soils. During several months of each year, the water in the river channels, now controlled by artificial levees, is 2 to 10 feet or more higher than the land surface. In order to obtain water for irrigation, siphons are built from the channels over the levees and the water is thus siphoned to the lands. It is distributed in a series of ditches from 2 to 3 feet deep and 1 foot wide having vertical sides. These ditches, spaced from

150 to 300 feet apart, provide adequate distribution of the water to irrigate small grains and root crops.

Some of these sub-irrigated lands have been injured and made less productive by saline and alkali conditions developed by the upward capillary water flow from the shallow water table. This reduction in productivity has made it necessary to discontinue sub-irrigation and irrigate large areas by the sprinkling method.

In the mountain states there are four notable areas on which natural sub-irrigation is practiced, namely: the Egin Bench area in upper Snake River Valley, Idaho; Cache Valley, Utah; Seedskadee Project, Wyoming; and San Luis Valley, Colorado. The conditions and procedure in the application of water are typified by the Egin Bench, Snake River Valley practice, which is described below.

Sub-Irrigation on Egin Bench, Idaho

The land slopes uniformly about 2 feet per 1000 feet. Surface loams and gravelly loams from 1.5 to 6 feet in depth are underlain by more

Fig. 14.7 Portable, gated aluminum pipe delivering water to furrows. (Courtesy W. R. Ames Company.)

permeable soil materials which rest on impervious lava rock at depths varying from a few feet to as much as 90 feet.

Clinton describes some of the noteworthy facts of this Egin Bench 28,000-acre sub-irrigated tract somewhat as follows.

The flow of water in the main canals is regulated by a ditchrider, whose job is very different from that of his co-worker on a conventional irrigation system. Essentially he is operating a huge ground-water reservoir, the water level of which is controlled by the rate of inflow and outflow. The inflow is from controlled water supplies and precipitation; the outflow, consumptive use by plants, evaporation from the soil, and seepage. Excess inflow means waterlogged soils, flooded fields, farmyards, and roads, while excess outflow results in wilted crops.

Irrigation does not interrupt cultivation of fields. Farm machinery can be operated in the fields at any time, even though the water table is near the ground surface.

The principal crops grown on the Egin Bench are, in order of percentage of cropped area, potatoes, alfalfa and clover hay, small grains, sugar beets, and field peas. Early in the agricultural development of the area, an attempt was made to irrigate the land by the usual flooding methods. Excessive deep percolation losses resulted, and frequent irrigation was found essential to ordinary crop yields. The gradual rise of the water table convinced the irrigators that smaller quantities of water would suffice under more favorable irrigation methods. Irrigation water is applied in shallow ditches about 3 feet wide and spaced from 100 to 300 feet apart. In general, these ditches do not exceed $\frac{1}{4}$ mile in length. A stream from $\frac{1}{4}$ to $\frac{1}{2}$ cfs is run into each ditch, from which it sinks to the ground water, causing the water table to rise high enough to moisten the root-zone soil by capillary action that sub-irrigation methods may be advantageously employed.

14.7 SUB-IRRIGATION AND DRAINAGE

In some localities, natural drainage is insufficient to carry away the excess water applied in sub-irrigation. It has been found necessary in the Lewiston Area, Cache Valley, Utah, to construct large open drains to prevent excessive waterlogging and salinity of soils. In certain favorable parts of large irrigated valleys such as the San Joaquin Valley, California, and the Salt River Valley, Arizona, where deep percolation from the flooding irrigation methods has resulted in the rise of the ground water to the extent that drainage by pumping has been found essential to maintain the soil productivity, it is possible that sub-irrigation methods may be advantageously employed.

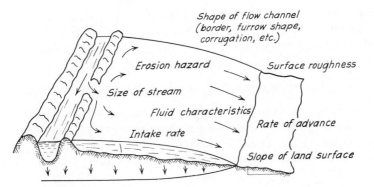

Fig. 14.8 Schematic view of border irrigation illustrating the basic variables involved in the hydraulics of surface irrigation.

14.8 ARTIFICIAL SUB-IRRIGATION

Under favorable soil conditions for the production of high priced crops on small areas a pipe distribution system is placed in the soil well beneath the surface. The process of applying water beneath the soil surface through various kinds of pipes or other conduits is designated artificial sub-irrigation. Favorable soil conditions which permit free lateral movement of water, relatively rapid capillary movement in the root-zone soil, and very slow downward movement in the subsoil, are essential to the mechanical success of artificial sub-irrigation. Persons not informed concerning the several methods of irrigation are sometimes prone to overvalue the advantages of artificial sub-irrigation and make expenditures far greater than economic results justify. The cost of this method of irrigation is usually prohibitive.

14.9 HYDRAULICS OF SURFACE IRRIGATION

The hydraulics of surface irrigation is rather complex and, consequently, not well understood. Nevertheless, good irrigation design depends upon understanding its principles. Figure 14.8 shows a schematic view of border irrigation, illustrating the basic variables involved in the hydraulics of surface irrigation. They are as follows:

1. Size of stream.
2. Rate of advance.
3. Length of run and time required.
4. Depth of flow.
5. Intake rate.

6. Slope of land surface.
7. Surface roughness.
8. Erosion hazard.
9. Shape of flow channel.
10. Depth of water to be applied.

It should be clearly understood that the hydraulics of surface irrigation has all the complexities of unsteady open-channel flow plus the added major complication of a variable intake. Hence, direct solution of this complex problem is not simple.

14.10 DESIGN OF SURFACE IRRIGATION SYSTEMS

At least ten principal criteria are important in the design of a surface irrigation system. All ten are in turn governed by the over-all economics of the entire farming operation. Thus the design of an irrigation system is complex and not readily subject to quantitative analysis.

1. *Store the required water in the root zone of the soil.* The amount of water to be stored varies with the crop and the month of the growing season. Development of an irrigation system flexible enough to meet the varying conditions is important, but sometimes difficult. A compromise design is usually the most economical design.

2. *Obtain reasonably uniform application of water.* It is almost impossible to design a surface irrigation system that will provide satisfactory control under all conditions. Size of stream may vary, and intake rate does vary from crop to crop. It varies also with seasons and the hour since irrigation was commenced. The more elaborate and costly the irrigation control devices, usually the more nearly a uniform irrigation can be obtained. For good distribution, the water should reach the end of the field in about one quarter of the total irrigation time.

3. *Minimize soil erosion.* Erosion cannot be eliminated, but it can be minimized. Observation of amount of erosion occurring with different size streams is about the only way to evaluate the permissible size stream. Observation will indicate the size stream which can be used without excessive erosion. Criddle has suggested that the maximum non-erosive furrow stream in gpm can be estimated by dividing 10 by the slope in percentage.

4. *Minimize runoff of irrigation water from the field.* Sizable quantities of water are wasted from the end of most fields during irrigation. This waste water is referred to as runoff. One of the most effective ways of minimizing runoff is to reduce the flow entering the field when the water nearly reaches the end of the field. The amount of wetted area will reduce somewhat, causing a slight decrease in intake, while the velocity will reduce considerably. Whenever possible, runoff should be re-used on lower lands.

5. *Provide for beneficial use of runoff water.* The beneficial use

of waste water should not be overlooked, since sizable quantities of runoff usually occur.

6. *Minimize labor requirement for irrigation.* Labor requirements should be held to a minimum. It is generally necessary to have well-trained and responsible labor for irrigation. Good preparation of the land, good control, and good irrigation layout will minimize labor requirements.

7. *Minimize land used for ditches and other controls to distribute the water.* Normally 5 percent of the land is used for surface irrigation ditches and controls. Thoughtful layout can minimize the amount of waste land. Sometimes it is advisable to use longer irrigation runs than are best for uniform water distribution to reduce the amount of waste land and to facilitate cultivation.

8. *Fit irrigation system to field boundaries.* The design of the irrigation system is often controlled by the size and layout of the farm. Frequently the length of an individual field is controlled by legal boundaries rather than desirable irrigation design.

9. *Adapt system to soil and topographic changes.* Considerable difficulty will result if a furrow traverses soils of markedly different texture, slope, or depth. Intake rate will be different in each portion and thus the flow rate cannot be adjusted to fit both types. Also the water-holding capacity will require a different irrigation frequency. Irrigation layout should be such that a minimum variation in these factors occurs in a given field.

10. *Facilitate use of machinery for land preparation, cultivating, furrowing, harvesting, etc.* Allowance must be made for machinery. As the degree of mechanization increases, wider borders and longer fields will be desirable.

REFERENCES

Christiansen, J. E., A. A. Bishop, and Yu-Si Fok, "The Intake Rate as Related to the Advance of Water in Surface Irrigation," Presented at the Winter Meetings of the *Am. Soc. Agr. Eng.,* Chicago, 1959.

Clinton, F. M., "Invisible Irrigation on Egin Bench," *Reclamation Era,* Vol. 34, No. 10, p. 182, 1948.

Criddle, Wayne D., "A Practical Method of Determining Proper Lengths of Runs, Sizes of Furrow Streams and Spacing of Furrows in Irrigated Lands," U.S.D.A. Soil Conservation Service, revised October 1950.

Criddle, Wayne D., and Sterling Davis, "A Method for Evaluating Border Irrigation Layouts," U.S.D.A. Soil Conservation Service, March 1951.

Criddle, Wayne D., Sterling Davis, Claude H. Pair, and Dell G. Shockley, "Methods for Evaluating Irrigation Systems," *U.S.D.A. Soil Conservation Service Agr. Handbook* 82, April 1956.

Hamman, A. J. and W. E. Code, "An Irrigation Guide for Colorado," *Agr. Ext. Ser., Agr. Exp. Sta., and Colo. Agr. and Mech. College, Bul.* 432-A, April 1954.

Hansen, Vaughn E., "Mathematical Relationships Expressing the Hydraulics of Surface Irrigation," *Proc. ARS, Soil Conservation Service,* Workshop on Hydraulics of Surface Irrigation, Denver, February 9–10, 1960.

Israelsen, O. W. and George D. Clyde, "Soil Erosion in Small Irrigation Furrows," *Utah Agr. Exp. Sta. Bul.* 320, 1946.

Jarvis, Joe W., "Irrigation Guide," Union Pacific Railroad Co., 1947.

Johnson, C. N., "Comparison Performances of Metallic and Plastic Siphons for Irrigation," *Agr. Eng.,* Vol. 27, pp. 469–471, 1946.

Kohler, Karl O., Jr., "Contour-Furrow Irrigation," *U.S.D.A. Soil Conservation Service Leaflet* 342, September 1953.

Lawrence, George A., "Furrow Irrigation," *U.S.D.A. Soil Conservation Service Leaflet* 344, December 1953.

Lewis, M. R., "Practical Irrigation," *U.S.D.A. Farmers' Bulletin,* 1922 (supersedes Bul. 864), 1943.

Marr, James C., "The Border Method of Irrigation," *Calif. Agr. Exp. Sta. Ext. Ser. Circ.* 408, March 1952.

Marr, James C., "Grading Land for Surface Irrigation," *Calif. Agr. Exp. Sta. Ext. Ser. Circ.* 438 (revised), August 1957.

Middleton, James E., "Water Management," *Agr. Ext. Ser. Univ. of Ariz. Circ.* 205 (revised), October 1953.

Quackenbush, T. H., G. M. Renfro, K. H. Beauchamp, L. F. Lawhon, and G. W. Eley, "Conservation Irrigation in Humid Areas," *Soil Conservation Service Agr. Handbook* 107, January 1957.

Soil Conservation Service National Engineering Handbook, Section 15, *Irrigation,* Chapter 12, "Land Leveling," U.S.D.A., March 1959.

Stanley, William R., "Corrugation Irrigation," *U.S.D.A. Soil Conservation Service Leaflet* 343, December 1954.

United States Department of Interior, Bureau of Reclamation, U.S.D.A. Soil Conservation Service, "Irrigation on Western Farms," *Agr. Information Bul.* 199, July 1959.

Wallinder, William O., "Development of Agricultural Land for Irrigation," *Agr. Eng.,* Vol. 25, p. 467, 1944.

CHAPTER 15 IRRIGATION IMPLEMENTS AND STRUCTURES

Extensive use of well-designed modern implements for leveling irrigated lands, making borders, construction and cleaning of ditches, and for making corrugations and furrows is contributing to irrigation advancement. It is essential to efficiency and economy that the lands of each irrigated farm be well prepared for irrigation and also be provided with structures that facilitate easy control and regulation of the irrigation water during its application to the land. It is essential also that the irrigator be able to control the stream at his disposal and to spread the water uniformly over the land surface in order to moisten the soil to the desired depth without sustaining excessive losses of water. Some of the modern implements used in irrigation farming and farm structures that facilitate the control of water are described in this chapter.

15.1 IMPLEMENTS

The farm implements of first importance in irrigation are plows, spike-tooth harrows, disc harrows, scrapers, and levelers. Good plows and good plowing contribute greatly to the possibility of uniformity in distributing irrigation water. Lands that are irrigated by the ordinary flooding methods especially require good plowing because, owing to the lack of specially prepared levees, there is no means of crowding water over the higher portions of poorly plowed fields. Careless plowing of lands that are to be irrigated by flooding, or plowing with dull, poorly kept plows, is followed by inefficient irrigation.

15.2 IMPLEMENTS FOR LEVELING LANDS

Scrapers and soil movers like the unit shown in Fig. 15.1 have contributed greatly to advancement in land leveling. After the large

Fig. 15.1 Farm tractor and carry-all scraper used for land leveling. (Courtesy Eversman Manufacturing Company.)

cuts and fills are accomplished with the scraper the lands are leveled and smoothed with large automatic levelers like the land plane shown in Fig. 15.2.

Fig. 15.2 Land plane for automatic leveling. (Courtesy The Farmhand Co.)

Fig. 15.3 Preparing land for the border method of irrigation. (Courtesy Union Pacific Railroad Company.)

Farmers who depend on the flooding and furrow methods are more and more using modern land leveling and tillage implements in order to obtain a smooth land surface in which there are few, if any, small depressions or elevations.

15.3 BORDER-MAKING IMPLEMENTS

In addition to the use of well-built scrapers and other land-leveling implements on the larger farms, special border-making scrapers drawn by tractors have proved to be economical. For smaller farms adjustable bordering machines like the one developed by Arizona farmers in the Salt River Valley, illustrated in Fig. 15.3, are very helpful. The frame attached to the rear end of the Arizona implement smooths and grades the top of the levee in the same operation which pushes the soil together to make a levee.

Fig. 15.4   Ditching machine, hydraulically operated for excavating ditches. (Courtesy Eversman Manufacturing Company.)

15.4   IMPLEMENTS FOR MAKING AND CLEANING DITCHES

The era of pioneer methods of digging irrigation ditches with hand labor and the pick and shovel is largely past in many countries. Power-drawn ditching machines, illustrated in Fig. 15.4, have greatly reduced the costs and the time required for construction of canals and ditches.   Where ditches with steep banks are feasible and desired, carry-type scrapers drawn by tractors are efficient and economical. In the construction of new canals as well as in the repairing of old ones it is often advantageous to compact the soils well in order to reduce seepage losses and also to add to the stability of the soil and thus reduce erosion of the bed and the banks of the canal.   The sheepsfoot roller, for many years always used in the construction of earth dams, is now found to be useful also on canals.

Ditchers of the type shown in Fig. 15.5 are especially useful for cleaning the banks of canals.

15.5   CORRUGATION IMPLEMENTS

Shallow furrows are designated as corrugations.   Several types of homemade corrugators are used: one, a roller around which collars of the desired thickness and depth are built; another, a drag having runners as corrugators. The roller type compresses and compacts the soil as a means of making furrows; the drag type crowds the soil to both sides of the runner. These corrugators are limited in use to newly plowed land. For old alfalfa land, clover land, or other land having a compact surface, heavy, well-constructed steel shovels are needed to make satisfactory furrows.

15.6 DEEP-FURROW IMPLEMENTS

Potatoes, corn, asparagus, celery, and orchards on some soils are best irrigated by using deep furrows, especially in heavy soils. Many orchard crops, such as apples, peaches, lemons, olives, and almonds, are also best irrigated by means of deep furrows. Common shovel plows are sometimes used for making the furrows. For orchard irrigation on land having a steep sidehill slope a standard moldboard plow may be used by throwing the soil downhill so as to avoid overflowing of the furrows. A two-way sulky plow saves time in making deep furrows on sidehill land.

15.7 FARM IRRIGATION STRUCTURES

Engineers apply the word structure to dams, head gates, sluices, flumes, inverted siphons, chutes, and drops which are built to divert

Fig. 15.5 This Gregerson ditcher with 18-foot blade for clearing canal banks weighs 3.2 tons and will operate in 5-foot-depth ditches having steep banks. (Courtesy Robinson Machinery Company.)

water from natural sources and convey it to the farms for irrigation. The devices and pieces of equipment used by the individual irrigator to divert water from a large canal into his ditch and convey it to the several parts of his farm are here designated farm irrigation structures. In some communities rather crude farm irrigation structures are made to suffice even though the labor cost required in the use of such structures is sometimes very high. As a rule it is economical, and it is always most satisfactory to the irrigator, to build structures that have the required capacity and the strength to control the water. Many irrigation canals are built along the rims of the valleys immediately above the irrigated lands so that each irrigator obtains water directly from the main canal which conveys water during the entire irrigation season. On such canals satisfactory take-out structures are especially necessary.

Farm irrigation structures include two general classes, namely, permanent and temporary structures. No structure is truly permanent, in the strict sense of the word, but the term is applied to those structures which remain in place during one or more irrigation seasons. Temporary structures are those that are moved from place to place during an irrigation, or those that are built for only one season's use. The irrigation devices here classed as "temporary farm irrigation structures" are described by some authorities as "irrigation equipment." A further desirable classification of structures is based on the function of the structure and includes: diversion, conveyance, and distribution structures.

15.8 PERMANENT DIVERSION STRUCTURES

A check gate is a gate placed across a stream from which it is desired to divert water. The function of the check gate is analogous to that of the dam or the diversion weir on the rivers at the heads of canal systems. Check gates are built across laterals and ditches for the purpose of diverting part or all of the stream. The take-out gate is a part of the farmer's "diversion works" and is analogous to the head gate of the main diversion works on the river system. Its function is to regulate the quantity of water flowing into the small lateral, the field ditch, or the furrow. Typical wooden check gates are illustrated in Figs. 15.6 and 15.7. These gates may also be used as take-out gates, although pipe or culvert take-outs are commonly used, especially to take small streams out of large canals. For check gates in large canals in which the quantity of water fluctuates appreciably, it is desirable to place flashboards on the bottom of the stream so that the water which passes the check gate and goes on down the canal is forced to flow over the check structure, not under it. A study of the

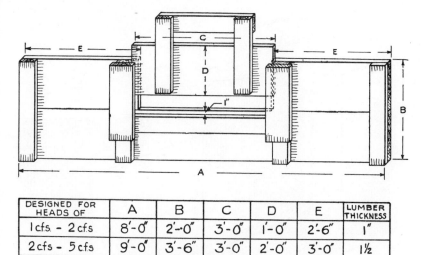

DESIGNED FOR HEADS OF	A	B	C	D	E	LUMBER THICKNESS
1 c.f.s. – 2 c.f.s.	8'-0"	2'-0"	3'-0"	1'-0"	2'-6"	1"
2 c.f.s. – 5 c.f.s.	9'-0"	3'-6"	3'-0"	2'-0"	3'-0"	1½
5 c.f.s. – 8 c.f.s.	10'-0"	3'-6"	4'-0"	2'-0"	3'-0"	1½

Fig. 15.6 Standard single-wing wooden check gate. (*U.S.D.A. Farmers' Bul.* 1243.)

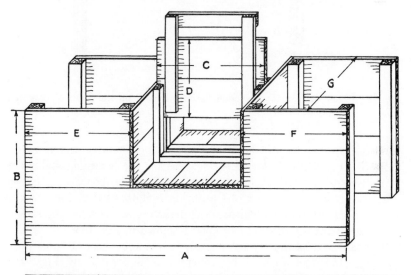

DESIGNED FOR HEADS OF	A	B	C	D	E	F	G
3 c.f.s. – 6 c.f.s.	9'-0"	3'-6"	3'-0"	2'-0"	3'-0"	3'-0"	2'-0"
6 c.f.s. – 10 c.f.s.	12'-0"	4'-0"	4'-0"	2'-0"	4'-0"	4'-0"	2'-6"
10 c.f.s. – AND UP	14'-0"	4'-0"	5'-0"	2'-0"	4'-6"	4'-6"	2'-6"

Fig. 15.7 Standard double-wing wooden check gate. (*U.S.D.A. Farmers' Bul.* 1243.)

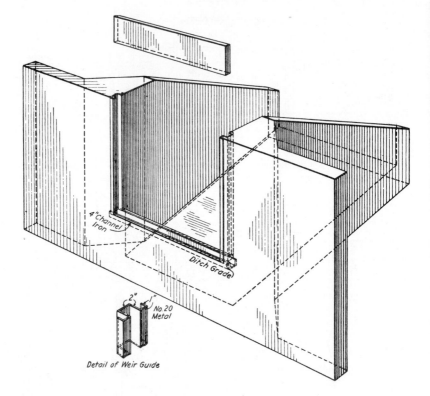

4"Channel Iron

Ditch Grade

2"

No.20 Metal

Detail of Weir Guide

Fig. 15.8 Concrete gate and wooden flashboard used on the Turlock Irrigation District, California.

hydraulic principles of check and take-out structures, given in the following article, will clarify the reasons for the foregoing statement. A well-made concrete check gate with wooden flashboard used by the Turlock Irrigation District of California is illustrated in Fig 15.8. A concrete structure that facilitates the diversion of the inflowing stream to either of three outflowing streams is shown in Fig. 15.9.

Permanent diversion structures can be made of wood as illustrated by Figs. 15.6 and 15.7, or by concrete as illustrated by Figs. 15.8 and 15.9. However, increasing use is being made of bricks and concrete blocks to form irrigation structures. Assembly costs are generally less than when concrete is used.

Prefabricated structures are beginning to be used widely. Mass production of simple shapes which can be combined into complex structures is reducing the cost of the material and the cost of installa-

Fig. 15.9 Concrete three-way outlet box on irrigation lateral. Water can be diverted to any one or more of the outlets. (Courtesy Soil Conservation Service.)

tion. A prefabricated low-cost steel structure developed by the Engineering Experiment Station at Utah State University is shown in Fig. 15.10.

15.9 HYDRAULIC PRINCIPLES OF DIVERSION STRUCTURES

In taking water out of a canal, it is desirable that the farmer obtain a flow as nearly constant as possible. Sudden increases in the quantity of water flowing in the canal, which occur as a result of storms or from closing off take-out gates, should be permitted to flow down the canal with as little obstruction as possible. These two conditions, i.e., approximately constant flow for the irrigator and a minimum of obstruction in the main canal, in general, may be provided by using submerged pipes or culverts as take-outs and overflow flashboards as checks in the main canal to cause the water to rise high enough to submerge the take-out gate and divert the desired quantity of water. To understand these principles clearly the reader should review Chapter 6 on the measurement of water and in particular Equations 6.4 and 6.8.

Equation 6.4 for an orifice shows that head h varies with the square of discharge q; hence to double the quantity of water flowing through

Fig. 15.10 Prefabricated low-cost steel water-control structure developed by the Engineering Experiment Station at Utah State University. (Courtesy United States Steel Company.)

a submerged culvert take-out the effective head, h, must be increased four times the original head. Thus, it is seen from Equation 6.8 for a weir that head H varies with the two-thirds power of discharge q; hence, to double the quantity of water flowing over flashboards as checks, the depth need be increased only 1.59 times. Therefore, streams through submerged take-outs are subject to much less variation than those through overpour take-outs.

15.10 TEMPORARY DIVERSION STRUCTURES

In order to divert water from the small ditches on the farm many irrigators use only temporary earth dams. They make each dam at the time and place desired by means of an ordinary shovel. In some soils that erode easily it is helpful to use a little partly rotted straw or weeds temporarily held in place by means of wooden stakes driven into the soil at the bottom of the ditch. Temporary earth dams are unsuited to streams of more than 2 cfs and in some soils are very difficult to maintain with a stream of 1 cf or more.

The labor requirement of temporary dams is greatly reduced by using portable dams of either metal, canvas, or plastic. Portable irrigation dams are illustrated in Figs. 15.11 and 15.12. Both dams can be used with fair accuracy to measure the flow since the opening through which the water passes has fixed dimensions. Portable metal dams are suited to streams smaller than those diverted by canvas dams. For streams of 5 cfs or more the metal dam required may be so large as to become burdensome to move or carry about the field. Well-built canvas or plastic dams are used to divert streams as large as 8 cfs or more, although streams of 2 to 3 cfs are more commonly diverted by using these portable dams. A heavy, durable, closely woven canvas is necessary to stand the water pressure and prevent excessive leaking.

An adjustable canvas or plastic dam is shown in Fig. 15.13. The dam stick hinges at the center and is held in place by a wing nut. Adjusting the wing nut will allow varying quantities of water to pass over the dam.

Fig. 15.11 Era-Gator portable irrigation-control gate. (Courtesy Page Metal Products Corporation.)

Fig. 15.12 Portable, rectangular weir made from canvas or plastic, utilizing sheet metal for the notch. (Photo by James R. Barker.)

15.11 WATER-CONVEYANCE STRUCTURES

The term structure as used herein applies quite as fully to ditches, levees, etc., which are made of earth as it does to devices built of wood, concrete, or metal. Most of the water-conveyance structures in the West are made of earth. Conveyance of water under pressure through underground concrete pipe to various points on the farm is becoming increasingly popular. The quantity of water that earth ditches, flumes, and pipes will convey may be estimated from the equations and the tables of Chapter 5. On excessive slopes caution must be exercised against erosion of farm ditches, and on small slopes it is important to guard against growth of weeds and grasses in order to maintain a satisfactory discharge capacity.

Fig. 15.13 Using an adjustable canvas-dam stick to control the amount of water passing the dam. (Photo by James R. Barker.)

Fig. 15.14 Conveyance flumes of concrete in the foreground, and aluminum to the left, to be removed for harvest of sugarcane in Hawaii. (Courtesy Hawaiian Commercial and Sugar Company.)

Fig. 15.15 Portable aluminum flumes used to irrigate sugarcane growing on steep land. (Courtesy Hawaiian Commercial and Sugar Company.)

15.12 REMOVABLE CONCRETE STRUCTURES

Several unique methods of water control have been developed on the large sugarcane plantations of Hawaii. Figure 15.14 shows thin-wall concrete sections with metal slide gates in the foreground and to the left an aluminum flume diverting water from the concrete flume. All concrete and aluminum sections shown will be removed from this field for harvesting and will be replaced after harvest. Figure 15.15 shows an aluminum flume conveying water down a steep hillside without loss of water or erosion. Note the plastic spill for discharging the water into the furrow. Several unique gates particularly adaptable for steep slopes have been developed by the Hawaii plantations. An experimental rubber pipe is being tried in Fig. 15.16.

15.13 CONCRETE PIPE UNDERGROUND

Irrigators are relying on concrete pipe more and more for distributing irrigation water. Thousands of miles of underground concrete pipe are used in the western states of the United States.

Fig. 15.16 Flexible rubber pipe with a thin-walled, removable concrete flume behind. (Courtesy Hawaiian Commercial and Sugar Company.)

Fig. 15.17 Fourteen furrows are being supplied controlled streams by this lockseam gated pipe. (Courtesy New Mexico State College.)

Fig. 15.18 A Washington farm owner operating a valve to release water from underground concrete pipe for irrigation of his orchard by the furrow method. (Courtesy Bureau of Reclamation.)

From the underground concrete pipe water flows upward through riser pipes, and through low-cost irrigation valves to an irrigation ditch, a check, or a basin. To prevent erosion of the soil around the riser pipe, top boxes of concrete or wood are sometimes used. Water flows from the concrete-pipe riser into the earth distribution ditch. Plastic or metal pipe sections about 3 feet long and ranging from 1 inch to 2 inches or more in diameter are often placed on the ditch bank to convey water from the ditch to the furrow. Some farmers prefer tubes made of lath.

Lightweight surface-gated pipe, mentioned in Chapter 14, is sometimes used with underground concrete pipe and connected to the concrete-pipe riser, thus at once solving the soil-erosion problem around the riser outlet and providing for complete control of the quantity of water flowing into each furrow. Figure 15.17 illustrates use of this pipe with a surface water supply. The convenience of the water control and the ease of regulating the quantity that is delivered to each furrow from the underground pipe are also illustrated in Fig. 15.18, showing the owner operating a valve to assure the best flow to the furrow.

15.14 SIPHON TUBES

Siphon tubes convey the water over the ditch bank into the furrow. They are especially useful with new ditch banks and are available in plastic, aluminum, galvanized iron, and rubber. Figure 15.19 shows water being conveyed in a concrete ditch and diverted by siphons to the furrows. Note that no gates are necessary and leakage from the ditch is thereby nullified.

The flow from the siphon is directly proportional to the size of the siphon and also proportional to the square root of the difference in elevation between the water in the ditch and the outlet, or the elevation of the downstream water surface if the siphon is submerged. Flow into borders or furrows is controlled by varying the number of siphons or the elevation of the discharge end of the siphon.

Fig. 15.19 Siphons diverting water from a concrete-lined ditch. (Courtesy Portland Cement Association.)

REFERENCES

Code, W. E., "Farm Irrigation Structures," *Colo. Agr. & Mech. College Agr. Exp. Sta. Bul.* 496-S, February 1957.

Currey, Albert S., "Irrigation Structures and Implements," *N. Mex. Agr. Ext. Ser. Circ.* 92, 1927 (revised) 1944.

Dusenberry, H. L. and O. W. Monson, "Irrigation Structures and Equipment," *Montana Ext. Ser., U.S.D.A. Bul.* 180 (revised), January 1951.

Givan, C. V., "Land Grading Calculations," *Agr. Eng.*, Vol. 21, p. 11, 1940.

Jensen, Max C., Mel A. Hagood, and Paul K. Fanning, "Irrigation Structures and Methods for Water Control," *Ext. Ser. Institute of Agr. Science State College of Washington Bul.* 491, September 1954.

Johnston, C. N., "Farm Irrigation Structures," *Calif. Agr. Exp. Sta. Circ.* 362, 1945.

Loving, M. W., "Concrete Pipe for Irrigation and Drainage," *Am. Concrete Pipe Assn. Bul.* 18, 1939 (reprinted) 1946–48.

Peikert, F. W., "Portable Pipe Irrigation Practices in Michigan," Reprinted from *Mich. Agr. Exp. Sta. Qtrly. Bul.*, Vol. 29, No 3, pp. 194–204, 1947.

Pillsbury, Arthur F., "Concrete Pipe for Irrigation," *Calif. Agr. Exp. Sta. Ext. Serv. Circ.* 418, November 1952.

CHAPTER 16 SPRINKLER IRRIGATION

The method of applying water to the surface of the soil in the form of a spray, somewhat as in ordinary rain, is known as sprinkling. This method of irrigation was started about 1900. The first agricultural sprinkler systems were an outgrowth of city lawn sprinkling. Before 1920 sprinkling was limited to truck crops, nurseries, and orchards.

It was used in humid regions as a supplemental method of irrigation. Most of these systems were stationary overhead perforated pipe installations, or stationary overtree systems with rotating sprinklers. These systems were expensive to install but often fairly inexpensive to operate. Portable sprinkler systems developed with the introduction of lightweight steel pipe and quick couplers in the early 1930's. This reduced the equipment cost for sprinkler systems and resulted in an increased number of sprinkler installations. The early development occurred principally in the Sacramento Valley, California. The number of sprinkler installations has increased rapidly since World War II owing to the development of more efficient sprinklers, lightweight aluminum pipe, more efficient pumps, and to the widespread distribution of low-cost electrical power and fuels for internal combustion engines. Sprinklers have been used on all soil types and on lands of widely different topography and slopes, and for many crops.

16.1 CONDITIONS FAVORING SPRINKLER IRRIGATION

When considering where sprinkler irrigation can be used to greatest advantage, the normal requirement of uniform distribution of water is of utmost importance. That method of irrigation which can most economically distribute the required water uniformly is generally the best method. Some of the conditions which favor sprinkler irrigation are as follows:

1. Soils too porous for good distribution by surface methods.
2. Shallow soils the topography of which prevents proper leveling for surface-irrigation methods.

3. Land having steep slopes and easily erodable soils.
4. Irrigation stream too small to distribute water efficiently by surface irrigation.
5. Undulating land too costly to level sufficiently for good surface irrigation.
6. Labor available for irrigating is either not experienced in surface methods of irrigation or is unreliable; good surface irrigation requires trained reliable labor.
7. Land needs to be brought into top production quickly. Sprinkler systems can be designed and installed quickly.

The following points should also be considered when comparing sprinkler and surface methods of irrigation.

1. Water measurement is easier with sprinkler than with surface methods.
2. Sprinkler systems can be designed so that less interference with cultivation and other farming operations occurs and less land is taken out of production than with surface methods.
3. Higher water-application efficiency can normally be obtained by sprinkler irrigation.
4. When water is already being pumped to the point of use, the pressure needed for sprinkling can be obtained with a minimum of additional capital investment.
5. When domestic and irrigation water comes from the same source, a common distribution line can frequently be used.
6. For areas requiring infrequent irrigation, sprinkler irrigation can be provided at a lower capital investment per acre of land irrigated than can surface irrigation.
7. Whenever water can be delivered to the field under gravity pressure, sprinkler irrigation is particularly attractive.
8. Frequent and small applications of water can be applied readily by sprinkler systems.

16.2   OTHER USES OF SPRINKLER SYSTEMS

Sprinkler systems have several secondary agricultural uses which are important in addition to the primary use for distributing irrigation water to be stored in the soil.

Light frequent irrigations, so easily managed by using sprinklers, are helpful in many situations, such as the following: shallow-rooted crops, germination of new plantings, control of soil temperature on certain crops such as lettuce, and control of humidity on such crops as tobacco. It is possible that light frequent irrigations may produce favorable results for coffee culture in lieu of shade trees which are commonly used in some parts of the world to control temperature and humidity.

Frost Protection

Frost protection can be provided by sprinklers. Sprinkler systems have been used for frost protection for several years on cranberry bogs, blueberries, strawberries, almonds, citrus, vegetables, and flowers. Crops have been saved in temperatures as low as 20°F by sprinkler irrigation. Fine misty sprays are desirable. Caution should be taken not to break off tree limbs with ice accumulation from overhead sprinklers. Sprinkling should be sufficient to keep the ice which forms wet and slushy. Water should be applied until the ice is melted. Spray should be started at or above 32°F for best protection.

Fertilizer Application

Several fertilizers and soil amendments can be applied quickly, economically, easily, and effectively through sprinkler-system spraying. The equipment is simple and the labor requirement is slight. By controlling the time of application most of the materials can be leached to the desired depth. Being in solution, these materials are distributed over the surface of the land quite uniformly. Fertilizer elements already in solution become available to the plant roots faster than when placed in the soil dry.

Fertilizer and amendment materials can be injected into the water in several ways. One of the simplest methods is to connect a barrel containing the dissolved material to the suction side of the pump. The barrel can also be connected to the throat of a venturi tube. Small high-pressure pumps are also used to inject the solution into the high-pressure side of the water line.

Preferred practice is to operate the sprinklers sufficiently before applying fertilizer to thoroughly wet the foliage and soil. Then the fertilizer should be applied for a sufficient time to provide good distribution over the land being irrigated. Application times of at least 30 to 60 minutes are required. Following application the sprinklers should be operated from 30 to 90 minutes depending upon the rate of application. Any corrosive material is thus washed out of the applicator pump and sprinkler system and toxic material is washed off the foliage of the plants. Also applying water after the application of fertilizer or soil amendment moves the material into the soil.

Ammonia nitrate, phosphoric acid, and ammonium sulfate are quite corrosive. Brass and bronze impellers in pumps are strongly attached

Fig. 16.1 Standard overhead sprinkling irrigation system. (Courtesy Skinner Irrigation Company.)

by solutions containing phosphorus especially when ammonium salts are also present. Hence, special care should be exercised to be sure that solutions entering the system on the suction side of the pump are fairly dilute and that the rate of application of the material is low.

16.3 TYPES OF SPRINKLERS

Three general types of sprinklers are used: fixed nozzles attached to the pipe, perforated pipe, and rotating sprinklers.

The earlier sprinkler systems were "fixed nozzle pipe" types, illustrated in Fig. 16.1. Parallel pipes are installed about 50 feet apart and supported on rows of posts. Water is discharged at right angles perpendicularly from the pipe line. The entire 50-foot width between pipe lines may be irrigated by turning the pipes through about 135°.

Perforated sprinkler lines illustrated in Fig. 16.2 are used more extensively in orchards and nurseries. Generally rates of application are in excess of 0.75 inch depth per hour and pressures are less than 35 psi, often as low as 10 psi. They do not cover a very wide strip. The rotating sprinklers illustrated in Fig. 16.3 are being used very extensively. The advantage of this sprinkler over other types is its ability to apply water at a slower rate while using relatively large nozzle openings. This factor is particularly favorable in waters containing silt and debris since less stoppage of sprinklers is experienced. Application rates less than 0.1 inch per hour are possible with these sprinklers. This slow rate is desirable on soils having low infiltration rates and advantageous to the small farmer doing his irrigation along with his field work. This low rate of application makes it necessary to move the sprinklers only once or twice a day on soils with low water-holding capacity. Pressures for rotating sprinklers normally range from 30 psi for the smaller sprinklers to over 100 psi for the large units.

16.4 PRESSURES

Sprinkler systems operate under a wide range of pressures from 5 psi to over 100 psi. The desirable pressure depends upon power costs, area to be covered, type of sprinkler used, sprinkler spacing, and

Fig. 16.2 Perforated sprinkler line in citrus orchard. (Courtesy W. R. Ames Co.)

Fig. 16.3 Sprinkler system on 400-acre field of tomatoes near Sacramento, California. (Courtesy Shur-Rane Irrigation Company and E. C. Olsen Company.)

crop being irrigated. Low pressures range from 5 to 15 psi; medium from 15 to 30, intermediate from 30 to 60, and high pressures from 60 to 100 psi. Giant sprinklers usually operate at pressure in excess of 80 psi. Sprinklers in the low-pressure range have small areas of coverage and relatively high sprinkling rates for the recommended spacings of the sprinklers. Their use is generally confined to soils having infiltration rates of more than $\frac{1}{2}$ inch per hour during the irrigation.

Medium-pressure sprinklers cover larger areas and have a wide range of precipitation rates, and water drops are well broken up.

High-pressure sprinklers cover large areas, and precipitation rates for recommended spacing are higher than for the medium pressures. Distribution patterns are usually good but are easily disrupted by winds because of higher water trajectories. They have high-application rates, above 0.75 inch per hour. The wetted diameter of the circle is from 200 to 400 feet. Distribution patterns are very good in calm air but are easily disturbed by winds.

In general, sprinklers can be obtained which will fit practically any operating condition in the field.

16.5 TYPES OF SPRINKLER SYSTEMS

The sprinkler system includes the sprinkler, the riser pipe, the lateral distribution pipe, the main line pipe, and, often, the pumping plant. Sprinkler systems may be classified as semi-permanent or portable according to the make-up of the components.

Semi-Permanent Installations (often referred to as solid systems)

There are semi-permanent installations wherein the main lines are buried and the lateral lines and sprinklers remain fixed during the irrigation season. Labor is reduced to a minimum, and quality and yield of crops are generally maximal, provided the system is operated properly. One man can irrigate 80 to 160 acres per day compared to approximately 40 acres with the movable system.

When light frequent irrigations are desirable and when the farmers are totally dependent upon irrigation to supply the crop moisture needed during the season, semi-permanent systems are finding favor. The most extensive use has been on potatoes which show marked increases in yield and quality with light frequent irrigations. Shallow-rooted crops, like radishes, are also responding well to permanent systems. For years nurseries have found them useful for maintaining a misty moist condition over plants.

Since permanent sprinkler systems are often used to apply small quantities of irrigation water, it is well to remember that wind may be a serious problem. Short periods of application do not allow for natural shifts in wind direction and variations in wind velocity which produce a more even distribution of water.

A variation of this system has sprinklers mounted on quick-coupling risers so that they may be moved along the buried lateral lines. This reduces the number of sprinklers and increases the labor needed for operation of the system.

Portable Systems

Semi-portable sprinkler systems consist of buried main pipe lines, portable lateral pipe lines and sprinklers, and a fixed pumping plant. The portable low-angle sprinkling system, illustrated in Fig. 16.4, is made up of a portable main pipe line, portable lateral pipe lines and sprinklers, and fixed pumping plant. The fully portable system has portable lateral pipe lines with sprinklers and portable pumping plant. A fully portable system can be moved readily from field to field, thus extending its utility.

Large rotating booms carrying combinations of rotating and fixed sprinklers are illustrated in Fig. 16.5. The size of pipe, length of the boom, nozzle sizes, and pressures can be varied to cover the desired acreage at each setting.

16.6 MOVING LATERAL LINES

Lateral lines containing sprinklers can be moved by several methods. The cheapest method in terms of capital equipment cost is the hand-

Fig. 16.4 Low-angle sprinklers in orchards keep water off soft fruit and leaves.
(Courtesy W. R. Ames Co.)

move system where the pipe is uncoupled, moved a length at a time
to the new location, and recoupled. Drag- or tow-type laterals are
being used wherein the entire line is pulled by a tractor or other

Fig. 16.5 Rotating boom-type sprinkler. (Courtesy W. R. Ames Co.)

vehicle to its new location on skids provided at each coupling. Wheels can be secured at the couplings in place of the skids, permitting easier transport. Both the skid and wheel arrangement allow movement parallel to the line with slight curvature being possible to displace the line laterally.

The wheel-move system uses the pipe as an axle and by mounting large diameter wheels the pipe line can be moved perpendicular to the direction of the lateral lines. The wheel-type units can either be hand powered or powered by a small auxiliary internal combustion engine.

The lateral system may also be moved by rotating the line in a circle about one end as a pivot. The rate of application is adjusted to compensate for different rates of travel along the line. The wheels are powered by hydraulic cylinders, designed to produce a rate of travel proportional to the distance from the pivot point.

16.7 GRAVITY SPRINKLER SYSTEMS

Often the most economical sprinkler system is one utilizing a source of water adjacent to and above the area to be irrigated. When the source of water is high enough on an adjacent hill to produce sufficient pressure at the sprinkler nozzles without use of a pump, the system is called a gravity sprinkler system.

There is no expense for a pump and motor, fuel or electricity. However, more main-line pipe must be purchased to convey the water from the source to the field than is required normally for a pumped sprinkler system. Larger main-line pipe produces more pressure at the sprinklers, but also increases the cost of pipe. The most economical design is obtained considering not only design requirements but also seasonal irrigation costs and expected production.

16.8 SOURCES OF WATER AND METHODS OF DEVELOPING PRESSURE

Sprinkler systems require sources of water free of debris that would clog the sprinklers. The best sources of debris-free water are wells and lakes. Screening devices are usually necessary to remove the debris when water is taken from irrigation canals or rivers.

Pressure for operation of sprinklers is obtained from gravity when practical and supplied by pumps when the source of water is located at such a level that gravity pressure will not operate the sprinklers. In some systems, combinations of both gravity and pump power are used to obtain the desired pressure.

When pumping from lakes, streams, and irrigation canals, centrifugal pumps are generally used to develop pressure. When pumping from wells either centrifugal, submersible, or turbine pumps are used. Turbine pumps or submersible pumps are better adapted to wells in areas having a variable water table during the irrigation season.

16.9   REMOVAL OF SILT AND DEBRIS

Silt and debris are very detrimental to sprinkler equipment. The abrasive action of silt causes excessive wear on pump impellers and sprinkler nozzles and bearings. Debris plugs nozzles and thus interferes with uniform application of water to the land. Clogged sprinkler nozzles cannot be tolerated. The method of removal of debris and silt will need to be adapted to site conditions, but a few general considerations will usually apply and should be carefully considered.

Silt is heavier than water and will settle out if given an opportunity. Hence, silt removal can best be accomplished by creating a quiet condition where silt can settle and subsequently be flushed from the container. Basins of various shapes can be used effectively.

Since silt is heavier than water it will tend to be carried along the bottom of the stream, provided the stream is not too turbulent. The intake of the pump should be located where the silt content is a minimum. Never place the intake in position where the silt will be carried directly into the intake. Frequently the approach conditions can be adjusted so that the portion of the flow with the greatest silt concentration is deflected past the intake. Of course screens can also be used to prevent silt from entering the intake to the pump.

In some regards debris presents more of a problem than does silt. Debris can be water soaked and float beneath the surface or can be carried on the bottom of the stream while lighter material will be carried on the surface. Screening devices clog easily. Sprinklers are easily clogged with debris and prevented from functioning properly. The structure used to remove debris from the water should be designed specifically to combat the type of debris present in the water.

Floating debris can be skimmed off the surface with a fixed or rotating screen and the sub-surface waters permitted to proceed into the pipeline. However, cleaning the screen presents a problem. If sufficient water is available to allow some to pass by the intake, the excess can be used to flush over the screen and remove the collected debris. In this case a horizontal screen just below the surface or a screen gently sloping up so that it is below the surface at the upstream end, and above or near the surface at the overflow, will clean reasonably well. Revolving screens have also been used with reasonable success.

Removal of debris such as moss and organic matter that is suspended in the flow presents more of a problem. Screens covering the entire cross section of flow are effective provided the size is small enough to remove the material and provided they are cleaned sufficiently often to prevent excessive resistance to flow. Two, or even three, screens used in sequence are helpful in reducing the tendency for clogging. The first should be coarser to retain the larger debris, and the final screen should be small enough to retain the smallest objectionable debris.

16.10 OPERATION OF SPRINKLER SYSTEMS

A sprinkler system may be well designed for the crop and field, but if it is not efficiently operated the result will be disappointing. A correctly designed sprinkler system will supply adequate water during periods of maximum water demand by the crop. Over-irrigation will result if the system is operated at full capacity when the water demand of the crop is less than maximum. Excessive application will cause leaching of soluble plant food, low water-application efficiencies, reduction in quality and quantity of crops, and, ultimately, a drainage problem.

16.11 COST ANALYSES

In considering the economics of any irrigation system, all costs and benefits should be included. Initial costs are important because the purchaser must finance this usually heavy capital investment. However, annual costs per acre compared to annual returns per acre are the best measure of the economics of an irrigation system. Annual costs per acre should include the following:

1. Annual depreciation of irrigation system.
2. Interest on investment for the system.
3. Water costs.
4. Power costs.
5. Repair, operation, and maintenance.
6. Labor.
7. Taxes.

When considering the return to be expected from a new irrigation system, credit should be shown for any savings resulting from the following:

1. Increased yield and quality.
2. Less land out of production.
3. Reduction in land preparation, tillage, and harvesting costs.

4. Saving in labor, repair and maintenance, and operating cost.

5. Saving in water and power costs.

Actual costs vary widely. The following figures are estimates based on normal sprinkler irrigation systems in use in the United States. The initial investment on a large sprinkler irrigation system with a simple layout may be as low as $50 per acre. Normal costs average about $125 per acre and will cost more as labor-saving devices are added. Permanent systems where the main and lateral lines are not moved may cost as much as $500 per acre. The average total cost per acre per year including initial investment and operating costs will vary from $15 to $35. The life expectancy of a sprinkler system varies with treatment, use, and storage but will average about 15 years.

Assume, as an example of how irrigation costs should be approached, that the average initial cost of a sprinkler system is $100 per acre. The annual cost of equipment for a system of this cost is normally 40 percent of the total annual cost. Fuel and repair would average about 30 percent and labor 30 percent. Considering 6 percent interest and a 15-year life for the system, recognizing that the life of individual components will vary, gives $10 per year on initial cost, $7.50 on fuel and repairs, and $7.50 on labor. Hence, the total annual cost of the sprinkler irrigation system will be $25 per acre per year. Similar analysis should be used when figuring the cost of any method of irrigation.

16.12 SPRINKLER SYSTEM DESIGN REQUIREMENTS

The important factors in the success of sprinkler irrigation systems are first, the correct design, and second, the efficient operation of the designed system. The basic information necessary for the design of a farm irrigation system is obtained from four sources, namely, the soil, the water supply, the crop to be irrigated, and the climate.

Information concerning soils includes the soil type, depth, texture, permeability, and available water-holding capacity of the root zone. Necessary water supply concerns the location of the water delivery point in relation to the fields to be irrigated, the quantity of water available, and the delivery schedule. The maximum consumptive use of water per day, the root-zone depth, and the peculiarities of irrigation necessary to be taken into account in the irrigation system are obtained from a knowledge of the requirements of the crop to be grown. Climatological information includes the natural precipitation and wind velocities and direction. All this information must be compiled in one form or another before starting to design a sprinkler system.

The performance requirements of a sprinkler system include applying water at a rate that will not cause runoff from the area irrigated. Water should be applied at such a rate that high water-application efficiency is obtained and in such a manner that high water-distribution efficiency results.

The sprinkler system must have the capacity to meet the peak water-use demands of each crop during the irrigation season. Allowance in capacity must be made for unavoidable water losses by evaporation, interception, and some deep percolation.

When a system is designed for supplemental irrigation or protective purposes, it should have a capacity to apply the necessary depth of water to the design area in a specified time. The cost of the system should be consistent with the insurance values involved.

There should not be more than a 20-percent variation in the depth of water applied to any part of the design area. To obtain reasonably uniform distribution the pressure in the lateral lines should not vary more than 20 percent. Variations in pressure occur as the result of friction loss in pipes and elevation changes in main lines or lateral lines. It may be necessary to control pressures with valves.

A sprinkler system must apply water so that it will not cause physical damage to the crop. In orchards, high-velocity streams of water from sprinkler nozzles have bruised growing apples when sprinklers have been placed too close to the trees. Also, in crops having fine seedling plants, a fine spray must be applied, or the plants will be beaten into the ground. Such a spray requires high pressures to break up the water drops at the nozzle.

Droplets must be small enough so that the soil is not damaged. Some soils will puddle under the impact of large droplets, causing the soil to crust. Smaller nozzles operated at higher pressures reduce drop size. Acceptable droplet size can be obtained using the following combination of nozzle sizes and pressures.

$$\frac{1}{8}'' \text{ to } \frac{3}{16}''\text{---}35 \text{ to } 50 \text{ psi}$$
$$\frac{3}{16}'' \text{ to } \frac{1}{4}''\text{---}45 \text{ to } 60 \text{ psi}$$
$$\frac{1}{4}'' \text{ to } \frac{3}{8}''\text{---}50 \text{ to } 70 \text{ psi}$$

Two nozzle sprinklers will produce acceptable breakup of drops at 5 psi less pressure than single nozzle sprinklers.

Wind can be critical, interfering excessively with distribution of the water. It is difficult to obtain good distribution of water in winds in excess of 10 miles per hour. Lateral lines should be set at an angle to the prevailing wind. It may be necessary to decrease the spacing between sprinklers and between lateral lines as much as 40 percent to obtain satisfactory distribution in the windy condition.

The sprinkler system should be designed to apply water at the lowest annual cost. A balance should be sought between pipe size and pumping costs in a system operated by pumping. Careful analysis should be made to arrive at a reasonable balance between equipment costs and power costs.

When used in practical field spacing with selected operating pressures, the sprinkler chosen must give satisfactory moisture distribution.

If a pump is necessary it must be picked on the basis of the maximum operating conditions of head and gallons per minute and must not overload at minimum operating conditions.

Careful attention should be given to the planning of the sprinkler system. To select the best combination of power plant, sprinkler system, and operating schedule requires skill and experience.

16.13 PLANNING FARM SPRINKLER SYSTEMS[1]

Adequate planning for sprinkler irrigation includes an inventory of the resources of the farm and integration of the operation of a system with other required farm or ranch operations. The system planner needs to consider the limitations of the farm resources as well as those factors of economics and farming operations which may influence the success or failure of the proposed irrigation enterprise.

In many cases, the initial installation of a sprinkler system is only a beginning step in the development of a complete farming sprinkler system. Adequate planning for sprinkler irrigation should recognize the need for future expansion. The initial system when installed should, insofar as is practical, fit with future expansions and, finally, into the ultimate farm irrigation system.

A factor often overlooked, both by the planner and the prospective user, is the cropping pattern which will develop from full use of the system. Often the system capacity, layout, and its operation are planned around one cash crop. Soil-building crop rotations will be required to maintain optimum production as the yields of this crop are stepped up with irrigation and the other necessary good farming practices. The requirements for moisture for the resulting crop rotation may be entirely different. With a crop rotation, it may be possible to reduce the required system capacity where peak moisture requirements for different crops occur at different times during the irrigation

[1] Abstracted from Chapter 10 of the second edition (1959) of "Sprinkler Irrigation," compiled and edited by Guy O. Woodward and published by the Sprinkler Irrigation Association.

season. With the development of a balanced crop rotation, mechanized equipment may be adapted for use on both row crops and close growing crops. Recognition of this feature may lead to reduced labor costs.

To prepare a successful sprinkler system plan there should be complete understanding between the planning technician and the prospective owner or user. When the inventory of farm or ranch resources is made, the planner and owner or user should work closely together. Differences in soils should be understood. Quite often the landowner or operator can give clues as to the intake rate of soils in the design area. If he has had previous irrigation experience, he has probably observed differences in ability of the same soils to hold water for crop use.

A complete understanding is needed concerning the labor requirements for irrigation operations. Without a full discussion of this phase of sprinkler irrigation, misunderstandings can arise, leading either to improper operation of the system or disappointment on the part of the user. Labor should be included in the annual irrigation costs.

The supplier-planner, who can reach these understandings and plan the irrigation facilities and operations accordingly, is going to be in a better position to maintain a profitable and reputable longtime career in sprinkler irrigation. In addition, he will have the satisfaction of playing an important role in building the agricultural production in his territory.

16.14 DESIGN OF A SPRINKLER SYSTEM

Proper sprinkler design involves careful consideration of essentially all of the concepts presented in earlier chapters. Soils, topography, crops, water supply, management, consumptive use, and hydraulics are involved in addition to labor, equipment, and power costs.

Table 16.1, modified from design sheets of W. R. Ames Co., shows how these factors can be organized in a systematic manner to arrive at a good sprinkler design. Technical data is available on sprinklers, pipe, pumps, and power units from the companies who sell sprinkler irrigation equipment.

Design should begin with the farm-resource inventory, items 1 to 7 inclusive of Table 16.1. Tabulation of location, water supply, crops, desired irrigation operations, power source, soils, and consumptive-use information, as well as a sketch of the farm, are essential before any detailed calculations can be made.

With this information the system capacity computations shown in item 8 can be made. Note that allowance is made for variation from

TABLE 16.1

Design and Selection of a Sprinkler Irrigation System

1. WATER SUPPLY   (a) Source _____

  (b) Estimated or measured quantity _____ gpm

  (c) Quality of water _____
                                    Good, Fair, Poor (Attach Test Results)

  (d) Delivery schedule _____

  (e) Seasonal variation in quantity _____ gpm to _____ gpm

  (f) Pressure available @ source _____ psi

  (g) Well data   (1) Total depth _____ ft.; diam. _____ in.

     (2) Water-bearing material _____

     (3) Water found @ _____ ft. to _____ ft. and _____ ft. to _____ ft.

     _____

     (4) Test—No. of hours _____ by _____

     Static W.L. @ _____ ft.

     Drawdown: _____ ft. @ _____ gpm

     _____ ft. @ _____ gpm

     _____ ft. @ _____ gpm

2. CROPS

Crop					Totals
Present ac.					
Initial plan					
Ultimate ac.					

3. DESIRED IRRIGATION OPERATIONS

  (a) Cover of area in _____ days   (b) No. of moves _____

  (c) To be moved by _____ men   (d) By owner only _____

  (e) Mechanical moving desired _____   (f) Type _____

4. POWER SOURCE _____ ; Unit cost at farm _____

5. BASIC SOILS INFORMATION

	Field Number and Acreage					System Design
	1	2	3	4	5	
(a) Surface texture						
(b) Subsoil texture						
(c) Effective depth of soil (ft.)						
(d) Moisture-holding capacity (inch/ft.)						
(e) Intake rate (inch/hour)						

  (f) Soil limitations _____
                    (depth, hardpan, gravel, low fertility, drainage, etc.)

TABLE 16.1

DESIGN AND SELECTION OF A SPRINKLER IRRIGATION SYSTEM (*Continued*)

6. GROSS AMOUNT AND FREQUENCY COMPUTATIONS

	Field Number and Acreage					System Design
	1	2	3	4	5	
(a) Kind of crop						
(b) Working root depths (ft.)						
(c) Net water applied per irrigation (in.)						
(d) Peak use rate (inches/day)						
(e) Irrigation interval (days)						
(f) Water application efficiency (percent)						
(g) Gross water applied per irrigation (in.)						

7. ANNUAL IRRIGATION REQUIREMENTS

(a) Net seasonal moisture req'd (in.)						
(b) Est. effective stored moist. (in.)						
(c) Est. effective rainfall (dry years)						
(d) Est. net irrigation requirement (in.)						
(e) Gross irrigation requirement (in.)						
(f) Est. number irrigations req'd (max.)						
(g) Est. number irrigations req'd (min.)						

8. SYSTEM CAPACITY COMPUTATIONS

(a) Application rate (inch/hour)						
(b) Time per set (hours)						
(c) Settings per day						
(d) Days of operation per interval						
(e) Systems capacity (gpm) (preliminary)						

9. FINAL DESIGN—SPRINKLER SYSTEM SPECIFICATIONS

(a) Final application rate _____ inch/hour   (b) Spacing _____' (line) x _____ (move)

(c) Sprinkler—model _____ ; nozzle _____" x _____"; _____ gpm @ _____ psi (av.)

(d) Lateral—total length _____'; _____' of _____" and _____' of _____" no. of spr. _____

(e) Head loss in lateral _____ psi   (f) Pressure @ head of lateral ___ psi _____ft.

(g) Total lateral capacity _____ gpm   (h) No. of laterals operating _____

(i) Total system capacity _____ gpm   (j) Acres covered per set _____

(k) Acres per day _____

TABLE 16.1

DESIGN AND SELECTION OF A SPRINKLER IRRIGATION SYSTEM (*Continued*)

10. MAIN LINE DESIGN—Capacity Requirements (Attach diagram and profile)

_____ gpm from source to _____ ft.; _____ gpm for _____ ft.

_____ gpm for _____ ft.; _____ gpm for _____ ft.

11. TOTAL DYNAMIC HEAD FOR MAIN LINE COMBINATIONS AND HP REQUIRED

Trials	Pipe Size and Length				Head Loss for Trial Pipe Combination			
	"	"	"	"	1	2	3	4
(a) 1								
2								
3								
4								
(b) Pressure @ lateral or gated inlet (ft.)								
(c) Maximum elevation difference (ft.)								
(d) Total suction lift (ft.)								
(e) Minor losses and fittings (ft.)								
(f) Total dynamic head (ft.)								
(g) Water horsepower req'd.								

12. ECONOMIC PIPE SIZE

(a) Approx. pipe comb. cost				
(b) Annual fixed cost				
(c) Annual fixed cost difference				
(d) Whp difference				
(e) Cost/whp per season				
(f) Annual power cost difference				

13. PUMP SELECTION

(a) Make, model and size _____

(b) Impeller diameter _____"

(c) Rpm @ design load _____

(d) Efficiency _____ %

(e) Required bhp @ design load _____

(f) Shutoff head _____ feet _____

(g) Required bhp @ minimum design load _____

(h) Rpm @ minimum load _____

TABLE 16.1

DESIGN AND SELECTION OF A SPRINKLER IRRIGATION SYSTEM (*Concluded*)

14. POWER UNIT SELECTION

(a) Make, model _____

(b) Cu. in. displ. _____

(c) Stroke in inches _____

(d) Rpm @ design load _____

(e) Piston speed @ design load _____

(f) Bmep @ design load _____ psi

(g) Type of power conversion _____

(h) Speed ratio _____

(i) Electric motor _____ hp; _____ phase; _____ rpm _____

15. CROP PRODUCTION COSTS

	Alt. 1	Alt. 2	Alt. 3
A. Seasonal irrigation costs			
(a) Labor costs	_____	_____	_____
(b) Equipment costs for moving	_____	_____	_____
(c) Power costs	_____	_____	_____
(d) Fixed costs	_____	_____	_____
(e) Maintenance	_____	_____	_____
(f) Total irrigation costs for season	_____	_____	_____
(g) Irrigation costs per acre	_____	_____	_____
B. Other crop production costs			
(a) Land preparation	_____	_____	_____
(b) Planting or seasonal preparation	_____	_____	_____
(c) Cultivating	_____	_____	_____
(d) Disease and insect control	_____	_____	_____
(e) Fertilizer	_____	_____	_____
(f) Harvesting or processing	_____	_____	_____
C. Total production costs	_____	_____	_____

16. EXPECTED OR POSSIBLE GROSS RETURNS (for comparison only) BASED ON GROSS ACREAGE AND YIELDS

	Alt. 1	Alt. 2	Alt. 3
(a) Total gross returns—farm produce	_____	_____	_____
(b) Deduction for quality grading	_____	_____	_____
(c) Deduction for land out of cultivation	_____	_____	_____
(d) Adjusted gross cash returns	_____	_____	_____

17. EXPECTED OR POSSIBLE PROFITS

_____   _____   _____

field to field, and yet one set of system-design values must be selected which will give the sprinkler system the capacity to meet the varying operating conditions.

Knowledge of the required capacity of the system and the operating conditions allows selection of an application rate, spacing, and distance between lateral lines. A particular sprinkler nozzle size and operating pressure can be selected and pipe lengths calculated. Power requirements and economic pipe sizes should be determined by studying several combinations. From these possibilities the most desirable pump and power unit can be selected.

The feasibility of irrigating by sprinklers cannot be fully ascertained until a cost analysis is completed showing the expected profits when production costs are subtracted from gross returns. Alternate designs should be considered in arriving at the best design.

The selection of sprinkler, nozzle size, pressure, and general layout should be based upon good engineering concepts tempered by field experience with sprinkler irrigation. Rate of application and droplet size are determined by considering several variables, but the feasible rate of application and droplet size will be restricted to a fairly narrow range of possibilities. Spacing of sprinklers along the lateral line, and spacing between lateral lines are also functions of sprinkler selection and pressure. Certain combinations work best for given field conditions of soil, crop, and wind. Much helpful information to guide designers is available from commercial sources as well as from public agencies.

REFERENCES

Christiansen, J. E., "Irrigation by Sprinkling," *Calif. Agr. Exp. Sta. Bul.* 670, 1942.

Christiansen, J. E., "Hydraulics of Sprinkling Systems for Irrigation," *Trans. Am. Soc. Civil Eng.*, Vol. 107, p. 221, 1942.

Christiansen, J. E., "Lawn Sprinkler Systems," *Calif. Agr. Exp. Sta. Circ.* 134, 1947.

Lewis, M. R., "Sprinkler or other Methods of Irrigation," *Agr. Eng.*, Vol. 30, No. 2, p. 86, 1949.

McCulloch, A. W., "Design Procedure for Portable Sprinkler Irrigation," *Agr. Eng.*, Vol. 30, No. 1, p. 23, 1949.

Woodward, Guy O., "Sprinkler Irrigation," second edition, Sprinkler Irrigation Assoc., 1959.

CHAPTER 17 DRAINAGE OF IRRIGATED LANDS

Adequate drainage of crop-producing lands requires a general lowering of shallow water tables. Some students, especially in humid regions, find it difficult to think of drainage as being essential in arid regions where irrigation is required for crop production. Experience has demonstrated fully the need for drainage of irrigated lands. In some valleys the higher lands never require drainage, but the need for drainage of the lower valley lands is frequently a result of the irrigation of the higher lands. From 20 to 30 percent of the irrigated lands in arid regions need drainage to perpetuate their productivity. The reclamation of saline and alkali soils has many important phases, but adequate lowering of the water table by drainage is a first and basic necessity.

Irrigation and drainage in arid regions are complementary practices, the necessity for drainage being greatly influenced by low efficiencies in the conveyance and application of irrigation water.

In humid regions drainage is of even greater necessity than in arid regions. Excess rainfall produces bogs in low-lying flat areas. In arid regions drainage usually follows irrigation. In humid regions drainage frequently must precede agricultural development and often must precede human habitation. And yet frequently the drained lands are the most productive agricultural lands. The challenge is to correctly evaluate the potential usefulness of the land, to design and install the most economical drainage on those lands justifying drainage, and to maintain a workable drainage system so that maximum and economic benefits will result.

17.1 BENEFITS OF DRAINAGE

Adequate drainage improves soil structure and increases and perpetuates the productivity of soils. Drainage is the first essential in reclamation of waterlogged saline and alkali soils. Even if only those lands which have been farmed are considered, drainage benefits irrigation agriculture and the public in many ways. For example, adequate drainage (1) facilitates early plowing and planting, (2) lengthens the crop-growing season, (3) provides more available soil moisture and plant food by increasing the depth of root-zone soil, (4) helps in soil ventilation, (5) decreases soil erosion and gullying, by increasing water infiltration into soils, (6) favors growth of soil bacteria, (7) leaches excess salts from soil, and (8) assures higher soil temperatures.

Drainage also improves sanitary and health conditions and makes rural life more attractive.

17.2 SOURCES OF EXCESS WATER

The major sources of excess water that make drainage necessary on parts of the irrigated lands are seepage losses from reservoirs or canals and deep percolation losses from irrigated lands. Efficient water application on the higher irrigated lands reduces the need for drainage of the lower lands. Flooding of low lands due to overflow of rivers and natural drainage channels during periods of maximum streamflow constitute important sources of excess water in certain low-valley areas. The flow of ground water toward waterlogged lands in arid regions may be in any direction. In some areas flow is largely downward through highly permeable surface soils to relatively impermeable subsoils.

In other areas unconfined or free ground water may flow under small hydraulic slopes. In still other areas the major source of excess water may be upward flow from an artesian aquifer. Two or more of these possible sources of excess water contribute to the maintenance of shallow water tables in some soils. Thorough ground-water investigations and subsoil studies are essential to intelligent design of drainage systems.

Extensive use of small-diameter pipes as piezometers, described in Chapter 9, enables engineers to develop ground-water contours, or flow patterns, as shown in Fig. 17.1, which indicate the directions of waterflow before irrigation and after. In April, before irrigation the highest water-table elevation was only 1.5 feet above the lowest level, and contours were widely spaced; whereas after irrigation in May, the highest elevation of the water table was 4.5 feet above the lowest level. Between these periods of measurement the water table rose 4.0 feet on part of the field and 1.5 feet in another. Ground-water studies illustrated by Fig. 17.1 are essential to intelligent design of drainage systems. Permeability measurements are very helpful because of the

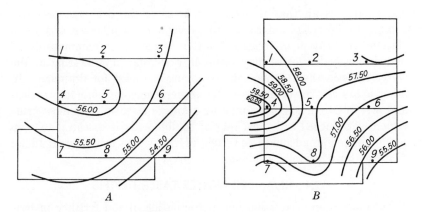

Fig. 17.1 Ground water contours. *A*. Before irrigation, April 14. *B*. After irrigation, May 12.

great range in soil permeability. The soil permeability is the dominant variable influencing the feasibility and the cost of drainage.

Gravelly and sandy soils, under natural conditions, are from 25,000 to 50,000 times as permeable as clay soils; and in some drainage studies permeability ratios of 100,000 to 1 have been measured. Subsoil formations and permeabilities thus influence sources of excess water in soils.

However, care should be exercised in interpreting the capacity of soils to transport water. Clays are well aggregated and may be quite permeable. Within the basin of the Great Salt Lake, the very fine-textured clays often transmit large quantities of water readily through cracks and fissures, apparently formed shortly after the clay was deposited.

17.3 CONTROL OF WATER SOURCES

In some areas, substantial progress toward a general lowering of the water table, and the solution of drainage problems, can be made by control of excess water sources. For example, as stated in Chapter 5, nearly 40 percent of the water taken into canals is lost during conveyance. Millions of acre-feet thus lost annually reach the water table and cause it to rise. Lining canals to reduce these losses should be encouraged.

Control of the source of excess water by percolation from the irrigated land also is a difficult and perplexing task. In a few areas, there is no excess salinity or alkali in the soil, and lowering of the water table is all that is necessary to solve the drainage problem. Lining of the irrigation canals to prevent seepage losses and more efficient application of irrigation water to reduce or eliminate deep-percolation water losses may result in a satisfactory lowering of the water table in these areas, thus removing the need for drainage. In other areas control of excess-water sources may be both impractical and prohibitive in cost, thus making drainage essential. In most arid-region waterlogged soils, leaching of soluble salts is essential to crop production, and adequate lowering of the water table by artificial drainage must precede leaching.

17.4 REQUIRED WATER-TABLE DEPTHS

Adequate crop production and perpetuation of soil fertility in irrigated areas generally require water-table depths of 6 feet or more. In many cases, even where drainage systems have been installed, the water table during part of the year may be less than 3 feet. A summary of required water-table depths approved by irrigation authorities and also by financial institutions that are interested in long-time loans for improving of irrigated land follows.

Classification	Range in Water-Table Depths
Good	Static water table below 7 ft; up to 6 ft for a period of about 30 days of the year.
Fair	Water table at 6 ft, up to about 4 ft for a period of 30 days. No general rise.
Poor	Some alkali on surface; water table 4 to 6 ft up to 3 ft for a period of 30 days.
Bad	Water table less than 4 ft and rising. Natural and artificial drains too far away to drain land.

Generous long-time loans are made by land banks and other lending agencies when good lands are mortgaged as security; restricted or short-time loans are made on fair lands, and no loans at all are made on poor and bad lands. Positive assurance of immediate and adequate lowering of the water table by thorough drainage may induce financial agencies to make limited loans on bad lands.

The above standards are applicable to large segments of irrigated agriculture and are accepted by agricultural authorities. However, proper management can produce good and even excellent results in farming land with a water-table depth between 3 and 6 feet. Special care is required, and careful application of irrigation water is essential. Sprinkler systems are well adapted to controlled applications. At least one adequate leaching irrigation must be applied each year, preferably during the dormant season when roots of perennial plants will not be seriously affected. When no crop is growing on the land, a heavier leaching irrigation should be applied at the beginning of the dormant season so that the water table will have the maximum time to return to its normal position before crops are planted in the spring.

In irrigated regions where irrigation water and application costs are high, another benefit will accrue from a water table higher than would normally be thought advisable. Crops may obtain a significant portion of their needed water from the saturated soils. As much as 50 percent can come from ground water without seriously reducing the rooting depth. Saving the cost of applying 20 to 50 percent of the normally needed irrigation water may more than offset the reduction in yield resulting from a somewhat higher water table.

Also by exercising care it may be possible to avoid the cost of drainage and still not seriously restrict the agricultural production. In general the water table, including the capillary fringe which is essentially saturated, should not be allowed to occupy more than the lower $\frac{1}{3}$ of the normal root zone of the crop. Since rooting depths vary with crops and vary during the season, no definite depth can be recommended. Furthermore, hazards are present and careful management is essential. Nevertheless, shallower water-table depths are economically feasible under excellent management.

17.5 LOWERING THE WATER TABLE

In addition to eliminating or controlling sources of excess water, improving natural drainage facilities and providing man-made drainage systems will also be of substantial aid in lowering the water table.

Proper maintenance of natural drainage systems, usually feasible at low costs, protects irrigated lands from excess percolation of water from rains and melting snows and also from flood damages.

In many arid regions artificial drainage also is required to provide adequate lowering of the ground-water table and is accomplished by one of three methods: (a) open channel drains; (b) covered clay or concrete pipe; and (c) pumping ground water.

In open-channel and tile drains, gravity pulls the excess water from the wet soils into the drains, and gravity also causes the flow in the drains. When gravity provides the mechanical power, the drains are designated gravity drains. Pumping of ground water in some valleys in arid regions provides for both irrigation and drainage.

17.6 FIELD INVESTIGATIONS

The four elements of primary concern in making a field investigation for drainage are topography, soil, water table, and water source.

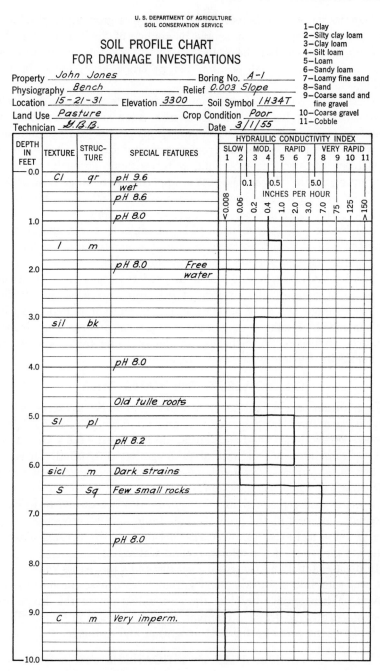

Table 17.1 Soil profile chart developed for drainage investigations by the Soil Conservation Service, U.S.D.A.

It is a generally accepted practice to classify soils with respect to texture and then to adjust the hydraulic conductivity, "HC" estimates with respect to soil structure, alkali, and other influencing factors. Soil texture refers to the relative proportions of the various size groups of individual soil grains in a mass of soil. Structure refers to the condition of the soil grains (clay, silt, sand, etc.), in the way they are arranged and bound together into aggregates with definite shape.

TEXTURE LEGEND			STRUCTURE LEGEND	
Clay.............	c	Loamy fine sand. lfs	Massive.........	m
Silty clay loam..	sicl	Sand........... s	Platy...........	pl
Clay loam.......	cl	Coarse sand..... cos	Prismlike	pr
Silt loam........	sil	Fine gravel...... fg	Blocky.........	bk
Loam...........	l	Coarse gravel... cg	Granular........	gr
Sandy loam.....	sl	Cobble......... co	Single grain......	sg

Aggregate length and thickness has an effect upon the "HC." The overlap of aggregates having a horizontal axis 3 or 4 times longer than the vertical can have a marked effect upon "HC." The "HC" of stratum can be changed by cracks, crevices and fractures. Some fractured (blocky structured) clay or shale strata are often much higher than sand or gravel strata. The "HC" of sandy stratum are generally higher when the grains are round and about the same size than when they are irregular in shape and of different sizes. Flat grains tend to overlap and reduce "HC" rates. The matric rather than the coarser material in a gravel or cobble stratum governs the "HC" and should be used as a basis in estimating the "HC."

The following two tables are suggested as a possible guide in rating the "HC" of various strata during a drainage soil survey. It is not anticipated that these tables will replace laboratory analysis and actual field "HC" measurements but that they will be used as estimates and bolstered or revised as additional experience and data become available.

HYDRAULIC CONDUCTIVITY INDEX NUMBER RANGE FOR TEXTURE-STRUCTURE CORRELATION

STRUCTURE / TEXTURE	MASSIVE	PLATY	PRISMLIKE	BLOCKY	GRANULAR	SINGLE GRAIN
		GENERAL INDEX NUMBER RANGE				
Clay	1	1–4	1–2	1–8	2–7	
Silty clay loam	2	1–4	1–4	2–7	3–7	
Clay loam	3	2–3	2–3	3–7	3–7	
Silt loam	4	3–4	2–4	3–6	4–6	
Loam	5	3–5	4–5	5–7	5–7	
Sandy loam		4–6	5–6	6–7	6–8	6
Loamy fine sand				6–7	7–8	7
Sand						8
Coarse sand						9

ESTIMATE OF ALKALI INFLUENCE ON HYDRAULIC CONDUCTIVITY BY pH COLOR INDICATORS

SOIL TEXTURE	pH ESTIMATES FROM COLOR INDICATORS							
	8.6		9.0		9.6		10.0	
	HC REDUCED %	INDEX NO.	HC REDUCED %	INDEX NO.	HC REDUCED %	INDEX NO.	HC REDUCED %	INDEX NO.
Clay	20	1	40	−1	70	−1	90	−1
Silty clay loam	15	2	30	2	60	1	80	1
Clay loam	15	3	25	3	50	2	70	2
Silt loam	15	4	25	4	50	3	70	2
Loam	10	5	20	5	40	4	50	4
Sandy loam	5	6	15	6	35	5	45	5
Loamy fine sand	5	7	15	7	30	6	40	6
Sand	5	8	10	8	25	8	40	7

U. S. DEPARTMENT OF AGRICULTURE
SOIL CONSERVATION SERVICE

FIELD HYDRAULIC CONDUCTIVITY TEST
PIEZOMETER METHOD
FOR DRAINAGE INVESTIGATIONS

Piezometer Number __A-1__

Estimated "HC" __0.2 in. hr__   Calculated "HC" __0.35 in. hr__

Location __John Jones 15-21-31__   Date __3/1/55__   Technician __L.B.B.__

Stratum Thickness __2.0 ft__   Texture __sil__   Structure __bk__

Depth Piez. __4.2 ft__   Auger Dia. __1 3/16 in.__   Piez. Dia. __1/4 in.__

Length Cavity __4 in.__   Sloughing __very slight__ Times Cleaned __4__

pH (Soil __8.0__ Water __8.1__) Salinity (Soil __2 x 10⁻³__   Water __250 x 10⁻⁶__   )

TIME	ELAPSED TIME	Δt	DISTANCE TO WATER SURFACE FROM REFERENCE POINT			Δh	RESIDUAL DRAWDOWN
			BEFORE PUMPING	AFTER PUMPING	DURING RECHARGING		
			B	A	R	A–R	R–B
	Minutes	Minutes	Feet	Feet	Feet	Feet	Feet
11:00			2.00				
:01	0			4.05			2.05
:06	5				3.70		1.70
:11	10				3.35		1.35
:16	15				3.00		1.00
:21	20	20			2.63	1.42	0.61
:26	25				2.35		0.35
:31	30				2.23		0.23
:36	35				2.15		0.15
:41	40				2.10		0.10

COMMENTS

The test was very good

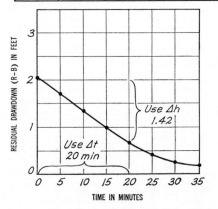

Table 17.2 Procedure and form used to obtain hydraulic conductivity in the field, developed by the Soil Conservation Service, U.S.D.A.

FIELD HYDRAULIC CONDUCTIVITY TEST

PIEZOMETER METHOD

The piezometer method is used to obtain the hydraulic conductivity of a given stratum or area in a soil profile. (Hyd. cond. is a permeability figure dependent on properties of the ground water as well as the soil profile.)   This is possible because the hole which is bored into the soil for the conductivity measurement is cased, except for a small cavity at its end. The rate of entry into this cavity is a measure of the hyd. cond. of the soil around the cavity.

EQUIPMENT: A 1½- to 2-inch worm auger with a square bit end is used for the test.   Electrical conduit 1¼- to 2- inch inside diameter, sharpened on one end, is used as the piez. The auger is ground to about 1/16 inch smaller than the inside dia. of the piez.   A driving head on the piez. top prevents damage during driving.   An electrical device sounding bell or blow tube can be used to measure the water level.   A soil tube jack or piez. removal equipment can be used to remove the piez.

METHOD: An auger hole is bored to a depth of 6 inches.   The piez. is then driven into the hole about 5 inches with light blows from a maul.   The hole is again augered to a depth of 6 inches below the piez.   This procedure is continued until the piez. reaches the desired depth.   A cavity 4 inches long is carefully augered below the end of the piez.   A stop on the auger handle helps make this length precise.   The auger should be removed very slowly to prevent sloughing of the cavity wall.   A hollow auger or small tube to the auger bit may be required to permit air to break the suction and prevent sloughing of the cavity.   The piez. is pumped or bailed out, with a pitcher pump or bail bucket, to permit the pores in the cavity wall to be flushed out.   Flushing is repeated until the rate of rise in the piez. is the same as a previous pumping.

TEST: The water level is lowered in the piez. a distance dependent upon the sloughing tendency of the profile. The water levels and times of observations are recorded and used in the following Kirkham Piezometer formula to calculate the hyd. cond.

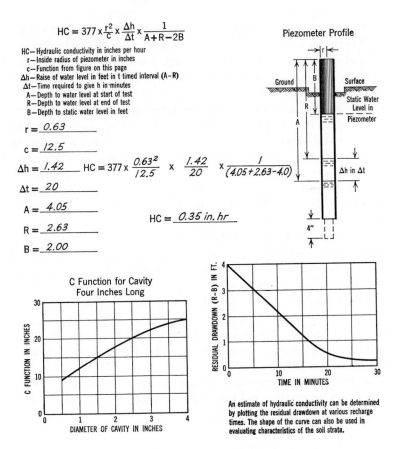

$$HC = 377 \times \frac{r^2}{c} \times \frac{\Delta h}{\Delta t} \times \frac{1}{A+R-2B}$$

HC—Hydraulic conductivity in inches per hour
r—Inside radius of piezometer in inches
c—Function from figure on this page
Δh—Raise of water level in feet in t timed interval (A–R)
Δt—Time required to give h in minutes
A—Depth to water level at start of test
R—Depth to water level at end of test
B—Depth to static water level in feet

r = 0.63

c = 12.5

Δh = 1.42 $HC = 377 \times \dfrac{0.63^2}{12.5} \times \dfrac{1.42}{20} \times \dfrac{1}{(4.05+2.63-4.0)}$

Δt = 20

A = 4.05

R = 2.63   HC = 0.35 in. hr

B = 2.00

Piezometer Profile

C Function for Cavity Four Inches Long

An estimate of hydraulic conductivity can be determined by plotting the residual drawdown at various recharge times. The shape of the curve can also be used in evaluating characteristics of the soil strata.

Each must be studied carefully and interpreted wisely. Existing data should be assembled and reviewed before proceeding with detailed field studies. The complexity and thoroughness of field investigation will depend upon the complexity of the drainage problems and the extent and usefulness of existing data.

The topography should be studied first by a visual field inspection. Aerial maps viewed under a stereoscope will frequently yield valuable information. Unless the problem can be solved readily, a topographic survey should be made. The accuracy needed will depend upon the problem and should be adjusted to give greater detail in critical areas.

The principal objective of the soils investigation is to ascertain the character of the soil profile and how it varies from location to location. Field investigations of the soil begin after reviewing existing data. Permeability is the major soil property of concern in a drainage study. Soil samples can be obtained and observations made with hand-powered augers, jetting rigs, and piezometers discussed in Chapters 8 and 9. Kind of drainage system and size, depth, and capacity of the system may be determined primarily from the soil survey. A soil profile chart developed for drainage investigations by George B. Bradshaw is shown in Table 17.1. A method and form for determining hydraulic conductivity in the field is shown in Table 17.2.

Observing the depth of the water table and its variations yields valuable data affecting the drainage design. Interpretations of differences of water-surface elevation in terms of direction and magnitude of the flow velocity of the ground water are discussed in Chapter 9. Frequently drainage can be made most effective at a minimum cost by intercepting flow before it enters the problem area. Also drainage lines should be constructed perpendicular, not parallel, to the flow to be most effective. Upward flowing water cannot generally be effectively drained by open or tiled drains unless the drains can be placed within a very permeable aquifer. Usually drainage wells relieving the pressure causing upward flow are the most effective methods of drainage. Hence, knowing the direction of ground-water movement can be very helpful and is generally essential.

Knowing the source of water often suggests ways of reducing the quantity at the source. Also when the source is known, it is often possible to estimate the quantity of water to be drained from the quantity measurement at the source.

The feasibility of intercepting the drainage water in a location where it can be readily used for irrigation is enhanced when the source and direction of flow are known.

17.7   DESIGN OF OPEN DRAINS

On some drainage projects open drains are used largely to convey water to distant outlets. Water may flow into open drains directly from the ground water and also from collecting tile lines. The designing engineer selects the drain outlet and determines appropriate elevation of bed of drain and water surface in the drain at times of maximum flow. Then he decides the bed slope, which ranges from ½ to 1½ feet per 1000 feet. The slopes of open drains in the lower nearly level lands should be as large as the ground-surface slope provided this will not cause excessive water-flow velocities and channel erosion. Uniformity of drain-bed slope is usually advantageous even though this causes some differences in depth of drain. Design of side slopes of open drains depends largely on the soil formation, with a range from the steep slope of ½ horizontal to 1 vertical in very stiff, compact clays, to flat slopes of 3 to 1 in loose, open sandy formations. Depths of open drains range from 6 to 12 feet or more.

The trend had been toward deeper drainage systems. Drains of 12 to 15 feet in depth were not uncommon. The high costs of very deeply constructed gravity drains and the value of land removed from production justify special effort toward pumping ground water for drainage in areas where the power costs and soil formations make pumping feasible or of installing tile drains so that the ground surface is not obstructed.

17.8   CONSTRUCTION METHODS AND COSTS

In modern construction of open drains, large power-driven drag-line excavators are used as illustrated in Fig. 17.2. Deep open drains require strips of land 75 to 100 feet or more in width, and this is one serious objection to the use of open drains in areas of high-priced

Fig. 17.2 Drag-line excavator in operation constructing an open drain.

Fig. 17.3 Typical open drain in New Mexico, bed width approximately 20 feet, capacity 125 cfs.

lands. A typical open drain of average depth is shown in Fig. 17.3.

Although costs of drains vary widely from time to time and from place to place, one typical example is presented. Consider an open drain 12 feet deep, having side slopes of 2 horizontal to 1 vertical, and bed width of 4 feet. The top width of channel is 52 feet, and the cross-sectional area is 336 square feet; nearly 38 square yards. The volume of excavation is 12⅔ cubic yards per foot of length. At 15 cents per cubic yard the excavation cost would be $1.90 per foot. If the excavated material is left on the land near the channel the total width of waste land is approximately 100 feet, and the area per mile of drain is 12 acres. At $200 per acre the cost of land is 45 cents per foot, thus making the drain cost $2.35 per foot exclusive of bridges and culverts.

Nitroglycerine dynamite is used in the construction of open drains in areas too wet for earth-moving machinery to work. When the soil is saturated and sticks of dynamite are uniformly spaced, the explosion will "propagate" from one charge to the next. This method requires only one blasting cap and is economical. Dynamiting affords an easy means of opening silted creeks and ditches as well as for making new open ditches. The soft bottom resulting from the blast is unsuitable for laying tile drains.

17.9 TILE DRAINAGE SYSTEMS

The two most common tile-drainage layouts are: (1) relief drains, nearly uniform in depth and spacing, for fairly level lands, and (2) intercepting drains on irregular slopes and near sidehill lands.

In the relief system, spacing of the drains is much influenced by the texture and permeability of the soil. In clay soils of low permeability and tile depths of 5 feet, close spacing of 200 feet may be essential to satisfactory drainage; in average loam soils, 400 to 600 feet is good spacing provided the tile is placed to depths of 6 feet or more; in sandy and gravelly soils, spacing at 800 feet or more represents the more general practice. Long main drains with short collecting laterals are called the herringbone system. The gridiron system consists of long parallel laterals connected to a short main drain. Manholes, sand traps, and observation wells at convenient points $\frac{1}{8}$ to $\frac{1}{4}$ mile along the lines facilitate essential inspection, cleaning, and maintenance of the lines.

The depths and locations of irregular cutoff drains to intercept seepage water and prevent it from flowing from sidehill lands toward the lower flat lands depend largely on the surface topography and soil formations. Intercepting drains must cut off the water flow in the sandy and gravelly soil strata because the rate of flow in these soils is very high compared with the flow in loams and clays.

Tile-drain sizes should be designed using the formulas presented in Chapter 5 for pipe flow. Allowance will need to be made for somewhat increased resistance due to the frequent joints between pieces of tile.

17.10 INSTALLATION OF TILE DRAINS

Trenching machines, like the ones shown in Figs. 17.4 and 17.5, have replaced hand labor on nearly all drainage projects that are large enough to warrant moving the heavy trenching machines to and from the fields that need drainage.

Either clay or concrete drain tile is hauled, usually direct from the factory to the field, and placed along the proposed drain lines. As the trenching machine moves forward in the field, the drain pipe is laid with each new length of pipe being placed securely against the one just laid. Some machines are equipped with hydraulic pressure devices to press the tile firmly together. Caving of soil into the trench near the excavator is prevented by a large two-walled steel shield. Water flows from the saturated soils into the tile through the pipe joints, not through the walls of clay or concrete pipe. To facilitate keeping soil materials from entering the tile it is good practice to place over the top of the joints a strip of tar paper and a screened-gravel envelope about 3 inches thick around the pipe. For soils of high stability, moderate moisture content, and no caving of banks at the time of placing the tile, the gravel may well be placed 10 to 15 feet behind the trencher. Immediately after the gravel is placed the tile should be carefully covered with soil to a depth of about 12 inches by

Fig. 17.4 Concrete tongue-and-groove joint pipe ready for placing in the drainage trench. (Photograph by J. R. Barker.)

hand labor. This process protects the pipe from displacement, and from damage when large volumes of earth material are forced into the trench by heavy machinery for backfilling.

The outlet end of a 3600-foot drain placed in sandy soil is shown in Fig. 17.6. The lower 1600 feet of the line is 8-inch concrete pipe, and the upper 2000 feet is 6-inch pipe. The trenching was done in January, with the equipment shown in Fig. 17.4. A gravel envelope was placed around all the joints. Shortly after completing the line the drain discharge was only 36 gpm, or 1 gpm per 100 feet of line. During the irrigation season the maximum discharge was 120 gpm.

Tile-laying machines can place about 1000 feet of tile a day when a tractor is used to load the gravel into the chute and backfill the trench.

17.11 MAINTENANCE OF DRAINS

The principal need in the maintenance of drainage systems is the removal of soil and vegetation from drains. Maintenance of open drains is the same as for other water conveyance channels and is discussed in Chapter 5.

Several types of trees and plants extend their root systems many feet to obtain water. Among these are greasewood, willows, and poplars. The roots enter the joints of closed drains and continue to grow inside the pipe, eventually obstructing the flow of drainage water. Partial obstruction by roots may retard the velocity of flow sufficiently to allow soil particles to settle in the pipe, thus gradually sealing the drains. Chemicals should be added to the drain water periodically when roots tend to penetrate the drains.

It is very difficult to construct drains in fine sandy soils with joints through which ground water may enter the pipe, and yet exclude all

Fig. 17.5 Laying shallow drainage tile in humid area. (Courtesy Iowa State College Extension Service.)

Fig. 17.6 The outlet-end of a 3600-foot, 8-inch-diameter concrete pipe drain immediately after installation during wintertime in Utah. (Photograph by J. R. Barker.)

soil particles. The volume of soil entering the drain is small when a suitable gravel envelope is provided for the joints.

There are large variations in the quantity and velocity of water flow in closed drains. The largest flows with the greatest velocities occur during or after storms and irrigations. Soil particles entering the drains during the periods of high-velocity flow may largely be carried in suspension to the sand boxes or to the drain outlet, but as the quantity and velocity of flowing water decreases the soil particles settle to the bottom of the pipe. They may be rolled along with the low-velocity flow of water, but with a further reduction in the velocity the soils come to rest. The soil may become sufficiently stable to resist the scouring action of subsequent high-velocity flows, after which more soil accumulates, thus gradually filling the pipe unless the soil is removed.

The term "wash-ins" refers to holes formed in the land where irrigation water flows downward through the backfilled trench, washing large volumes of soil into the closed drain through the joints. Wash-ins have occurred most frequently soon after construction, and less frequently as the backfill material settled in the trench, thus becoming more compact.

Wash-ins have been attributed to careless irrigation methods. Irrigation water applied excessively and allowed to pond over the drains is especially conducive to wash-ins. The damage caused by wash-ins is twofold: the soil washed in obstructs the drains, and the holes formed in the land surface render that part of the land unproductive until repaired.

In sandy soils drain pipes have been found shifted as much as 90° from alignment. Loosely placed or unmatched joints cause this condition. Drainage water flowing over a wide or unmatched joint erodes the soil from under the joint, thus allowing it to settle. As the joint settles and the pipe ends separate the eddies in the hole forming under the joint become greater, thus increasing the eroding effect of the water. This continues until the flow of water in the drain is completely blocked by soil entering the joint or by silt collecting in the lower reaches of the drain.

17.12 NEED FOR A GOOD QUALITY TILE

Good quality tile is essential if drainage systems are to operate over extended periods of time. Tile must be highly resistant to deterioration from freezing and thawing, resistant to attack by acids and sulfates, and dense enough to be relatively impermeable. Strong dense tiles are desirable. Modern processes of manufacture and availability of new high quality cements remove any excuse for using poor quality tile.

17.13 GRAVEL ENVELOPES FOR TILE DRAINS

Gravel envelopes are placed around tile drains for two reasons: (1) to prevent soil particles from moving into the drain, which may clog the line and also cause "wash-ins" as discussed above and (2) to provide a more permeable material adjacent to the tile, thus increasing the effective diameter of the tile.

Design of gravel envelopes for drains follows the same procedure outlined in Chapter 3 for wells. Because of the rather large quantities of gravel frequently used on a given drainage project, it becomes rather costly to adjust the gravel mix to fit the theoretical considerations. In practice, naturally occurring gravels which are properly graded have been used successfully. Gravels used for filters should contain a predominance of coarse- and medium-sized sand.

17.14 DRAIN DEPTHS, SPACING, AND GROUND-WATER FLOW

Two types of soil profile are considered to illustrate the influence of drain depth, spacing, and other factors on the quantity of flow of ground water toward and into drains.

In highly permeable sandy soils, underlain by compact clay of low permeability 6 to 10 feet below the land surface, as illustrated in Fig. 17.7, the ground-water flow is essentially horizontal toward the drains.

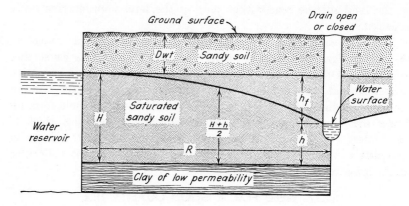

Fig. 17.7 Illustrating for sandy soils over clay the linear flow of ground water toward drains spaced $2R$ feet in which the water table midway between drains is $H - h$ feet above the water surface in the drain.

To simplify the illustration, the source of water flowing toward the drain is considered a reservoir as illustrated on the left of Fig. 17.7. The water surface is maintained in the reservoir and adjoining soil a distance of H feet above the clay. Flow from the reservoir to the drain is steady, it being assumed for simplicity that the reservoir is the only source of water.

Ground water actually flows to the drain from both sides. Let $2q$ represent the flow into a drain in length L. Then the ground-water flow from one side to the drain is

$$q = av$$

and from Darcy's law,

$$v = k\,\frac{h_f}{R} = k\left(\frac{H - h}{R}\right)$$

Consider the depth of saturated sand about midway between the reservoir and the drain as average; then the average area of saturated soil, in drain length L, through which the ground water flows is:

$$a = \left(\frac{H + h}{2}\right) L$$

and the quantity of flow from the reservoir to the drain

$$q = \left(\frac{H + h}{2}\right) L \times k\left(\frac{H - h}{R}\right) = \frac{kL(H^2 - h^2)}{2R} \qquad (17.1)$$

The quantity of flow to the drain from reservoirs on both sides would be

$$Q = 2q = \frac{kL(H^2 - h^2)}{R} \tag{17.2}$$

from which

$$R = \frac{kL(H^2 - h^2)}{Q} \tag{17.3}$$

For example, assuming that the reservoir is the only source of ground-water flow in a 15-foot depth of waterlogged sand, what spacing of drains will draw, from the soil on both sides of a drain, a stream Q of 1 cfs in a 2500-foot length of a 15-foot depth drain in which the water is 2 feet deep, when the water table is 5 feet below ground midway between the drains?

Referring to Fig. 17.7, for this example:

$$H = 10 \text{ ft} \qquad L = 2500 \text{ ft}$$

$$h = 2 \text{ ft} \qquad q = 1 \text{ cfs}$$

The average permeability measured in field soils in this case is

$$k = 2 \times 10^{-3} \text{ ft/sec}$$

Then since the drain spacing S equals $2R$ it follows that

$$S = \frac{2 \times 2 \times 2500 \times (100 - 4)}{1000 \times 1} = 960 \text{ ft}$$

The lengths H, h, and L can be accurately measured at any time, and q can be measured within 5 percent accuracy. However, computations of drain spacing using Equation 17.3 must be regarded as only approximations because of the fact that permeability k varies greatly in field soils.

Rainfall and irrigation water which percolate below the water table may be the source of drainage water rather than essentially horizontal flow from a distant source such as a reservoir. Since the flow entering the ground water is essentially vertical, a modification must be made in the above analysis.

In the previous development, flow midway between the reservoir and the drain was assumed to be q. However, with uniform vertical flow the discharge past this midsection will be only $q/2$. Hence Equation (17.1) becomes

$$q = \frac{kL(H^2 - h^2)}{R} \tag{17.4}$$

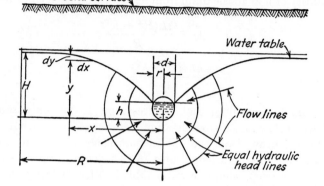

Fig. 17.8 Illustrating for deep, uniform soils the radial flow of ground water toward drains spaced $2R$ feet, in which the water table midway between drains is $H - h$ feet above the water surface in the drain.

and the quantity of flow into the drain will be

$$Q = \frac{2kL(H^2 - h^2)}{R} \tag{17.5}$$

from which

$$R = \frac{2kL(H^2 - h^2)}{Q} \tag{17.6}$$

For soils of great depth and approximately uniform permeability, as illustrated in Fig. 17.8, ground water flows radially toward the drain from all directions. With these soil conditions and designating radial flow through a semi-circular area (a little less than the actual area) it is essential to use the calculus to derive the rational equation

$$Q = \frac{\pi kL(H - h)}{2.3 \log_{10} R/r} = \frac{\pi kL(H - h)}{2.3 \log_{10} S/d} \tag{17.7}$$

in which all the symbols but S and d have the same meaning as in Equation 17.2. S is the spacing between drains, and d is the diameter of the drain.

The basic differences of the two flow conditions of special importance are: for the first condition, if h is small as compared to H, the flow to the drain is proportional approximately to the square of the effective depth $(H - h)$, whereas in the second condition the flow is proportional to the first power of the effective depth.

17.15 LIMITATIONS OF EQUATIONS

Of necessity Equations 17.1 to 17.8 are based on uniform soil conditions and a uniform source of water. Also, the capillary range above the water table and the seepage surface at the drain have been neglected. Hence, the formulas shown are useful as approximations, to be modified by experience and applied with wisdom. The method of developing the equations and the assumptions made will be helpful in visualizing the physical processes involved in drainage and the degree to which variation from assumed conditions will influence results. Further discussions of these principles are contained in Chapters 3 and 9.

17.16 PUMPING FOR DRAINAGE

The main physical defect of gravity drainage systems is failure to lower the water table to an adequate depth. Many gravity drains are either too shallow, spaced too far apart, or both. Pumping ground water in some areas is a more effective means of lowering the water table.

The influences on well discharge of the soil permeability, effective well depth, and diameter of well when pumping from ground water are considered in Chapter 3. In pumping from confined ground water in an artesian aquifer of depth D the water flows radially to the well through cylindrical surfaces having a vertical axis and the following well-discharge formula applies:

$$Q = \frac{2\pi k D (H - h)}{2.3 \log_{10} R/r} \qquad (17.8)$$

Comparisions of Equations 17.7 and 17.8 show that if the length of drain L in Equation 17.7 is equal to the depth of aquifer D in Equation 17.8, and all other items are the same, then the flow to the well through "vertical cylindrical" surfaces would be twice that to the drain through "horizontal semi-cylindrical" surfaces.

Pumping ground water for drainage is influenced favorably by adequate depths and permeabilities of the water-bearing formations, by high values of pumped water for irrigation, and by low power costs. The experiences of one California irrigation district and of the Salt River Valley Water Users Association in Arizona indicate the feasibility of ground-water pumping where subsoils are favorable and the pumped water can be used for irrigation.

From 1907 to 1922, the Modesto Irrigation District in California spent $356,000 for construction and maintenance of gravity drains for 45,000 acres. Sub-irrigation had prevailed for several years. In many locations where the rich soil had previously produced abundant crops, the yields decreased, orchard trees died, and vines withered for the alkali salts had become sufficiently concentrated to render the soil unfit for plant growth. In 1922, the Modesto group drilled the first drainage well, and by 1939 had put into operation 77 pump wells, reaching a combined capacity of 207 cfs. On 50,000 acres subject to Modesto's high water table in a 17-year period the drainage-pump cost per acre was $12.24, counting $4.38 for construction, maintenance, and operation, and $7.86 for power cost. This is a third more than the per-acre cost for gravity drains. During the period in which the district operated the pumps, a total of 602,000 acre-feet of water was pumped and about 75 percent of that water was utilized for irrigation. At the rate of $1.36 per acre-foot, the 1940 evaluation of water in the Modesto district, the pumped irrigation water had a value of $612,050, entirely offsetting all drainage-pumping costs. The Modesto experience leads to the conclusion that the operation of deep-well pumps is not only a most satisfactory method of sub-surface drainage but also a self-liquidating method when physical conditions are favorable.

In the Salt River Valley, Arizona, irrigation was greatly advanced in 1911 by completion of the Bureau of Reclamation Roosevelt dam and reservoir. Drainage did not become a problem there until about 1918; then the Water Users Association decided to pump ground water. The association pumped 50,000 acre-feet in 1920 and the same amount in 1921. In 1922, it increased the drain to 100,000 acre-feet, and the menacing water table began to go down. Since then, the volumes of water pumped annually have increased, and the water-table depth has also greatly increased, thus solving the drainage problem. In 1960 the association operated 250 pumped wells, drawing 500,000 acre-feet of water, nearly one-half of its irrigation water supply, and the average depth of the water table was greater than 170 feet.

17.17 RE-USE OF DRAINAGE WATER

With the increasing shortage of water, drainage designs should include provision for maximum re-use of drainage water for irrigation purposes. The practice by Modesto Irrigation District and Salt River Valley Water Users, discussed in the preceding paragraphs, shows what can be done. Many other irrigated areas having drainage problems could do likewise. Others may not be able to drain by pumping, but drainage water could be combined with surface water to advantage. Public and company officials have an obligation to see that drainage water is re-used to its fullest potential.

REFERENCES

Aronovici, V. S. and W. W. Donnan, "Soil-Permeability as a Criterion for Drainage-Design" (reprint), *Trans. Am. Geophys. Union,* pp. 95–102, No. 1, February 1946.

Donnan, W. W., "Model Tests of a Tile-Spacing Formula," *Proc. Soil Sci. Soc. Am.,* Vol. 11, pp. 131–136, 1946.

Donnan, W. W., George B. Bradshaw, and Harry F. Blaney, "Drainage Investigation in Imperial Valley, California, 1941–51 (A 10-year summary)," *U.S.D.A. Soil Conservation Service Tech. Pub.* 120, 1954.

Gardner, Willard and O. W. Israelsen, "Design of Drainage Wells," *Utah Eng. Exp. Sta. Bul.* 1, 1940.

Jessup, L. T., "Drainage of Irrigated Land by Pumping from Wells often Advisable," *U.S.D.A.* Yearbook of Agriculture, pp. 229–231, 1930.

Luthin, James N. (editor), "Drainage of Agricultural Lands," *Am. Soc. of Agron.,* Madison, Wisconsin, 1957.

Maughan, J. Howard, Orson W. Israelsen, and Eldon G. Hansen, "Drainage Districts in Utah—their Activities and Needs," *Utah Agr. Exp. Sta. Bul.* 333, January 1949.

Schwab, Glen O., et al., "Elementary Soil and Water Engineering," John Wiley and Sons, New York, 1957.

Sutton, John G., "How to Plan a Tile Drainage System," U.S.D.A., Soil Conservation Service, 17 pp. Fig. 10, 1948.

"Water," *U.S.D.A.* Yearbook of Agriculture, pp. 478–564, 1957.

CHAPTER 18 LEGAL AND ADMINISTRATIVE ASPECTS OF IRRIGATION AND DRAINAGE[1]

Since legal and administrative aspects are so diverse, it is impractical to present more than a brief summary of the important principles. To illustrate these principles, legal and administrative patterns in the United States and irrigation practices in the arid West will be used. The individual interested in irrigation should become intimately acquainted with legal and administrative structures used in his own area and then do his utmost to improve the existing patterns.

18.1 IRRIGATION AND DRAINAGE ENTERPRISES

The type of organization formed to deliver irrigation water and/or to drain land is vitally important. The legal powers and the relation between the organization and the water users are governed by the kind of organization. Good irrigation projects have failed because the wrong organization was adopted. The most successful enterprises have been those owned and operated by the water users. When the users are involved directly, their interest is keen and their services are rendered at a nominal cost. Most private associations formed to make a profit through the conveyance and delivery of irrigation water have become insolvent, or have been replaced by cooperative organizations formed by the water users. Most successful irrigation and drainage enterprises are operating on a non-profit basis. Three types of organizations are used: private, quasi-public, and public.

Private enterprises include individual projects, and mutual and commercial company organizations.

Quasi-public enterprises operate under public laws which prescribe

[1] In the previous edition, this chapter on "Social and Administrative Aspects of Irrigation" was written by J. Howard Maughn, Assistant to Dean, Division of Agricultural Science, Utah State University.

procedure and public agencies participate in the organization and management without assuming direct financial responsibility.

Public irrigation enterprises are organized under public laws, administered by public agencies, and financed with public funds.

18.2 PRIVATE ENTERPRISES

Commercial Companies

The irrigation enterprise which supplies water for compensation to irrigators who have no direct financial interest in the irrigation works, or who hold an equity which has not yet ripened into ownership and control, is designated a commercial irrigation company. Some commercial companies furnish water on an annual rental basis; others sell the prospective irrigator a water right and in addition charge an annual rental; and some sell a water right which carries with it a perpetual interest in the irrigation system. Commercial companies of the last type ultimately become mutual companies in which the irrigation works are owned and operated by the irrigators. Service rates of companies furnishing water on an annual rental basis are generally subject to public regulation as a result of dedication of the water to public use. The annual rentals charged by companies which sell water rights are not subject to public regulation if the contracts for sale of rights and charging of rentals are held to be private contracts.

Individual and Partnership Enterprises

Comparatively few large, but many small, irrigation projects are built by individuals working alone or in partnerships. Small streams closely adjacent to arable land in isolated sections favor individual effort in irrigation. Also, where ground water is available for pumping, or where other water sources may be best developed by small pumping plants, individuals build and operate their own irrigation projects. The advantages claimed for individual enterprises are that they permit the farmer to irrigate at any time he desires, so that he can regulate his own practices, and that he is independent of the assessments, rules, regulations, and irrigation practices of his neighbors.

Individual irrigation activity is usually more expensive than the combined activity of groups who need irrigation water, and, moreover, it is rigorously restricted by nature since it is quite impossible, as a rule, for the farmer to build the storage and diversion works, and conveyance canals necessary to provide water for lands which are at great distances from the sources of water supply.

Cooperative Enterprises

There are water resources and arable lands which can be brought together by small groups of individuals voluntarily forming an association for the purpose of constructing and operating irrigation systems. Cooperative irrigation enterprise is locally designated by a variety of names. All such enterprises are included in two main types of organization, incorporated and unincorporated, termed herein, respectively, mutual irrigation companies and mutual associations. The incorporated group units are far more numerous and extensive.

The success of an unincorporated mutual association, sometimes designated as a mutual company, rests largely on the fairness and congeniality of each member, because the association provides no means of legally enforcing the payments of dues or of enforcing contributions to the maintenance, betterment, renewal, and operation or expansion of the project. The major asset of the association is the labor of its members. Its activities are limited to small projects which require no difficult construction and but little capital.

Mutual Irrigation Company

A corporate body of irrigators, voluntarily organized for the purpose of supplying water to its stockholders, is known as a mutual company. It is a non-profit organization for the delivery of water to its members only. It obtains its revenues by stock assessments, and its dividends consist of water delivered in proportion to the stock owned by each irrigator. It enforces payment of assessments by the sale of stock if necessary. The stockholders delegate the responsibility of management to a board of directors, from three to seven or more, elected by ballot. Each stockholder has as many votes as he owns shares of stock. The tenure of office of directors, fixed by the articles of incorporation, ranges from 1 to 3 or more years. The directors elect one of their members as president and appoint a secretary, treasurer, and watermaster, any of whom may or may not be directors. In the small irrigation companies the watermaster has charge of the project operation and maintenance, including the distribution of water to stockholders. The watermaster, with the aid of check gates, takeout gates, and diversion structures, is expected to distribute equitably a valuable commodity to numerous claimants.

The larger mutual companies sometimes employ an engineer-manager who is assigned the responsibility of water distribution and to whom the watermasters, one for each of several districts, are instructed to report. The mutual irrigation company has wide flexibility.

It is especially suited to maintenance and operation of irrigation projects and is the dominant type of operating organization in many of the western United States irrigated areas.

Mutual companies are exempt from general taxation as long as they are used for the service of their own stockholders only. Some States relieve mutual irrigation companies from license taxes assessed against corporations.

18.3 QUASI-PUBLIC ENTERPRISES

Irrigation Districts

An irrigation district is a quasi-public corporation for providing water for lands within its boundaries. The fundamentals attributed to an irrigation district are authoritatively given by Hutchins as follows:

It is a public corporation, a political subdivision of a State with defined geographical boundaries. It is created under authority of the State legislature through designated public officials or courts at the instance and with the consent of a designated fraction of the landowners or of the citizens, as the case may be, of the particular territory involved. Being public and political, the formation of a district is not dependent upon the consent of all persons concerned, but may be brought against the wishes of the minority. In this respect the district differs fundamentally from the voluntary mutual company and the commercial irrigation company.

It is a cooperative undertaking, a self-governing institution, managed and operated by the landowners or citizens within the district. Supervision by State officials is provided for to the extent of seeing that the laws are enforced, and in most States is extended in greater or less degree over organization, plans, and estimates prior to bond issues, and construction of works.

It may issue bonds for the construction or acquisition of irrigation works, which bonds are payable from the proceeds of assessments levied upon the land.

Hence, it has the taxing power. Each assessment becomes a lien upon the land. While the ultimate source of revenue, therefore, is the assessment, an additional source frequently provided for is the toll charged for water. Other revenue may in some cases be obtained from the sale or rental of water or power to lands or persons outside the district.

Finally, the purpose of the irrigation district is to obtain a water supply and to distribute the water for the irrigation of lands within the district. Additional authority is granted irrigation districts, almost without exception, to provide for drainage. In some States districts may also develop electric power. These additional powers, however, are subsidiary and are intended to make more effective the principal function of the organization, which is to provide irrigation water.

During the sixty-odd years of its history the irrigation district has become an increasingly important irrigation agency. It is adapted to large-scale irrigation enterprise. Many of the larger irrigation systems of the West are managed by this type of organization.

18.4 PUBLIC ENTERPRISES

United States Reclamation Projects

The enactment of the Reclamation Law in 1902 was noteworthy in first providing direct use of federal funds without interest for construction of large irrigation projects. The Bureau of Reclamation has made an outstanding contribution in the design and building of engineering structures of large magnitude.

Ultimately all federal irrigation projects will be owned and operated by the irrigators as some now are. Either a mutual irrigation company or an irrigation district is created by the irrigators when they assume control; and through this organization they conduct their affairs with the government.

The salient features of projects constructed by the United States under the Reclamation Law are:

(a) The settler has the use of public non-interest-bearing money and normally a period of 40 years in which to repay construction costs.

(b) Public lands for which the development of a water supply was so costly as to be in general unattractive to private capital were included in many federal projects.

(c) Until a substantial part of the construction charges are paid the project is under complete control of the Bureau of Reclamation.

(d) Annual payments of construction charges and operation and maintenance costs are fixed by the Bureau and paid by the settler to its representatives.

(e) The construction costs of multiple-purpose projects are prorated according to the benefits to irrigation, municipal water development, power development, flood control, navigation, and other purposes.

The essentials of success, as on irrigation districts and other irrigation enterprises, have proved to be productive land, sufficient water, reasonable construction costs, and adequate land settlement.

18.5 LEGAL ASPECTS OF IRRIGATION AND DRAINAGE

The legal as well as the administrative structure under which irrigation and drainage is practiced will often insure the success or seriously retard, if not destroy, the undertaking. Therefore the discussion of technical aspects alone is not sufficient. The student of irrigation and drainage must understand and appreciate the importance of legal and administrative aspects. The legal aspects focus upon two conflicting doctrines, riparian rights and appropriation.

Doctrine of Riparian Rights

Settlers along streams in humid regions used the water as needed under the common-law doctrine of riparian rights, each owner along the stream being entitled to have the water flow in its natural channel "undiminished in quantity and unpolluted in quality." Use does not create nor does disuse destroy the riparian right.

Since the doctrine of riparian rights does not contemplate or allow for consumptive use of water or pollution, changes in concept in arid regions were inevitable. Thus in some areas modification of doctrine has been made to allow each proprietor to divert water for use on his adjoining lands for "domestic" and "natural purposes." In some areas diversion for these purposes is permitted even though it may consume the water previously used by a lower riparian owner. Thus no right to the water is secured by use.

Use of water for irrigation is considered as "artificial" and is not normally allowed. Modifications have been made which result in the "natural flow" doctrine which permits a riparian owner to divert water for irrigation only if there is a surplus to the needs of lower riparian owners, and if such diversion does not appreciably lower the water level or quality. A "reasonable use" doctrine permits a more liberal removal of water for irrigation.

Under any of the riparian doctrines, beneficial use does not create priority. Usually land must be adjacent to the stream, and diversion as well as return flow must be made upon the same land to be irrigated.

Riparian and modifications of the riparian doctrine are used in the more humid regions. However, the increasing use of water for irrigation in humid areas is creating serious water-right problems which can only be solved by a modified doctrine approaching the doctrine of appropriation.

Doctrine of Appropriation

When irrigation was begun in the arid West of the United States, the doctrine of riparian rights was readily seen to be inadequate. Agricultural and community developments depended upon a reliable supply of water. Frequently the existing waters were used fully. Development, therefore, depended upon establishing a right to the water by prior and by beneficial use. Out of this need came the doctrine of appropriation which asserts that all water rights are founded upon priority of use: that use creates the right, and that disuse destroys or forfeits the right. Beneficial use is declared "the basis, the measure, and the limit of the right."

18.6 GROUND-WATER LAWS

Laws governing surface water are generally applied in their essence to ground water. However, because of the difficulty of defining ground water, considerable confusion exists. The courts have attempted to distinguish between different underground streams; "underflow of surface streams," "percolating waters," "artesian water," etc. Since these distinctions cannot be supported by sound physical facts, much confusion has resulted.

Vigorous attempts are being made to establish good, sound laws which will permit maximum utilization of underground water reservoirs and resources.

The common law doctrine of riparian rights is still used in most of the states of the U.S. However, at least fifteen states have some form of ground-water laws based on the doctrine of appropriation, and about six apply the reasonable-use concept to the riparian doctrine.

18.7 DRAINAGE ENTERPRISES

Group action is essential in the drainage of irrigated land. For one landowner to be able to provide adequate drainage for his land without cooperating with his neighbors is the exception rather than the rule. Group action in some areas is obtained by organizing drainage districts which are quasi-public corporations provided for by state laws and which have authority to tax irrigated land for drainage purposes.

Two principal purposes influence the organization of drainage districts, namely: (1) to consolidate into one drainage agency the lands of an area in need of drainage and contributing to that need, and (2) to provide the authority and the procedure, and to assign the responsibility for the design, financing, construction, and maintenance of drainage systems.

Noteworthy powers of a drainage district are: (1) the power to include in the district all lands to be benefited by the drainage system, and thus assure strength of the district and ability to promote equity in its dealings with landowners; (2) the power to tax lands of the district and enforce tax collections. The second power carries the authority to foreclose on tax-delinquent land and sell it if necessary.

Drainage districts have legal authority to carry out all the functions and activities pertaining to drainage of farm land including financing, design, and installation of drains, and their operation and maintenance.

There must be a need for drainage in the area proposed for the district, and it must be shown that the benefits to the included lands will exceed the costs; the desires of a majority of the landowners to participate must be expressed; the specified organizing procedure must be carried out. Land ownership within a drainage district is the usual requirement for membership and participation in district activities.

After a district has been organized, its taxing procedure set in operation, its capital financing provided, and its drainage system installed, its problems concern largely management, operation, and maintenance, and the discharging of financial obligations.

In Utah, a board of three supervisors constitutes the governing body of the drainage district and carries the responsibility for all the district affairs. Supervisors are appointed by the county commissioners, usually upon nomination by the district landowners. The supervisors elect from their number a president, secretary, and treasurer. They must ascertain the needs for drainage in the area and determine the nature, extent, and probable cost of a drainage system designed to meet those needs. It is also their responsibility to provide capital financing for the installation of the drainage system, usually involving the issue and sale of district bonds, and to set up an equitable taxing system to provide adequate and dependable finances to meet the annual revenue requirements. The ability and willingness of supervisors to serve the district may determine the success or failure of the enterprise.

The large irrigation enterprises, irrigation districts, and mutual water users' associations are developing more and more interest in the drainage of irrigated lands. The trend is toward combining responsibilities for irrigation and drainage systems into one organization, particularly where legal authority is adequate for an irrigation company to assume responsibility for irrigation and drainage.

REFERENCES

Hayden, Senator Carl, *National Irrigation Policy—Its Development and Significance,* U.S. Senate Document 36 of the 76th Congress, 1939.

Hibbard, Benjamin Horace, "A History of the Public Land Policies," New York, Peter Smith, 1939.

Hutchins, Wells A., "Selected Problems in the Law of Water Rights in the West," *U.S.D.A. Misc. Pub.* 418, 1942.

Israelsen, O. W., J. Howard Maughan, and George P. South, "Irrigation Companies in Utah, their Activities and Needs," *Utah Agr. Exp. Sta. Bul.* 322, March 1946.

Ryker, Rodney, "A State Ground Water Code," *Agr. Eng.,* Vol. 27, p. 571, December 1946.

METRIC CONVERSIONS,
EQUATIONS, FIGURES,
AND TABLES

AMERICAN-ENGLISH-METRIC CONVERSION CONSTANTS

TABLES

American-English-Metric Conversion Constants

Length and Distance	Inches	Feet	Yards	Statute Miles	Nautical Miles	Centimeters	Meters	Kilometers
one inch equals	1	.083	.027	—	—	2.54	—	—
one foot	12	1	.333	—	—	30.48	.305	—
one yard	36	3.0	1	—	—	91.44	.914	—
1 statute mile	—	5280	1760	1	.865	—	1609	1.61
1 nautical mile	—	6080	2027	1.15	1	—	1853	1.85
1 centimeter	.394	.0328	.0109	—	—	1	.01	—
1 meter	39.37	3.281	1.094	—	—	100	1	.001
1 kilometer	—	3281	1094	.6214	.5396	—	1000	1

USEFUL APPROXIMATIONS: 5 cm. = 2 in.; 10 cm. = 4 in.; 30 cm. = 1 ft.; 10 meters = 33 ft.; 10 km. = 6 miles; 16 km. = 10 miles; 1 fathom = 6 ft.; 120 fathoms = 1 cable length.

Areas	Square Inches	Square Feet	Square Yards	Acres	Square Centimeters	Square Meters	Hectares
1 sq. inch	1	.007	—	—	6.45	.00064	—
1 sq. ft.	144	1	.1111	—	—	.0929	—
1 sq. yd.	1296	9	1	—	—	.8361	—
1 acre	—	43560	4840	1	—	4050	.405
1 sq. cm.	.155	—	—	—	1	.0001	—
1 sq. m.	1550	10.76	1.20	—	10.000	1	.0001
1 hectare	—	107.600	11.955	2.47	—	10.000	1

USEFUL APPROXIMATIONS: 1 hectare = 2½ acres; 1 sq. in. = 6½ sq. cm.; 1 sq. m. = 11 sq. ft.; a hectare is 330 feet on each side.

```
Centimeters
0   1   2   3   4   5   6   7   8   9   10   11   12   13
|   |   |   |   |   |   |   |   |   |   |    |    |    |
|     |       |       |       |       |
0     1       2       3       4       5
              Inches
```

Volume and Capacity	Cubic Inches	Cubic Feet	American Gallons	Imperial Gallons	Cubic Meters	Liters	Quarts
1 cu. inch	1	.00057	.0043	.0035	—	.0163	.0173
1 cu. foot	1728	1	7.48	6.23	.0283	28.3	29.9
1 Am. gallon	231	.134	1	.833	.0038	3.78	4
1 Imp. gallon	277.4	.161	1.201	1	.0045	4.55	4.8
1 cu. meter	61030	35.3	264.2	220	1	1000	1057
1 liter	61.03	.0353	.264	.220	.001	1	1.057

USEFUL APPROXIMATIONS: 16.4 cc. = 1 cu. in.; 35 cu. ft. in a cu. m.; 1 register ton = 100 cu. ft.; 1 U.S. shipping ton = 40 cu. ft. or 32.14 U.S. bushels or 31.14 Imp. bu.

Weights	Grains	Grams	Ounces (avoir.)	Pounds (avoir.)	Kilos	Short Ton	Long Ton	Metric Ton
1 grain	1	.0648	.0023	.00014	—	—	—	—
1 gram	15.4	1	.035	—	—	—	—	—
1 ounce	437.5	28.35	1	.0625	.0283	—	—	—
1 pound	7000.—	454.—	16	1	.4536			
1 kilogram	15432.	1000.	35.27	2.205	1			
1 short ton	—	—	—	2000.—	907.	1	.893	.907
1 long ton	—	—	—	2240	1016	1.12	1	1.016
1 metric ton	—	—	—	2205	1000	1.102	.984	1

AIR PRESSURE: In terms of atmospheres—One atmosphere equals 14.7 pounds per sq. in. or 1.033 kilo per square centimeter. Normal tire pressure is 2 atmospheres = 29 lbs./sq. in.

	Temperature			Weight				Volume	
	To convert Centigrade into Fahrenheit multiply by 1.8 and add 32		One kilo equals 2.205 lbs. One Pound " .4536 kg. One Stone " 14.00 lbs.				A liter is slightly more than a quart		
Réaumur	Centi-grade	Fahren-heit	Kilo-grams	Pounds	Stone	Liters		Am. Gals.	Imp. Gals.
−12	−15	5	.45	1	.07	1		.26	.22
−8	−10	14	.91	2	.14	5		1.32	1.10
−4	−5	23	1.00	2.205	.16	10		2.64	2.20
0	0	32	1.36	3	.21	15		3.96	3.30
4	5	41	1.82	4	.29	20		5.28	4.40
8	10	50	2.00	4.410	.32	25		6.60	5.50
12	15	59	2.27	5.00	.36	30		7.93	6.60
16	20	68	3.00	6.63	.47	35		9.25	7.70
20	25	77	4.00	8.82	.63	40		10.50	8.80
24	30	86	4.54	10.—	.71	45		11.90	9.90
25.6	32	89.6	5.00	11.02	.79	50		13.20	11.00
27.2	34	93.2	6.34	14.00	1.00	55		14.00	12.10
28.8	36	96.8	7	15.50	1.10	60		15.80	13.20
29.6	37	98.6	8	12.70	1.27	65		17.20	14.30
30.4	38	100.4	9	19.90	1.43	70		18.50	15.40
31.2	39	102.2	10	22.05	1.58	75		19.80	16.50
32	40	104.0	45.36	100	7.14	80		21.10	17.60
32.8	41	105.8	68.06	150	10.71	85		22.00	18.70
33.6	42	107.6	453.60	1000	71.4	90		23.70	19.80
80	100	212.0	1000	2205	158.00	100		26.40	22.00

EQUATIONS IN METRIC UNITS

Horsepower

$$\text{WHP} = \frac{Qh}{273} \tag{4.3}$$

where WHP = water horsepower
Q = discharge in cubic meters per hour
h = vertical lift in meters

or

$$\text{WHP} = \frac{Qh}{76}$$

where WHP = water horsepower
Q = discharge in liters per second
h = vertical lift in meters

Specific Speed

$$N_s = 0.86 \frac{\text{RPM}\sqrt{Q}}{H^{3/4}} \tag{4.6}$$

where N_s = specific speed
Q = discharge in cubic meters per hour
H = vertical lift in meters

Chezy Equation

$$V = \frac{C}{\sqrt{3.28}} \sqrt{RS} \tag{5.7}$$

where V = velocity in meters per second
C = coefficient of roughness, as tabulated in English units
R = hydraulic radius in meters
S = slope of water surface or piezometric-head line

Manning Equation

$$V = \frac{1.00}{n} R^{2/3} S^{1/2} \tag{5.8}$$

where V = velocity in meters per second
n = Horton's values of n as tabulated in Table 5.2
R = hydraulic radius in meters
S = slope of water surface or piezometric-head line

Submerged Orifice

$$Q = 0.61 \times 10^{-3} A \sqrt{2gh} \tag{6.5}$$

where Q = discharge in liters per second
A = area of orifice in square centimeters
g = acceleration—981 centimeters per second
h = head in centimeters

Suppressed Weir (Francis Formula)

$$Q = 0.0184 L H^{3/2} \tag{6.9}$$

where Q = discharge in liters per second
L = length of weir crest in centimeters
H = total head in centimeters

Trapezoidal Weir

$$Q = 0.0186 L H^{3/2} \tag{6.11}$$

where Q = discharge in liters per second
H = total head in centimeters
L = length of weir crest in centimeters

90° Triangular Weir

$$Q = 0.0138 H^{5/2} \tag{6.12}$$

where Q = discharge in liters per second
H = total head in meters

Water Applied

$$cqt = ad \tag{7.9}$$

where c = constant depending upon units of discharge q
q = the size of stream
t = time in hours required to irrigate the area
a = area irrigated in hectares
d = depth in centimeters that the volume of water used would cover the land irrigated if quickly spread uniformly over its surface

When q = flow in cubic meters   When q = flow in liters
per hour   per second

then $c = 100$ then $c = 27.8$

Time of Application

$$t = \frac{P_w A_s D a}{100cq} \tag{7.10}$$

where t = time of application in hours
   P_w = moisture percentage, dry-weight basis
   A_s = apparent specific gravity of soil
   D = depth of soil in centimeters
   a = area irrigated in hectares
   q = size of stream in cubic meters per hour
   c = constant depending upon units of discharge q

When q = flow in cubic meters   When q = flow in liters
   per hour   per second

then $c = 100$ then $c = 27.8$

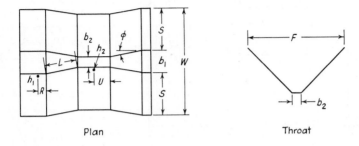

Plan   Throat

Profile   End view

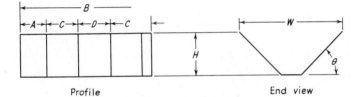

Flume no.	Description	b_1	b_2	A	L	C	D	E	F	H	B	R	S	U	W	θ	ϕ
1	Large 60°-v	5.1	0	17.8	17.8	17.6	17.8	7.6	20.0	17.1	78.4	3.8	10.2	8.9	25.4	60°	8.25°
2	Small 60°-v	5.1	0	12.7	10.5	10.3	12.7	5.1	12.1	10.2	51.4	2.5	6.0	6.4	17.2	60°	11.20°
3	5.1 cm-60° wsc	12.4	5.1	20.3	21.6	21.4	21.6	7.6	35.6	34.3	92.2	3.8	15.2	10.8	42.9	60°	9.22°
4	5.1 cm-45° wsc	12.4	5.1	20.3	21.6	21.3	21.6	7.6	56.4	26.9	92.1	3.8	26.9	10.8	66.2	45°	9.91°
5	5.1 cm-30° wsc	12.4	5.1	20.3	21.6	21.3	21.6	7.6	93.1	25.4	92.1	3.8	44.0	10.8	100.4	30°	9.80°
6	10.2 cm-60° wsc	20.3	10.2	22.9	25.4	23.9	25.4	7.6	50.8	35.2	105.7	3.8	20.3	12.7	61.0	60°	11.20°
7	5.1 cm-30° csu	25.4	5.1	25.4	27.3	25.4	25.4	7.6	90.1	24.5	109.2	3.8	42.5	12.7	110.4	30°	21.80°

Note: Dimensions in centimeters except as shown. See Fig. 6-17 for the relationship of Q to depth in each of the 7 flumes.

Fig. 6.16 Details of designs of trapezoidal flumes.

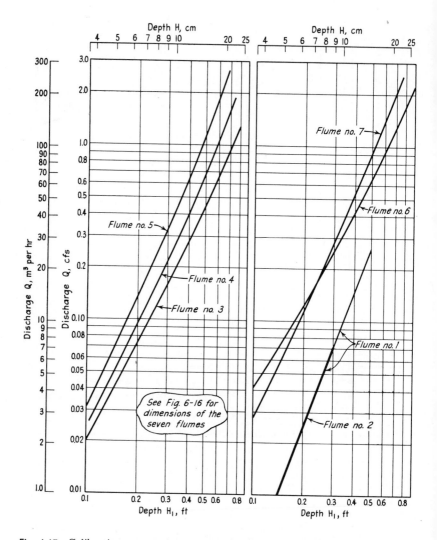

Fig. 6.17 Calibration curves for trapezoidal flumes under free-flow conditions. (Directions for use are same as for Fig. 12.8.)

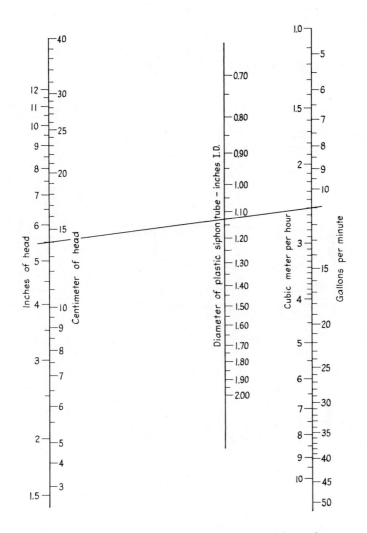

Fig. 6.24 Discharge related to head for plastic siphon tubes.

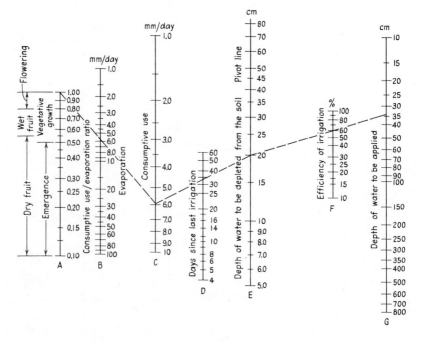

Fig. 12.7 Calculating depth of water to be applied per irrigation starting with the stage of growth of the crop.

To use: 1. Select appropriate values on scales A, B, D and F.
2. Place a ruler _from_ the point on scale A, _through_ the point on scale B, to the first pivot line, scale C.
3. Place the point of a sharp pencil against the ruler on the pivot line, slide the rule on the pencil _through_ the point on scale D _to_ the second pivot line, scale E.
4. Repeat step 3, pivoting on E, passing _through_ the point on F _to_ G.
5. Read the answer on G.

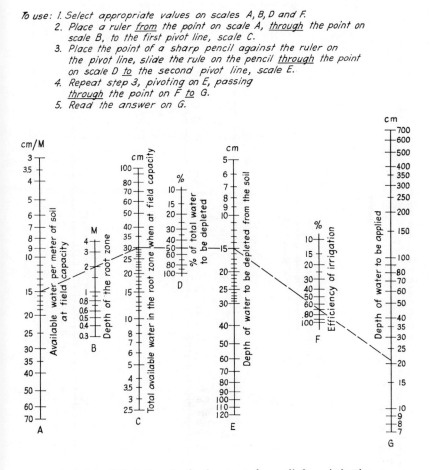

Fig. 12.8 Calculating depth of water to be applied per irrigation.

To use: /. Select appropriate values on scales A, B, and D
          2. Lay a ruler from the point on scale A *through* the point on scale B.
             to the pivot line.
          3. Place the point of a sharp pencil against the ruler *on* the pivot line
          4. Slide the ruler on the pencil *to* the point on scale D
          5. The answer appears where the ruler intersects scale E.

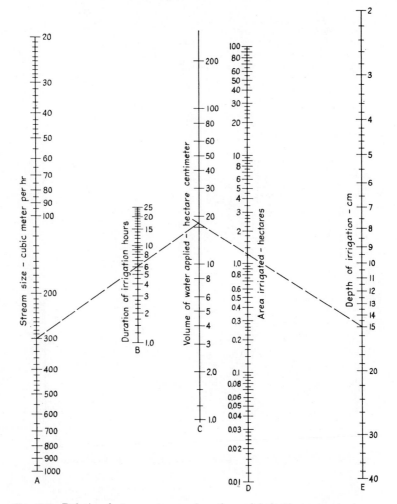

Fig. 12.9 Relation between stream size, time of irrigation, area to be covered, and depth of water to be applied, $qt = ad$.

TABLE 4.1

HORSEPOWER REQUIRED TO LIFT DIFFERENT QUANTITIES OF WATER
TO ELEVATIONS OF 10 TO 90 METERS

(Efficiency of pumping plant 50 percent of theoretical.
Use for estimating only.)

Discharge		Horsepower Required for Elevations of								
Cubic Meters per Hour	Liters per Second	10 meters	20 meters	30 meters	40 meters	50 meters	60 meters	70 meters	80 meters	90 meters
10	2.78	0.7	1.5	2.2	2.9	3.7	4.4	5.1	5.9	6.6
20	5.56	1.5	2.9	4.4	5.9	7.3	8.8	10.3	11.7	13.2
30	8.34	2.2	4.4	6.6	8.8	11.0	13.2	15.4	17.6	19.8
40	11.1	2.9	5.9	8.8	11.7	14.7	17.6	20.5	23.4	26.4
50	13.9	3.7	7.3	11.0	14.7	18.3	22.0	25.6	29.3	33.0
60	16.7	4.4	8.8	13.2	17.6	22.0	26.4	30.8	35.2	39.6
70	19.5	5.1	10.3	15.4	20.5	25.6	30.8	35.9	41.0	46.2
80	22.2	5.9	11.7	17.6	23.4	29.3	35.2	41.0	46.9	52.7
90	25.0	6.6	13.2	19.8	26.4	33.0	39.6	46.2	52.4	59.3
100	27.8	7.3	14.7	22.0	29.3	36.6	44.0	51.3	58.6	65.9
125	34.8	9.2	18.3	27.5	36.6	45.8	54.9	64.1	73.3	82.4
150	41.7	11.0	22.0	33.0	44.0	54.9	65.9	76.9	87.9	98.9
175	48.6	12.8	25.6	38.5	51.3	64.1	76.9	89.7	102.6	115.4
200	55.6	14.7	29.3	44.0	58.6	73.3	87.9	102.6	117.2	131.9
250	69.5	18.3	36.6	54.9	73.3	91.6	109.9	128.2	146.5	164.8
300	83.4	22.0	44.0	65.9	87.9	109.9	139.9	153.8	175.8	197.8

Computed from the formula:

$$HP = \frac{QH}{273n}$$

TABLE 4.2

LOSS OF HEAD IN METERS DUE TO FRICTION PER 1000 METERS OF
15-YEAR-OLD ORDINARY STEEL PIPE

(BASED ON ORIGINAL DATA BY WILLIAMS AND HAZEN)

$$V = CR^{.63}S^{.54} \text{ where } c = 100$$

Rated Flow in Cubic Meters per Hour	Liters per Second	Nominal Diameter of Pipe									
		¾" (1.90 cm)	1" (2.54 cm)	2" (5.08 cm)	2½" (6.35 cm)	3" (7.62 cm)	4" (10.61 cm)	5" (12.70 cm)	6" (15.24 cm)	8" (20.32 cm)	10" (25.40 cm)
.25	0.069	10	2.4								
.5	0.139	36	9								
.75	0.208	82	18								
1.00	0.278	145	33	1.2	0.4						
1.5	0.417	300	68	2.5	0.8						
2	0.556		110	4.0	1.5	0.6	0.1				
3	0.834			8.9	3.0	1.2	0.3				
4	1.11			15	5.0	2.0	0.5	0.2			
5	1.39			22	7.5	3.0	0.7	0.3			
6	1.67			31	10	4.2	1.0	0.4			
8	1.95			54	18	7.3	1.8	0.6	0.3		
10	2.78			80	27	11	2.7	0.9	0.4		
12	3.34			110	32	15	3.8	1.1	0.5		
14	3.89			150	38	20	5.0	1.3	0.7	0.2	
16	4.45			190	65	26	6.5	2.2	0.9	0.2	
18	5.00			240	80	32	8.0	2.7	1.1	0.3	
20	5.56			265	95	38	9.0	3.2	1.3	0.3	
25	6.95			430	150	60	14	5.0	2.0	0.5	0.2
30	8.34					80	20	7.0	2.8	0.7	0.2
35	9.73					108	27	9.5	3.7	1.0	0.3
40	11.1						34	12	4.8	1.2	0.4
45	12.5						43	14	5.8	1.5	0.5
50	13.9						50	18	7.2	1.8	0.6
55	15.3						60	22	8.8	2.4	0.8
60	16.7						70	25	10	2.6	0.8
65	18.1						84	30	12	3.0	1.0
70	19.5						98	34	13	3.4	1.2
80	22.2							42	17	4.3	1.5
90	25.0							54	22	5.5	1.9
100	27.8							64	26	6.6	2.3
110	30.6							76	30	8.0	2.7
120	33.4							90	36	9.4	3.0
130	36.2							103	42	11	3.7
140	38.9							120	49	12	4.2
150	41.7							135	55	14	4.8
160	44.5								64	16	5.4
170	47.2								70	18	6.0
180	50.0								80	20	6.6
190	52.8								88	22	7.5
200	55.6								95	24	8.0
225	62.5									30	10
250	69.5									34	12
275	76.5									43	15
300	83.4									50	17
350	97.3									68	22
400	101									88	29
450	125									110	36
500	139									130	44
600	167										60
700	195										80
800	222										103
900	250										130

TABLE 4.4

Performance Standards for New Deep-Well Pumping Plants
(Computed from Schleusener et al.)

Energy Source	Rated Load hp-hr/liter for Representative Power Units	Performance Standard in whp-hr/liter[1]
Diesel	3.90[2]	2.92
Gasoline	2.98[2]	2.24
Tractor fuel	2.92[2]	2.20
Propane	2.36[2]	1.76
Natural gas	28.92[3] per 1000 liters	21.68 per 1000 liters
Electric	88 percent efficient	0.885[4] per kw-hr

[1] Based on pump efficiency of 75 percent.

[2] Taken from Test D of Nebraska Tractor Test Reports. Drive loss for flat belt is included.

[3] Manufacturer's data corrected for 5 percent drive loss.

[4] Not corrected for drive loss. Assume a direct connection.

TABLE 4.5

Maximum Fuel Requirements for a Good Pumping Plant[1]
(Computed from Schleusener et al.)

Flow of Water in Cubic Meters per Hour	Lift in Meters[2]	Water Horse-power	Fuel Required in Liters per Hour				
			Pro-pane	Die-sel	Gasoline or Tractor Fuel	Natural Gas	Elec-tricity[3]
	20	7.5	4.2	2.7	3.5	350	8.5
100	50	18.5	10.5	6.2	8.5	860	21
	70	26.0	14.7	9.0	11.7	1200	29
	20	11	6.2	3.7	5.2	510	12.5
150	50	28	15.7	9.5	13.0	1290	32
	70	39	22.0	13.5	18.2	1800	44
	20	15	8.5	5.2	6.7	690	17
200	50	37	21.0	12.5	16.5	1710	42
	70	52	29.5	17.7	23.5	2400	59
	20	19	10.7	6.5	8.5	880	22
250	50	46.5	26.5	16.0	21.0	2150	53
	70	65	36.7	22.2	20.2	3000	73

[1] Based on standards in Table 4.4.

[2] If you have a pressure at the pump discharge, add the number of meters shown in Table 4.6 to the distance the water must be lifted while pumping.

[3] Kilowatt-hours per hour.

TABLE 4.6

RELATIONSHIP BETWEEN DISCHARGE PRESSURE AND LIFT
(COMPUTED FROM SCHLEUSENER ET AL.)

Discharge Pressure in Atmospheres	Equivalent "Lift" in Meters
1	10.3
2	20.7
3	31.0
4	41.4
5	51.7
6	62.0
7	72.4

TABLE 5.1

VALUE OF ROUGHNESS e FOR VARIOUS CONDUIT MATERIALS
(COMPUTED FROM ALBERTSON, BARTON, SIMONS)

	Relative Roughness (e in centimeters)
Glass, drawn brass, copper, and lead	Smooth–0.00015
Commercial steel or wrought iron	0.005
Asphalted cast iron	0.012
Galvanized iron	0.015
Cast iron	0.026
Wood stave	0.02–0.09
Cement-lined steel	0.04
Concrete	0.03–0.3
Riveted steel	0.09–0.9
Corrugated metal pipe	3.0–6.0
Large tunnels, concrete or steel lined	0.06–0.12
Blasted-rock tunnels	30–60

TABLE 5.3

LIMITING VELOCITIES FOR ESSENTIALLY STRAIGHT CANALS AFTER AGING
(COMPUTED FROM FORTIER AND SCOBEY)

Material	Value of Manning's n	Velocity in Meters per Second	
		Clear Water	Water Transporting Colloidal Silts
Fine sand, colloidal	0.020	0.4	0.8
Sandy loam, non-colloidal	0.020	0.5	0.8
Silt loam, non-colloidal	0.020	0.6	0.9
Alluvial silts, non-colloidal	0.020	0.6	1.1
Ordinary firm loam	0.020	0.8	1.1
Volcanic ash	0.020	0.8	1.1
Stiff clay, very colloidal	0.025	1.1	1.5
Alluvial silts, colloidal	0.025	1.1	1.5
Shales and hardpans	0.025	1.8	1.8
Fine gravel	0.020	0.8	1.5
Graded loam to cobbles when non-colloidal	0.030	1.1	1.5
Graded silts to cobbles when colloidal	0.030	1.2	1.7
Coarse gravel, non-colloidal	0.025	1.2	1.8
Cobbles and shingles	0.035	1.5	1.7

TABLE 6.1

Discharge in Liters per Second from a Submerged Rectangular Orifice[1]

Head H Centimeters	Cross-sectional Area A of Orifice, square centimeters							
	100	250	500	1000	1250	1500	1750	200)
1.0	2.71	6.77	13.6	27.1	33.9	40.6	47.4	54.2
1.5	3.32	8.30	16.6	33.2	41.5	49.8	58.1	66.4
2.0	3.83	9.58	19.2	38.3	47.9	57.4	67.0	76.6
2.5	4.28	10.7	21.4	42.8	53.5	64.2	74.9	85 6
3.0	4.69	11.7	23.5	46.9	58.6	70.4	82.1	92.8
3.5	5.07	12.7	25.4	50.7	63.4	76.1	88.7	101
4.0	5.42	13.6	27.1	54.2	67.8	81.3	94.9	108
4.5	5.75	14.4	28.8	57.5	71.9	86.3	101	115
5.0	6.06	15.2	30.3	60.6	75.8	90.9	106	121
5.5	6.35	15.9	31.8	63.5	79.4	95.3	111	127
6.0	6.64	16.6	33.2	66.4	83.0	99.6	116	133
6.5	6.91	17.3	34.6	69.1	86.4	104	121	138
7.0	7.17	17.9	35.9	71.7	89.6	108	125	143
7.5	7.42	18.6	37.1	74.2	92.8	111	130	148
8.0	7.66	19.2	38.3	76.6	95.8	115	134	153
8.5	7.90	19.8	39.5	79.0	98.8	119	138	158
9.0	8.13	20.3	40.7	81.3	102	122	142	163
9.5	8.35	20.9	41.8	83.5	104	125	146	167
10.0	8.57	21.4	42.9	85.7	107	129	150	171
10.5	8.78	22.0	43.9	87.8	110	132	154	176
11.0	8.99	22.5	45.0	89.9	112	135	157	180
11.5	9.19	23.0	46.0	91.9	115	139	161	184
12.0	9.39	23.5	47.0	93.9	117	141	164	188
12.5	9.58	24.0	48.0	95.8	120	144	168	192
13.0	9.77	24.4	48.9	97.7	122	147	171	195
13.5	9.96	24.9	47.8	99.6	125	149	174	199
14.0	10.1	25.3	50.6	101	126	152	177	202
14.5	10.3	25.8	51.5	103	129	155	181	206
15.0	10.5	26.3	52.5	105	131	158	184	210
15.5	10.7	26.8	53.5	107	134	161	187	214
16.0	10.8	27.0	54.0	108	135	162	189	216
16.5	11.0	27.5	55.0	110	138	165	193	220
17.0	11.2	28.0	56.0	112	140	168	196	224
17.5	11.3	28.3	56.5	113	141	170	198	226
18.0	11.5	28.8	57.5	115	144	173	201	230
18.5	11.7	29.3	58.5	117	146	176	205	234
19.0	11.8	29.5	59.0	118	148	177	207	236
19.5	12.0	30.0	60.0	120	150	180	210	240
20.0	12.1	30.3	60.5	121	151	182	212	242

[1] Computed from the formula $Q = 0.61 \times 10^{-3} A \sqrt{2gh}$.

TABLE 6.2

DISCHARGE IN LITERS PER SECOND, PER METER OF LENGTH OF
WEIR CREST BY THE FRANCIS FORMULA: $Q = 0.0184LH^{3/2}$

Head in Centimeters	Discharge in Liters per Second	Head in Centimeters	Discharge in Liters per Second	Head in Centimeters	Discharge in Liters per Second
1.0	1.84	15.0	107	29.0	288
1.5	3.38	15.5	112	29.5	295
2.0	5.20	16.0	118	30.0	302
2.5	7.27	16.5	123	30.5	310
3.0	9.56	17.0	129	31.0	318
3.5	12.0	17.5	135	31.5	325
4.0	14.7	18.0	141	32.0	333
4.5	17.6	18.5	146	32.5	341
5.0	20.6	19.0	152	33.0	349
5.5	23.7	19.5	158	33.5	357
6.0	27.0	20.0	165	34.0	365
6.5	30.5	20.5	171	34.5	373
7.0	34.1	21.0	177	35.0	381
7.5	37.8	21.5	183	35.5	389
8.0	41.6	22.0	190	36.0	397
8.5	45.6	22.5	196	36.5	406
9.0	49.7	23.0	203	37.0	414
9.5	53.9	23.5	210	37.5	422
10.0	58.2	24.0	216	38.0	431
10.5	62.6	24.5	223	38.5	440
11.0	67.1	25.0	230	39.0	448
11.5	71.8	25.5	237	39.5	457
12.0	76.5	26.0	244	40.0	466
12.5	81.3	26.5	251	40.5	475
13.0	86.2	27.0	258	41.0	483
13.5	91.3	27.5	265	41.5	492
14.0	96.4	28.0	273	42.0	501
14.5	102	28.5	280	42.5	510

TABLE 6.3

DISCHARGE OVER CIPOLLETTI'S TRAPEZOIDAL WEIR FOR VARIOUS LENGTHS AND HEADS[1]

Head in Centimeters	Length of Weir Crest in Meters									
	0.25	0.50	0.75	1.0	1.5	2.0	3.0	4.0	5.0	6.0
	Discharge in Liters per Second									
1	0.5	0.9	1.4	1.9	2.8	3.7	5.6	7.4	9.3	11.2
2	1.3	2.6	3.9	5.3	7.9	10.1	15.8	21.0	26.3	31.6
3	2.4	4.8	7.2	9.7	14.5	19.3	29.0	38.7	48.3	58.0
4	3.7	7.4	11.2	14.9	22.3	29.8	44.6	59.5	74.4	89.3
5	5.2	10.4	15.6	20.8	31.2	41.6	62.4	83.2	104	125
6	6.8	13.7	20.5	27.3	41.0	54.7	82.0	109	137	164
7	8.6	17.2	25.8	34.4	51.7	68.9	103	138	172	207
8	10.5	21.0	31.6	42.1	63.1	84.2	126	168	210	252
9	12.6	25.1	37.7	50.2	75.3	100	151	201	251	301
10	14.7	29.4	44.1	58.8	88.2	118	176	235	294	353
11	17.0	33.9	50.9	67.8	102	136	204	271	339	407
12	19.3	38.7	58.0	77.3	116	155	232	309	387	464
13	21.8	43.6	65.4	87.2	131	174	262	349	436	523
14	24.4	48.7	73.1	97.4	146	195	292	390	487	585
15	27.0	54.1	81.1	108	162	216	324	432	540	649
16	29.8	59.5	89.2	119	178	238	357	476	595	714
17	32.6	65.2	97.8	130	196	261	391	522	652	782
18	35.5	71.0	107	142	213	284	426	568	710	852
19	38.5	77.0	116	154	231	308	462	616	770	924
20	41.6	83.2	125	166	250	333	499	666	832	998
21	44.8	89.5	134	179	268	358	537	716	895	1074
22	48.0	96.0	144	192	288	384	576	768	960	1152
23	51.3	103	154	205	308	410	616	821	1026	1231
24	54.7	109	164	219	328	437	656	875	1094	1312
25	58.1	116	174	232	349	465	698	930	1162	1395
26	61.6	123	185	247	370	493	740	986	1233	1480
27	65.2	131	196	261	392	522	783	1044	1305	1566
28	68.9	138	207	276	414	551	827	1103	1378	1654
29	72.6	145	218	291	436	581	872	1162	1452	1743
30	76.4	153	229	306	458	611	917	1222	1528	1834

[1] Computed from the equation $Q = 0.0186LH^{3/2}$.

TABLE 6.4

DISCHARGE TABLE FOR 90° TRIANGULAR WEIR[1]

Head in Centimeters	Discharge in Liters per Second	Head in Centimeters	Discharge in Liters per Second	Head in Centimeters	Discharge in Liters per Second
1.0	0.014	15.5	13.1	30.0	68.0
1.5	0.038	16.0	14.1	30.5	70.9
2.0	0.078	16.5	15.3	31.0	73.8
2.5	0.136	17.0	16.4	31.5	76.9
3.0	0.215	17.5	17.7	32.0	79.9
3.5	0.316	18.0	18.9	32.5	83.1
4.0	0.441	18.5	20.3	33.0	86.4
4.5	0.592	19.0	21.7	33.5	89.7
5.0	0.731	19.5	23.2	34.0	93.0
5.5	0.977	20.0	24.7	34.5	96.5
6.0	1.21	20.5	26.2	35.0	100
6.5	1.49	21.0	27.9	35.5	104
7.0	1.79	21.5	29.5	36.0	107
7.5	2.11	22.0	31.3	36.5	111
8.0	2.49	22.5	33.1	37.0	115
8.5	2.90	23.0	35.1	37.5	117
9.0	3.34	23.5	37.0	38.0	123
9.5	3.85	24.0	38.9	38.5	127
10.0	4.36	24.5	41.0	39.0	131
10.5	4.92	25.0	43.1	39.5	135
11.0	5.54	25.5	45.3	40.0	140
11.5	6.20	26.0	47.6	40.5	144
12.0	6.91	26.5	49.9	41.0	148
12.5	7.65	27.0	52.3	41.5	153
13.0	8.41	27.5	54.8	42.0	158
13.5	9.27	28.0	57.3	42.5	163
14.0	10.2	28.5	59.9	43.0	167
14.5	11.0	29.0	62.5	43.5	172
15.0	12.0	29.5	65.3	44.0	177

[1] Computed from $Q = 0.0138H^{5/2}$.

TABLE 6.7

WEIR-BOX DIMENSIONS FOR RECTANGULAR, CIPOLLETTI, AND
90° TRIANGULAR NOTCH WEIR

(All dimensions are in meters.   The letters at the heads of the columns in this table refer to Fig. 6.10.)

Rectangular and Trapezoidal Weirs with End Contractions

Flow (Cubic meters per hour)	H Maximum Head	L Length of Weir Crest	A Length of Box above Weir Notch	K Length of Box below Weir Notch	B Total Width of Box	E Total Depth of Box	C End of Crest to Side	D Crest to Bottom	F Hook Gage Distance Upstream	G Hook Gage Distance Across Stream
50 to 300	0.30	0.30	1.83	0.61	1.68	1.07	0.69	0.61	1.22	0.61
200 to 500	0.34	0.46	2.13	0.91	2.13	1.22	0.84	0.76	1.37	0.61
400 to 800	0.36	0.61	2.44	1.22	2.59	1.37	0.99	0.84	1.52	0.76
600 to 1400	0.40	0.91	2.74	1.52	3.66	1.52	1.37	0.99	1.68	0.91
1000 to 2200	0.46	1.22	3.05	1.83	4.27	1.68	1.52	1.07	1.83	0.91

90° Triangular Notch Weir

Flow (Cubic meters per hour)	H Maximum Head	L Length of Weir Crest	A Length of Box above Weir Notch	K Length of Box below Weir Notch	B Total Width of Box	E Total Depth of Box	C End of Crest to Side	D Crest to Bottom	F Hook Gage Distance Upstream	G Hook Gage Distance Across Stream
50 to 250	0.30	..	1.83	0.61	1.52	0.91	0.76	0.46	1.22	0.61
200 to 450	0.38	..	1.98	2.59	1.98	0.99	0.99	0.46	1.52	0.76

TABLE 6.9

Free Flow Through Parshall Flume*

Upper Head, H_a in cm	Throat Widths in Centimeters									
	7.5	15	23	30	61	91	122	152	183	244
	Flow in Liters per Second									
1										
2										
3	.78	1.4	2.5							
4	1.2	2.3	4.0							
5	1.6	3.1	5.4							
6	2.3	4.5	7.3	9.8	18	27	35			
7	2.9	5.7	9.0	12	23	34	45			
8	3.5	7.1	11	15	28	41	54	67	79	
9	4.3	8.6	13	18	35	51	67	83	99	130
10	5.0	10	15	21	41	60	79	97	116	152
11	5.8	12	18	24	46	69	91	112	133	176
12	6.5	13	21	27	52	78	103	127	152	200
13	7.5	15	24	31	61	90	119	148	176	233
14	8.5	17	27	35	68	101	133	165	190	260
15	9.6	19	29	38	75	111	147	183	218	288
16	10.3	21	32	42	82	122	162	201	240	316
17	11	23	35	47	92	137	181	225	270	356
18	12	25	38	51	100	149	197	246	295	388
19	13	27	42	55	108	161	213	266	317	421
20	14	29	45	59	117	173	230	286	342	454
21	16	32	49	64	127	190	252	312	374	495
22	17	35	52	69	136	204	270	336	404	531
23	18	37	56	73	145	216	288	358	430	570
24	19	40	60	78	155	234	310	383	464	615
25	21	43	64	84	166	248	332	414	495	660
26	22	45	68	89	176	264	350	440	525	698
27	23	48	72	94	186	278	370	463	555	730
28	25	51	76	100	199	298	398	496	595	780
29	26	54	80	105	209	313	418	522	625	835
30	27	57	84	110	220	330	440	550	660	880
32	30	63	93	122	244	368	488	612	734	980
34		70	103	134	270	400	540	680	810	1060
36		76	110	146	290	440	590	740	880	1180
38		83	121	157	320	480	640	810	970	1300
40			131	170	350	520	690	880	1050	1400
42			142	184	380	560	750	940	1140	1520
44			152	198	400	600	810	1010	1210	1630
46			163	210	430	650	870	1090	1310	1750
48			174	230	460	690	920	1160	1400	1870
50				240	490	740	990	1240	1490	2000
52				260	520	790	1060	1320	1590	2130
54				270	550	830	1110	1400	1690	2270
56				290	580	880	1180	1490	1790	2410
58				300	610	930	1260	1580	1890	2540
60				320	640	980	1320	1660	2000	2690
62				340	680	1030	1390	1750	2120	2840
64				350	710	1080	1460	1840	2220	2980
66				370	740	1130	1510	1920	2320	3120
68				390	780	1190	1580	2020	2440	3280
70				400	820	1250	1670	2100	2560	3440
72				420	850	1290	1740	2180	2680	3580
74				440	890	1350	1820	2300	2780	3760
76				460	940	1420	1920	2420	2920	3940

* Source: Water Measurement Manual, U.S. Dept. of the Interior, Bureau of Reclamation, 1953.

TABLE 6.11

STANDARD DIMENSIONS OF PARSHALL MEASURING FLUMES FROM
7.5 CENTIMETERS TO 23 CENTIMETERS THROAT WIDTH

Dimension Letter (Fig. 6.13A)	Dimensions in Meters for Throat Widths (W) of				
	2.5 cm	5 cm	7.5 cm	15 cm	23 cm
A	0.36	0.42	0.47	0.62	0.87
2/3A	0.24	0.28	0.31	0.41	0.60
B	0.36	0.41	0.46	0.61	0.86
2/3B	0.24	0.27	0.30	0.40	0.57
C	0.09	0.14	0.18	0.39	0.38
D	0.17	0.21	0.26	0.39	0.57
E	0.23	0.25	0.38	0.45	0.61
F	0.08	0.11	0.15	0.30	0.30
G	0.20	0.25	0.03	0.61	0.45
K	0.02	0.02	0.03	0.08	0.08
N	0.003	0.017	0.06	0.12	0.12
X	0.008	0.016	0.03	0.05	0.05
Y	0.01	0.025	0.04	0.08	0.08

TABLE 6.12

STANDARD DIMENSIONS OF PARSHALL MEASURING FLUMES
FROM 0.30 METER TO 3.0 METERS

Throat Width, W, in Meters	Dimensions in Meters[1]					
	A	2/3A	B	2/3B	C	D
0.30	1.37	0.9	1.32	0.9	0.61	0.84
0.61	1.52	1.0	1.47	1.0	0.91	1.21
0.91	1.65	1.1	1.62	1.12	1.22	1.57
1.22	1.83	1.22	1.77	1.22	1.52	1.94
1.52	1.98	1.32	1.93	1.30	1.83	2.29
1.83	2.13	1.42	2.08	1.40	2.13	2.67
2.13	2.28	1.53	2.25	1.50	2.44	3.04
2.44	2.44	1.63	2.39	1.60	2.75	3.40
3.05	2.75	1.83	2.65	1.79	3.35	4.12

[1] Letters refer to Fig. 5.13B in which other dimensions for these flumes are shown.

TABLE 7.2

DEPTH OF IRRIGATION WATER IN CENTIMETERS REQUIRED TO ADD
DIFFERENT AMOUNTS OF FIELD-MOISTURE TO ONE METER OF SOIL
FOR SOILS HAVING DIFFERENT APPARENT SPECIFIC GRAVITIES.
BASED ON EQUATION 7.8,

$$d = \frac{P_w A_s D}{100}$$

Available Field Moisture Capacity (P_w)	Apparent Specific Gravity (A_s)							
	1.2	1.3	1.4	1.5	1.6	1.7	1.8	1.9
4.0	4.8	5.2	5.6	6.0	6.4	6.8	7.2	7.6
4.2	5.1	5.5	5.9	6.3	6.7	7.1	7.6	8.0
4.4	5.3	5.7	6.2	6.6	7.0	7.5	7.9	8.4
4.6	5.5	6.0	6.4	6.9	7.4	7.8	8.3	8.7
4.8	5.8	6.2	6.7	7.2	7.7	8.2	8.6	9.1
5.0	6.0	6.5	7.0	7.5	8.0	8.5	9.0	9.5
5.2	6.2	6.8	7.3	7.8	8.3	8.8	9.4	9.9
5.4	6.5	7.0	7.6	8.1	8.6	9.2	9.7	10.3
5.6	6.7	7.3	7.8	8.4	9.0	9.5	10.1	10.6
5.8	7.0	7.5	8.1	8.7	9.3	9.9	10.4	11.0
6.0	7.2	7.8	8.4	9.0	9.6	10.2	10.8	11.4
6.2	7.4	8.1	8.7	9.3	9.9	10.5	11.2	11.8
6.4	7.7	8.3	9.0	9.6	10.3	10.9	11.5	12.2
6.6	7.9	8.6	9.2	9.9	10.6	11.2	11.9	12.5
6.8	8.2	8.8	9.5	10.2	10.9	11.6	12.2	12.9
7.0	8.4	9.1	9.8	10.5	11.2	11.9	12.6	13.3
7.2	8.6	9.4	10.1	10.8	11.5	12.2	13.0	13.7
7.4	8.9	9.6	10.4	11.1	11.8	12.6	13.3	14.1
7.6	9.1	9.9	10.6	11.4	12.2	12.9	13.7	14.4
7.8	9.4	10.1	10.9	11.7	12.5	13.3	14.0	14.8
8.0	9.6	10.4	11.2	12.0	12.8	13.6	14.4	15.2

TABLE 7.3

TIME IN HOURS REQUIRED WITH A STREAM q TO ADD VARIOUS
PERCENTAGES OF AVAILABLE MOISTURE P_w TO 1.0 HECTARE-METER
OF SOIL, THE APPARENT SPECIFIC GRAVITY A_s OF WHICH IS 1.4,

$$\text{BASED ON } t = \frac{P_w A_s D a}{100 c q} \text{ WHERE } c = 1.00$$

Col. No.		1	2	3	4	5	6	7	8	9	10
Line No.	Available Capacity P_w	Size of Stream q in Cubic Meters per Hour									
		50	100	150	200	250	300	350	400	450	500
1	4.0	11.2	5.6	3.7	2.8	2.2	1.86	1.60	1.40	1.24	1.12
2	4.2	11.8	5.9	3.9	2.9	2.4	1.96	1.68	1.47	1.31	1.18
3	4.4	12.3	6.2	4.1	3.1	2.5	2.05	1.76	1.54	1.37	1.23
4	4.6	12.9	6.4	4.3	3.2	2.6	2.15	1.84	1.61	1.43	1.29
5	4.8	13.4	6.7	4.5	3.4	2.7	2.24	1.92	1.68	1.49	1.34
6	5.0	14.0	7.0	4.7	3.5	2.8	2.34	2.00	1.75	1.56	1.40
7	5.2	14.6	7.3	4.9	3.6	2.9	2.43	2.08	1.82	1.62	1.46
8	5.4	15.1	7.6	5.0	3.8	3.0	2.52	2.16	1.89	1.68	1.51
9	5.6	15.7	7.8	5.2	3.9	3.1	2.62	2.24	1.96	1.74	1.57
10	5.8	16.2	8.1	5.4	4.1	3.2	2.71	2.32	2.03	1.80	1.62
11	6.0	16.8	8.4	5.6	4.2	3.4	2.80	2.40	2.10	1.87	1.68
12	6.2	17.4	8.7	5.8	4.3	3.5	2.90	2.48	2.17	1.93	1.74
13	6.4	17.9	9.0	6.0	4.5	3.6	2.99	2.56	2.24	1.99	1.79
14	6.6	18.5	9.2	6.2	4.6	3.7	3.08	2.64	2.31	2.05	1.85
15	6.8	19.0	9.5	6.3	4.8	3.8	3.18	2.72	2.38	2.11	1.90
16	7.0	19.6	9.8	6.5	4.9	3.9	3.27	2.80	2.45	2.18	1.96
17	7.2	20.2	10.1	6.7	5.0	4.0	3.36	2.88	2.52	2.24	2.02
18	7.4	20.7	10.4	6.9	5.2	4.1	3.46	2.96	2.59	2.30	2.07
19	7.6	21.3	10.6	7.1	5.3	4.3	3.55	3.04	2.66	2.36	2.13
20	7.8	21.8	10.9	7.3	5.5	4.4	3.64	3.12	2.73	2.43	2.18
21	8.0	22.4	11.2	7.5	5.6	4.5	3.74	3.20	2.80	2.49	2.24

TABLE 7.4

REPRESENTATIVE PHYSICAL PROPERTIES OF SOILS

Soil Texture	Infiltration[1] and Permeability, Centimeters per Hour I_f	Total Pore Space % N	Apparent Specific Gravity A_s	Field Capacity % FC	Permanent Wilting % PW	Total Available Moisture[2]		
						Dry Weight % $P_w = FC - PW$	Volume % $P_v = P_w A_s$	Centimeters per Meter $d = \frac{P_w}{100} A_s D$
Sandy	5 (2.5–25)	38 (32–42)	1.65 (1.55–1.80)	6 (6–12)	4 (2–6)	5 (4–6)	8 (6–10)	8 (7–10)
Sandy Loam	2.5 (1.3–7.6)	43 (40–47)	1.50 (1.40–1.60)	14 (10–18)	6 (4–8)	8 (6–10)	12 (9–15)	12 (9–15)
Loam	1.3 (0.8–2.0)	47 (43–49)	1.40 (1.35–1.50)	22 (18–26)	10 (8–12)	12 (10–14)	17 (14–20)	17 (14–19)
Clay Loam	0.8 (0.25–1.5)	49 (47–51)	1.35 (1.30–1.40)	27 (23–31)	13 (11–15)	14 (12–16)	19 (16–22)	19 (17–22)
Silty Clay	.25 (.03–0.5)	51 (49–53)	1.30 (1.30–1.40)	31 (27–35)	15 (13–17)	16 (14–18)	21 (18–23)	21 (18–23)
Clay	0.5 (.01–0.1)	53 (51–55)	1.25 (1.20–1.30)	35 (31–39)	17 (15–19)	18 (16–20)	23 (20–25)	23 (20–25)

Note: Normal ranges are shown in parentheses.

[1] Intake rates vary greatly with soil structure and structural stability, even beyond the normal ranges shown above.

[2] Readily available moisture is approximately 75% of the total available moisture.

TABLE 8.1

GUIDE FOR JUDGING HOW MUCH OF THE AVAILABLE MOISTURE HAS BEEN REMOVED FROM THE SOIL

Soil Moisture Deficiency	Feel or Appearance of Soil and Moisture Deficiency in Centimeters of Water per Meter of Soil			
	Coarse Texture	Moderately Coarse Texture	Medium Texture	Fine and Very Fine Texture
0% (Field capacity)	Upon squeezing, no free water appears on soil but wet outline of ball is left on hand. 0.0	Upon squeezing, no free water appears on soil but wet outline of ball is left on hand. 0.0	Upon squeezing, no free water appears on soil but wet outline of ball is left on hand. 0.0	Upon squeezing, no free water appears on soil but wet outline of ball is left on hand. 0.0
0–25%	Tends to stick together slightly, sometimes forms a very weak ball under pressure. 0.0 to 1.7	Forms weak ball, breaks easily, will not slick. 0.0 to 3.4	Forms a ball, is very pliable, slicks readily if relatively high in clay. 0.0 to 4.2	Easily ribbons out between fingers, has slick feeling. 0.0 to 5.0
25–50%	Appears to be dry, will not form a ball with pressure. 1.7 to 4.2	Tends to ball under pressure but seldom holds together. 3.4 to 6.7	Forms a ball somewhat plastic, will sometimes slick slightly with pressure. 4.2 to 8.3	Forms a ball, ribbons out between thumb and forefinger. 5.0 to 10.0
50–75%	Appears to be dry, will not form a ball with pressure.[1] 4.2 to 6.7	Appears to be dry, will not form a ball.[1] 6.7 to 10.0	Somewhat crumbly but holds together from pressure. 8.3 to 12.5	Somewhat pliable, will ball under pressure.[1] 10.0 to 15.8
75–100% (100% is permanent wilting)	Dry, loose, single-grained, flows through fingers. 6.7 to 8.3	Dry, loose, flows through fingers. 10.0 to 12.5	Powdery, dry, sometimes slightly crusted but easily broken down into powdery condition. 12.5 to 16.7	Hard, baked, cracked, sometimes has loose crumbs on surface. 15.8 to 20.8

[1] Ball is formed by squeezing a handful of soil very firmly.

TABLE 9.1

THE FLOW OF WATER IN SOILS. COMPARISON OF PERMEABILITY
OF SOIL TO WATER STATED IN DIFFERENT UNITS

	1	2	3
		Permeability in	
Line No.	Cu M per Sq M per 24 Hr	Surface Cm per Hr	Liters per Sec per Hectare
1	0.003	0.013	0.35
2	0.006	0.025	0.70
3	0.009	0.038	1.05
4	0.012	0.051	1.40
5	0.015	0.064	1.75
6	0.018	0.076	2.10
7	0.021	0.089	2.45
8	0.024	0.102	2.80
9	0.027	0.114	3.14
10	0.030	0.127	3.50
11	0.061	0.254	6.99
12	0.091	0.381	10.5
13	0.122	0.508	14.0
14	0.152	0.635	17.5
15	0.183	0.762	21.0
16	0.213	0.889	24.5
17	0.244	1.02	28.0
18	0.274	1.14	31.4
19	0.305	1.27	34.9
20	0.366	1.52	41.9
21	0.427	1.78	48.9
22	0.488	2.03	55.9
23	0.549	2.29	62.9
24	0.610	2.54	69.9
25	0.671	2.79	76.9
26	0.732	3.05	83.9
27	0.792	3.30	90.9
28	0.854	3.56	97.8
29	0.915	3.81	105
30	0.976	4.06	112
31	1.04	4.32	119
32	1.10	4.57	126
33	1.16	4.83	133
34	1.22	5.08	140
35	1.37	5.72	157
36	1.52	6.35	175
37	1.83	7.62	210
38	2.14	8.89	245
39	2.44	10.2	280
40	2.75	11.4	314
41	3.05	12.7	349
42	4.57	19.1	524
43	6.10	25.4	699
44	7.62	31.8	874
45	9.14	38.1	1050
46	10.7	44.4	1220
47	12.2	50.8	1400
48	13.7	57.2	1570
49	15.2	63.5	1750

TABLE 11.13. TOTAL CONSUMPTIVE USE AND PEAK DAILY USE, WESTERN UNITED STATES (COMPUTED FROM WOODWARD)

| Crops | Southern Coastal* (1) | | | | South Pacific Coastal Interior and North Coastal (2) | | | | | | | |
| | 300 Days Plus | | 250–300 Days | | 250–300 Days | | 210–250 Days | | 180–210 Days | | 150–180 Days | |
	Season Use cm	Daily Use cm/day	Season Use cm	Daily Use cm/day	Season Use cm	Daily Use cm/day	Season Use cm	Daily Use cm/day	Season Use cm	Daily Use cm/day	Season Use cm	Daily Use cm/day
Alfalfa	91	0.51	76	0.43	94	0.69	81	0.56	66	0.51	56	0.46
Pasture	85	0.51	71	0.43	84	0.69	76	0.56	61	0.51	51	0.46
Grain—small	41	0.46	36	0.41	43	0.56	37	0.51	31	0.51	25	0.46
Beets—sugar	74	0.51	64	0.46	76	0.69	66	0.56	..	..	..	..
Beans—field	31	0.46	25	0.41	33	0.56	28	0.46	..	..	..	..
Corn—field	..	..	..	..	..	0.51	48	0.64	41	0.56	36	0.51
Potatoes	..	..	..	..	61	0.51	51	0.46	..	0.46	..	..
Peas—green	25	0.46	21	0.41	28	0.64	23	0.46	20	0.41	18	0.41
Legume seed	64	0.51	56	0.46	66	0.51	56	0.56	51	0.51	46	0.46
Tomatoes	46	0.41	38	0.41	48	0.46	41	0.46	33	0.41	28	0.41
Vegetable seed	41	0.41	36	0.41	46	0.46	41	0.46	36	0.41	31	0.41
Beans—pole	..	..	..	..	46	0.51	41	0.51	36	0.46	31	0.41
Corn—sweet	41	0.46	36	0.41	41	0.51	36	0.46	31	0.46	28	0.41
Apples	..	..	..	..	61	0.51	56	0.51	51	0.46	46	0.46
Cherries	..	..	..	..	61	0.51	56	0.51	..	..	..	..
Peaches	..	..	..	..	61	0.51	56	0.51	..	..	..	..
Prunes	..	..	..	..	56	0.51	51	0.51	..	..	..	..
Apricots	..	..	..	..	56	0.51	51	0.51	..	..	..	..
Oranges	51	0.41	46	0.41	56	0.46	..	..	..	..	..	..
Avocados	46	0.41	..	..	..	..	..	..	..	..	..	..
Walnuts	56	0.51	46	0.46	61	0.64	56	0.56	51	0.51	..	..
Strawberries	56	0.51	46	0.46	58	0.64	46	0.56	41	0.51	36	0.46
Lettuce	10	0.41	10	0.41	15	0.46	13	0.46	48	0.51	..	..
Mint	..	..	..	..	58	0.61	53	0.56	..	..	43	0.46
Hops	..	..	..	..	51	0.56	51	0.51	..	..	..	..

* Fog Belt.

TABLE 11.13. TOTAL CONSUMPTIVE USE AND PEAK DAILY USE, WESTERN UNITED STATES (*Continued*)

Central Valley—California and Valleys East Side of Cascade Mountains (3)

Crops	250–300 Days		210–250 Days		180–210 Days		150–180 Days		120–150 Days		90–120 Days	
	Season Use cm	Daily Use cm/day	Season Use cm	Daily Use cm/day	Season Use cm	Daily Use cm/day	Season Use cm	Daily Use cm/day	Season Use cm	Daily Use cm/day	Season Use cm	Daily Use cm/day
Alfalfa	102	0.76	86	0.71	76	0.64	66	0.56	51	0.51	36	0.46
Pasture	92	0.76	76	0.71	71	0.64	61	0.56	46	0.51	33	0.46
Grain—small	46	0.56	41	0.56	38	0.51	36	0.46	33	0.46	31	0.41
Beets—sugar	84	0.76	71	0.64	61	0.56	51	0.51	46	0.46	..	..
Beans—field	43	0.56	33	0.51	33	0.51	31	0.46	31	0.46	..	..
Corn—field	66	0.89	56	0.81	56	0.76	51	0.64	46	0.56	43	0.51
Potatoes—summer	31	0.41										
Potatoes—fall	..	..	48	0.71	46	0.56	46	0.51	43	0.46	41	0.41
Peas—green	..	..	20	0.46	18	0.46	18	0.46	18	0.41	15	0.38
Peas—field	..	..	25	0.46	23	0.46	22	0.46	20	0.41	20	0.38
Tomatoes	51	0.51	46	0.46	46	0.46	43	0.43	41	0.41	..	..
Cotton	66	0.76	56	0.71	..	..	..	..	..	..	20	0.38
Grain—sorghum	38	0.51	33	0.46	31	0.43	25	0.41	23	0.38	..	..
Apples	..	..	66	0.51	58	0.51	53	0.46	..	..	..	..
Cherries	..	..	61	0.51	53	0.51	48	0.46	..	..	..	..
Peaches	56	0.56	56	0.51	51	0.51	46	0.46				
Apricots	51	0.56	43	0.51	38	0.51						
Oranges	71	0.46										
Strawberries	61	0.51	51	0.51	46	0.46						
Lettuce—winter	102	0.51	..	..	..	..						
Mint	..	..	51	0.56	46	0.51						
Hops	..	0.64	46	0.51	41	0.46						
Grapes	76	..	64	0.56	56	0.51						
Walnuts	61	0.56	51	0.51	..	..						
Almonds	56	0.64	51	0.56	..	..						

TABLE 11.13. TOTAL CONSUMPTIVE USE AND PEAK DAILY USE, WESTERN UNITED STATES (*Concluded*)

Intermountain, Desert, and Western High Plains (4)

Crops	250–300 Days		210–250 Days		180–210 Days		150–180 Days		120–150 Days		90–120 Days	
	Season Use cm	Daily Use cm/day	Season Use cm	Daily Use cm/day	Season Use cm	Daily Use cm/day	Season Use cm	Daily Use cm/day	Season Use cm	Daily Use cm/day	Season Use cm	Daily Use cm/day
Alfalfa	132	1.02	112	0.81	91	0.74	76	0.66	76	0.66	48	0.51
Pasture	121	1.02	102	0.76	84	0.71	71	0.64	71	0.64	41	0.51
Grain—small	53	0.64	46	0.56	41	0.51	41	0.51	41	0.51	36	0.46
Beets—sugar	94	0.76	81	0.76	76	0.71	66	0.64	66	0.64	46	0.51
Beans—field	56	0.64	43	0.51	36	0.51	36	0.46	36	0.46	31	0.38
Corn—field	..	..	76	0.89	66	0.76	61	0.71	61	0.71	..	..
Potatoes—fall	..	..	58	0.76	53	0.71	51	0.64	51	0.64	43	0.51
Peas—field	..	..	..	..	25	0.48	25	0.46	25	0.46	23	0.38
Tomatoes	..	..	51	0.56	46	0.51	43	0.46	43	0.46	..	..
Cotton	81	0.76	76	0.71	..	..	..	..	..	..	..	..
Grain—sorghum	48	0.64	46	0.51	41	0.51	36	0.46	36	0.46	..	..
Apples	..	..	..	..	71	0.56	61	0.51	61	0.51	..	..
Cherries	..	..	..	..	66	0.56	..	..	..	..	..	..
Peaches	..	..	74	0.64	69	0.56	..	..	..	..	..	..
Apricots	66	0.64	61	0.64	64	0.51	..	..	..	..	..	..
Almonds	56	0.64	51	0.64	..	..	..	..	..	..	..	..
Vineyards	102	0.69	81	0.64	66	0.56	..	..	..	..	..	..
Legume seed	..	..	..	..	..	..	..	..	..	..	36	0.41
Grass seed	..	..	..	..	..	..	..	..	..	..	31	0.36
Potatoes—seed	..	..	..	..	..	..	..	..	..	..	36	0.38
Grapefruit	114	0.51	..	..	..	..	..	..	..	..	..	..
Oranges	91	0.46	..	..	..	..	..	..	..	..	..	..
Lettuce—winter	15	0.46	51	0.56	..	..	..	..	..	..	..	..
Melons	56	0.64	..	..	46	0.51	41	0.46	41	0.46	..	..
Palm dates	152	0.76	..	..	..	..	..	..	..	..	..	..
Truck crops	51	0.64	46	0.56	36	0.51	31	0.46	31	0.46	25	0.38

APPENDIX B PROBLEMS AND QUESTIONS

CHAPTER 1

1. Define irrigation.
2. List six purposes for applying irrigation water to the soil.
3. How is irrigation accomplished?
4. What are the four major sources of irrigation water? Define and explain each.
5. What precautions should be kept in mind in determining the amount of water available from precipitation?
6. Is all of the precipitation that falls on cropland available to the crops? Defend your answer.
7. What atmospheric conditions, other than precipitation, will produce significant amounts of supplemental water?
8. Does ground water contribute directly to the water needs of plants? When? When is it harmful?
9. What are some of the advantages of irrigation in humid areas?
10. How does the need for irrigation in humid areas differ from that in arid regions?
11. What are the major factors which should be considered in the development of an irrigation project?
12. Answer the following questions in relation to the agricultural area most familiar to you.
    (a) What is the present status of irrigation?
    (b) How do the climate, water supply, and soils influence irrigation practices?
    (c) In what manner will the agricultural economy be improved by more extensive and improved irrigation?

CHAPTER 2

1. Why is accurate information regarding water supply important to the development of an irrigation system?
2. What is a snow survey?
3. What are the advantages of small ponds and reservoirs for irrigation? What are some of the disadvantages?

4. What factors should be considered when selecting a site for a reservoir?
5. How can reservoir silting be reduced or eliminated? List four ways and describe the operation of each.
6. What can be done to reduce loss of irrigation water by evaporation?
7. What are phreatophytes? Are they beneficial? Why?
8. What are the two principal methods of "rainmaking"? Explain each.
9. Discuss the feasibility of saline water conversion for irrigation purposes.
10. What are some of the sources of water that can be used for recharging ground water reservoirs?
11. What are the methods used in ground water recharge operations?
12. What does "safe-yield" refer to?
13. List three approaches to ground water investigations and describe the techniques used in each.

CHAPTER 3

1. List the three principal methods of drilling wells and explain each. Discuss the advantages and disadvantages of each.
2. What are the purposes of well screens and of gravel packing a well? When would you use each? Should they be used together? If so, when? If not, why?
3. Why is the development of a well so important?
4. List the three principal methods of well development and explain the operation of each.
5. What is the difference between a confined and an unconfined well?
6. Is a confined or artesian well necessarily a flowing well? Explain.
7. What are some of the assumptions made in the development of well equations?
8. What discharge can be expected from an unconfined well 15 inches in diameter when the draw-down is 20 feet in an aquifer saturated to a depth of 50 feet? Assume the coefficient of permeability to be 70 feet per day and the radius of influence to be 600 feet.
9. What diameter well is required to deliver a discharge of 2 cfs from a confined aquifer 50 feet thick with a draw-down of 70 feet? Assume a radius of influence of 400 feet and a permeability of 50 feet per day.
10. Determine the height of the seepage face for an unconfined well 1 foot in diameter, a permeability of 0.02 fps, a discharge of 800 gpm, and a depth of water in the well of 15 feet.

CHAPTER 4

1. Define horsepower, theoretical horsepower, brake horsepower, water horsepower, and pumping plant efficiency.
2. Define static head and draw-down.
3. What are the characteristic curves of a pump? Of what value are they?
4. Define specific speed. What types of pumps are characterized by a high specific speed? By a low specific speed?

5. Under what conditions would you use a centrifugal pump? What are some of the problems connected with its use?

6. Define total dynamic head.

7. An irrigator desires to lift a stream of 500 gpm a vertical height of 40 feet. If the loss of head in the casing and pump results in a 62 percent over-all pump efficiency and the electric motor has an efficiency of 91 percent, how many horsepower will his motor need? How many kilowatts will it use while pumping? Assume 1 hp = 0.746 kw.

8. With the same height of lift and the same efficiencies as given in Problem 7, how many kilowatts would a motor require in order to deliver a stream that would supply enough water in 30 hours to cover a 10-acre tract to a depth of 6 inches? Assume 1 hp = 0.746 kw.

9. The gross water requirement for orchard irrigation is assumed to be 3 acre-feet per acre. If a pumping plant (motor and pump) operates at an efficiency of 57 percent, how many kilowatt-hours of energy will be required to lift water 30 feet for each acre?

10. Compute the horsepower required to pump:
    (a) A stream of 2 cfs against a head of 40 feet, assuming 100 percent efficiency.
    (b) Actual pumping plant efficiency is 59 percent. What is the horsepower requirement?
    (c) Using the "Electric Service Schedules and Costs," Chapter 4, what would the charge be per month (30 days) if the motor is run continuously? Assume that voltage and term discount are received. Assume 1 hp = 0.746 kw.

11. Would it be advisable to use the centrifugal pump whose characteristics are shown in Fig. 4.2 when it is desired that 1600 gpm be pumped against a head of 42 feet? What is the efficiency? Why is a high efficiency desirable? What horsepower would be used?

12. (a) A farmer pumps 1 cfs for 24 hours each day of the irrigation season. The static head is 20 feet and the draw-down while pumping 1 cfs is 5 feet. If his pumping plant is 60 percent efficient, what does it cost him per month if he pays the rates as given in the article "Electric Service Schedules and Costs"?

CHAPTER 5

1. Discuss the advantages and disadvantages of earth canals.

2. Why should amortized annual cost of lining a canal be considered instead of the initial cost?

3. How might canal lining be related to drainage?

4. Consider a farm irrigation ditch in a loam soil having the following dimensions:
    (a) Bottom width, 2.00 feet.
    (b) Total depth, 1.75 feet.
    (c) Side slopes of 1 horizontal to 1 vertical.

(d) Depth of water, 1 foot.

Find the following properties:

(a) Cross-sectional area of the stream.

(b) Wetted perimeter.

(c) Hydraulic radius of the stream.

5. If the bottom of the ditch described in Problem 1 has a uniform slope of 5.28 feet/mile (1 foot/1000), and if the bottom and sides are kept smooth and free from weeds, what will be the mean velocity of flow and the discharge? Use $n = 0.02$.

6. If the canal described in Problem 1 were permitted to grow weeds on the sides and bottom, what would be the velocity and discharge? Use $n = 0.04$.

7. For a canal of the same dimensions as in Problem 1 built in earth on a slope of 10.56 feet/mile (2.00 feet/1000), determine the velocity and discharge.

8. How do you account for the fact that the velocity and discharge are not doubled when the slope is doubled?

9. For a concrete-lined ditch in good condition having the same dimensions and slope as the ditch in Problem 4, determine the velocity and discharge. Use $n = 0.014$.

10. Re-work Problem 6 using the Chezy equation and Fig. 5.2. Use an e value of 0.000 and consider the water temperature to be 60°F. (Alternative—Consider ν to be 1.21×10^{-5} ft²/sec.)

11. Compute C and f for the conditions of Problem 6.

12. Compute C and f for the conditions of Problem 3.

13. Show that a square, covered box flowing 0.95 full carries more water than the same box flowing full. Explain.

14. Given a canal with 1 to 1 side slopes and cross-section area of 112 square feet, find the dimensions for the best hydraulic cross section.

15. List the principal materials used as canal linings. Discuss the advantages and disadvantages of each.

16. Describe the "inflow-outflow" method for determining seepage losses from a canal.

CHAPTER 6

1. What are the principal advantages and disadvantages of the following water measuring devices?

(a) Orifice

(b) Weir

(c) Parshall flume

(d) Trapezoidal flume

(e) Current meter

2. When using a current meter at 0.6 of the depth, do you measure 0.6 from the water surface or from the bottom of the stream?

3. Explain why more flow will occur over a dull-crested weir than over a

sharp-crested weir, even though the depth of water over the crest is the same.

4. (a) Find the theoretical velocity of a jet of water flowing out of a square orifice in a large tank if the center of the orifice is 2 feet below the water surface.

   (b) If the orifice opening considered in Problem 1a is $\frac{1}{2}$ foot by $\frac{1}{2}$ foot, what is the theoretical discharge in cubic feet per second?

   (c) What are the probable maximum and probable minimum actual discharges in cubic feet per second?

5. (a) In measuring the water that flows through a submerged orifice, is it necessary to know the vertical distance from the upstream water surface to the center of the orifice? Explain.

   (b) Find the discharge in cubic feet per second through a rectangular standard submerged orifice 18 inches long (horizontal dimension) by 8 inches deep (vertical dimension) if the upstream surface is 7 inches vertically above the downstream water surface. First use the appropriate equation and check your result by use of a table.

6. (a) In using weirs to measure water, is it essential to make direct measurement of the velocity of the water as it flows through the weir notch? Explain.

   (b) By means of the appropriate equation, compute the cubic feet per second over a rectangular weir having suppressed end contractions if the weir crest is 24 inches long and the water surface at a point 8 feet upstream from the weir is $5\frac{1}{2}$ inches vertically above the weir crest. Check your result with a weir table.

   (c) If the weir described in Problem 3b has complete end contractions, would the discharge be more or less than your computed result? How much?

7. For the same length of weir crest and depth of water over the crest as in Problem 3b, compute the discharge over a trapezoidal weir. Check your results with a table.

8. For a right-angle triangular notch weir, what is the discharge when the depth of water vertically above the apex of the weir notch is 0.6 feet at a point 5 feet upstream from the weir?

9. Show that doubling the effective head, causing discharge through a submerged orifice, increases the discharge approximately 41 percent.

10. Show that doubling the head over a rectangular or trapezoidal weir makes the discharge 2.8 times greater.

CHAPTER 7

1. Is it practicable for the irrigation farmer to greatly modify the texture of his soils? Why?

2. What soil structure is best suited to irrigation and crop production? Describe ways in which the farmer can maintain a favorable structure in his soil.

3. Distinguish between the *real* and the *apparent* specific gravity of a soil. Is it possible for the apparent specific gravity to be equal to or larger than the real specific gravity? Explain.

4. What substances occupy the pore space of a soil? Is the percentage pore space of a field soil influenced by its water content?

5. Why is the rate of water-flow into soils of importance in irrigation practice?

6. For a soil of given texture and structure, will a 4-foot depth of well-drained root-zone soil hold twice as much irrigation water as one of 2-foot depth? Assume that the water table is 30 feet or more below the land surface. Give reasons for your answers.

7. What properties of the soil determine the percentages of these three classes of moisture in the soil: hygroscopic, capillary, and gravitational?

8. Are irrigated soils that are naturally well drained ever completely saturated? Explain.

9. Why are the actual heights to which water will rise by capillary action in a soil usually less than the theoretical heights computed from Equation 7.5?

10. Why is moisture content expressed in percent volume more useful in the metric system of measurement than in the English system?

11. How can the concept of field capacity be determined and used even though there is no point on the moisture drainage curve that uniquely defines field capacity?

12. Why is the moisture content at which a crop permanently wilts a function of consumptive-use rate as well as soil texture?

13. A sharp-edged cylinder 6 inches in diameter is carefully driven into the soil so that negligible compaction occurs. An 8-inch column of soil is secured. The wet weight is 5780 grams and the dry weight is 5180 grams.
(a) What is the percent moisture on a dry weight basis?
(b) What is the apparent specific gravity of the soil?

14. A cylinder was carefully pushed into the soil without compressing or disturbing the soil. The cross-sectional area of the cylinder was 0.25 square feet. The length of the column of soil within the cylinder was 12 inches. The weight of the soil within the cylinder was 21 pounds when it was dried. The weight of the soil before drying was 25.2 pounds. Determine P_w, A_s, P_v.

15. A stream of 4 cfs is used to apply 5 acre-inches of water per acre to an 8-acre field. How long will it take?

16. An irrigator uses a stream of 3.5 cfs for two days (48 hours) to irrigate 30 acres of sugar beets. What is the average depth of water applied?

17. A farmer desires to irrigate a border which is 40 feet wide and 500 feet long. He wants to apply an average of 3 inches depth of water to the area with a stream of 1000 gpm. How long will it take him to irrigate this border?

18. The soil moisture at field capacity is 27.2 percent and the moisture content at the time of irrigating is 19.0 percent. The apparent specific gravity is 1.3 and the depth of soil to be wetted is 3 feet.

(a) How many acre-inches per acre of water must be applied?

(b) How long will it take to irrigate the 12 acres with a 4 cfs stream if the water application efficiency is 65 percent?

19. Soil samples indicate an average moisture and apparent specific gravity in the soil as follows:

Depth	P_w (dry wt)	A_s
0–1 foot	14.7	1.34
1–2 feet	15.3	1.39
2–3 feet	17.6	1.32
3–4 feet	18.2	1.30

Compute the depth of water held in the first 4 feet.

20. A farmer owning "bench" land, in which the sandy loam soil averaged about 4 feet in depth and was underlain by gravel and coarse sand to a depth of 30 feet or more, discovered in March by borings with a soil auger that the light winter precipitation had penetrated the soil to a depth of only 6 inches. He at once applied a 5-cfs stream of flood water to his 20-acre tract and kept the stream well spread out on the land for a period of four 24-hour days in order to give the soil a good soaking. There was no surface runoff. Find approximately what percentage of the water applied was lost by deep percolation.

CHAPTER 8

1. Why is measurement of soil moisture important?
2. What characteristics should be considered when deciding whether to use a spiral-bit auger or a posthole auger to obtain a soil moisture sample?
3. Concerning the Oakfield probe:
   (a) Under what conditions does it work best?
   (b) What conditions make it difficult to use?
4. For what purposes would you consider feel and appearance of the soil sufficiently accurate as an indication of soil moisture content?
5. What are the principal advantages and disadvantages of adsorption blocks for determining soil moisture?
6. What characteristics should be considered when deciding whether resistance blocks or tensiometers should be used?
7. Describe the principle of operation of the neutron method of measuring soil moisture.
8. List 8 methods of estimating soil moisture.

CHAPTER 9

1. Show by computation that kinetic energy can usually be ignored for flow-through soils.
2. Show that permeability k in Equation 9.3 is dependent upon size of pores by comparing Equations 5.3 and 9.3.
3. From Equation 9.5 and the definition following, it is seen that A is the gross area at right angles to the flow direction. What approximate change

would occur in the computed magnitude of permeability k for flow-through loam soil if A were taken as net cross-sectional area?

4. Imagine a soil column of unit cross-sectional area at right angles to the direction of flow of water and state whether or not it is practicable to measure accurately the net cross-sectional area of the channels through which flow occurs: (a) for a saturated soil; (b) for an unsaturated soil. Give reasons for your answer.

5. Consider a vertical soil column of 1 square foot cross-sectional area and 4 feet long. If 5 cubic feet of water percolate through the column in 36 hours from a supply pipe which permits the water to flow onto the soil just fast enough to keep the soil surface covered, what is the permeability in feet per 24-hour day?

6. Measurement of the permeability of a 50-foot stratum of saturated clay soil overlying a water-bearing gravel shows that $k = 2 \times 10^{-2}$ inches per hour. If the pressure head in the gravel is 75 feet of water (measured at the lower surface of the clay) and 0 feet near the soil surface, water is flowing vertically upward through the clay. Compute the flow in cubic feet per second through a block of clay 50 feet thick and 640 acres in area.

7. A contour map of water pressures overlying an artesian basin shows an average fall in pressure head of 30 feet per mile. Assume a mean thickness of water-bearing gravel of 26 feet and that $k = 2 \times 10^{-4}$ feet/second. Compute the underground flow in cubic feet per second through a section of gravel 1000 feet long at right angles to the direction of flow.

8. Assume that a 40-acre tract of land is irrigated frequently and given enough water to keep the soil practically saturated below the 6-foot depth but that the water table is 100 feet deep. If the average $k = \frac{1}{2}$ cfs per acre, compute the number of acre-feet of water that flow vertically downward to the water table each month.

9. A soil sample weighing 455 grams when brought from the field in a cylinder of 300 cc volume, weighed 375 grams when dried. The soil particles displaced 145 cc of water. Compute the following:
(a) Proportion of the total volume filled with water.
(b) Percentage of the pore space filled with water.
(c) Inches of water in 2.5 feet depth of soil at the given moisture content.

10. The following intake data were taken prior to the designing of an irrigation system:

Elapsed Time (minutes)	Accumulated Intake (inches)
2	0.28
4	0.42
6	0.47
10	0.60
30	1.10

Determine the values of a and n in Equation 9.9.

11. Intake data obtained using cylinders and buffers will most closely approximate the actual intake resulting from what method of irrigation?

12. Show that Equation 9.13, $I_{ave} = \dfrac{\text{gpm}}{\text{spacing}}$, can be derived from Equation 7.9, $qt = ad$.

13. Discuss the characteristics of the three basic zones in the phenomena of water movement through soil during irrigation.

CHAPTER 10

1. Explain why humid-region soils do not contain excessive amounts of alkali.

2. How does the amount of water moving through the soil affect the salinity or lack of salinity in the soil?

3. Water having a conductivity of 500 micromhos per cm has how many tons of salt per acre-foot? What is the concentration in milliequivalents per liter and in parts per million?

4. Is sodium carbonate a black salt? What is black alkali? What salts give rise to the occurrence of black alkali?

5. Why is exchangeable-sodium percentage important in reclamation of soils?

6. Provided one-half of the 831 ppm of alkali salts in lower Sevier River irrigation water were deposited in the upper 3 feet of soil each year, how many years would it take to add 0.5 percent total salts to the soil provided 2 feet of water are applied to the soil each year? Assume $A_s = 1.40$.

7. A drain tile main outlet from a 1000-acre tract discharges an average of 1 cfs during each of the 12 months of the year. If the average salt content of the drainage water is 1200 ppm, the irrigation water applied to the tract is practically free from salinity, and the mean depth of drains is 6 feet, what is the annual reduction in salt content of the soil in terms of the percentage of the weight of the dry soil?

8. What added precautions are necessary in reclaiming an alkali soil that need not be considered when reclaiming a saline soil?

9. Explain why it may be easier to leach an alkali soil with salty water than with pure water.

10. Explain fully why lowering of the water table is helpful in the prevention of accumulations of soluble salts on the surface of the soil.

11. In addition to the lowering of the water table, describe other means of preventing, or at least decreasing, the accumulations on the surface of the soil.

12. Under what conditions, if any, is it advisable to use, for irrigation purposes, water that contains appreciable percentages of soluble salts? What precautions are necessary to minimize the danger of using saline irrigation water?

13. Are the texture and the structure of soils related to the salinity and alkali problems? If so, explain.

14. What is the primary reason for adding chemical amendments to the soil?

15. What is the most important reclamation procedure for saline soil?

16. Of what value is water-quality data?
17. Define an alkali soil. Define a saline soil.
18. Describe an agricultural area where in your judgment proper management of a high water table would be a better solution economically than drainage.
19. Discuss the practicality of each of the three methods for temporarily controlling salts on irrigated land.
20. In the permanent reclamation of saline and alkali lands, what four conditions should be maintained? Why are all four necessary?
21. What management precautions must be taken when using saline waters for irrigation?
22. Is the following statement true? If the irrigation water is salty, less water should be applied because this will result in less salt being applied to the soil. Discuss.

CHAPTER 11

1. Are studies of the consumptive use of water in irrigation likely to increase in importance as time advances? Why?
2. Are the terms "consumptive use" and "evapotranspiration" the same? Explain.
3. Draw a graph showing how both transpiration and evaporation vary during the growing season for a particular crop. Explain briefly the trends shown.
4. List all of the factors that affect consumptive use.
5. List those factors affecting the evaporation component of consumptive use and state which factors management can alter, thereby changing evaporation.
6. How is the wilting of a plant related to the amount of available energy in the atmosphere?
7. Will a light rainfall reduce the consumptive use? Explain.
8. Calculate the number of tons of coal that would be required to evaporate an amount of water equivalent to the daily consumptive use from one acre. Assume 17,900 BTU per pound of bituminous coal, a boiler efficiency of 65 percent, and a daily consumptive use of 0.25 inches per day.
9. What are the principal liabilities and limitations of the following methods of determining consumptive use:
   (a) Tanks and lysimeters?
   (b) Soil-moisture studies?
10. What kind of valley is best suited for determining consumptive use by the inflow-outflow method?
11. The Penman and Thornthwaite methods of determining consumptive use were developed in humid climates, whereas the Lowry-Johnson and Blaney-Criddle methods were developed in arid climates. Does this fact influence the accuracy and utility of these methods in estimating consumptive use on either arid or humid lands?

12. What climatic data is required for each of the following formulas: Penman, Thornthwaite, Lowry-Johnson, and Blaney-Criddle?

13. Would you expect a Piche evaporometer and an atmometer to indicate a higher evaporation ratio than an evaporation pan? Why?

14. Explain why consumptive use rates will vary from day to day because of the stage of growth of the plants.

15. Estimate the value of the consumptive use-evaporation ratio for the following stages of crop growth:
    (a) Tasseling of corn.
    (b) Flowering of potatoes.
    (c) Filling of pea pods.
    (d) Ripening of tomatoes.
    (e) Cutting of alfalfa.
    (f) Digging of sugar beets.
    (g) Cutting of sugar cane.
    (h) Harvesting of lettuce.
    (i) Picking of cotton.

16. A field of corn about 5 feet tall is growing vigorously, and tasseling has just started. What would you estimate the consumptive use–evaporation ratio to be? If records indicate the evaporation to be 0.20 inches per day, what would the consumptive use be?

17. Using the Penman method, compute the consumptive use in mm of water per day for the following conditions:
    Air temperature $= 80°$ F.
    Relative humidity $= 50$ percent.
    Sunshine, $n/N = 85$ percent.
    Wind speed, $U_2 = 120$ miles per day at 2 meters.
    Radiation rate, $R_A = 19$ mm of water per day.
    Radiation coefficient $= 30$ percent.

18. Using the Thornthwaite method, compute the average daily consumptive use in inches of water per day for the following conditions:
    Heat index—30.
    Temperature—$20°$C.
    Latitude—$35°$N.

19. Using the Lowry-Johnson method, compute the effective heat-day-degrees if the annual consumptive use is 22 inches depth of water per acre. If the mean maximum temperature in degrees Fahrenheit during August is 75, what percentage of the annual effective heat-day-degrees occur during August?

20. Using the Blaney-Criddle method, calculate the annual consumptive use for cotton growing, with the temperatures and percent daytime hours given in Table 11.10. Assume an average $k = 0.70$.

21. Find a graph showing the average mean daily temperature in your locality and compute the seasonal heat available to alfalfa in day-degrees. Specify assumptions you consider necessary to this computation.

22. Select a crop from Table 11.11 which is commonly grown in your area. Using recorded local temperatures, compute by the Blaney-Criddle method the annual consumptive use.
23. For an average July temperature of 77°F in the Intermountain Area of arid western U.S.A., what would be the average number of frost-free days and the seasonal and daily use of water by small grains?

CHAPTER 12

1. List the three principal factors which influence time of irrigation and amount of water to apply.
2. Outline the reasons for reduced yield when the soil is too wet or too dry.
3. List several crops which show by their appearance that soil moisture is deficient. Describe the appearance of each crop listed.
4. Why is it poor practice to wait until a crop shows signs of drought before irrigating?
5. By early spring snow surveys show that irrigation water supplies will be critically short during the latter part of the growing season. The crops normally grown are alfalfa, sorghum, and cotton. Outline the adjustment in cropping practice that you would recommend to meet the impending drought.
6. What irrigation and fertility practices should be followed when vegetative growth is desired?
7. What are the major purposes of irrigating soils during the non-growing or dormant season?
8. Under what conditions, if any, is it justifiable to divert water from partly filled storage reservoirs during the non-growing or dormant season for irrigation purposes?
9. Discuss the feasibility and need for fall and winter irrigation in your area.
10. Why does early spring irrigation often retard growth of plants?
11. Describe the probable difference in root systems of crops which have been irrigated by frequent, light irrigations and widely-spaced, heavy irrigations.
12. A crop of alfalfa is being produced on a net water use of 24 inches. Each of four irrigations applies 6 inches of water over the surface.
    (a) Assuming the soil moisture extraction pattern shown in Fig. 12.6, calculate the average depth of water stored in each foot of the 4 feet of root zone.
    (b) If the total water-holding capacity of the soil is 2.0 inches per foot, what percentage is stored in each foot during each irrigation?
13. What fertilizer element is most essential during (a) vegetative growth? (b) fruiting?
14. How should irrigation practice be modified during vegetative, wet fruit, and dry fruit stages of plant growth?

15. Explain how irrigation practice may influence the availability to plants of nitrogen and of phosphate fertilizer in the soil.

16. A rather limited water supply had been applied to a field by sprinkler irrigation for several years, allowing only 1 inch of water to enter the soil per irrigation. Explain why the fertile, deep-loam soil exhibited droughty characteristics and was considered to be shallow and low in fertility under these conditions of management.

17. Calculate the consumptive use rate of a crop that is just entering the wet fruit stage of growth when the evaporation is 0.30 inches per day. If 9 days have elapsed since the last irrigation, how much water has been depleted from the soil? What depth of water should be applied if the irrigation efficiency is to be 60 percent?

CHAPTER 13

1. Enumerate the conditions in the order of their importance, which you consider most essential to the attainment of community economical use of irrigation water.

2. (a) State three major conditions which tend to satisfy irrigation farmers with a low water-application efficiency.
   (b) State three major conditions which tend to stimulate irrigators to attain a high water-application efficiency.

3. The soil of an irrigated farm is a clay loam of comparatively uniform texture to a depth of 6 feet, below which there is a coarse gravel to a great depth. Moisture determinations before irrigation and again 48 hours after irrigation showed an average of 4.5 acre-inches per acre irrigation water stored in the soil from an irrigation in which the irrigator used a stream of 3 cfs continuously for 24 hours on a 10-acre tract of alfalfa. Neglecting consumptive use between completion of irrigation and the taking of samples for moisture determinations, what was the water-application efficiency? What was the efficiency considering that 0.3 inch per day was used on each of the two days between irrigation and sampling?

4. The average apparent specific gravity of the soil of the tract considered in Problem 2 is 1.3. Provided the mean increase in moisture content to a depth of 6 feet equals 5.35 percent, what is the water-application efficiency?

5. One hundred cfs are diverted from a river into a canal. Of this amount 75 cfs are delivered to the farms. The surface runoff from the irrigated area averages 15 cfs. The contribution to the ground water is 10 cfs.
   (a) What is the water-conveyance efficiency?
   (b) What is the water-application efficiency?

6. A farmer irrigates 5 acres of wheat the first week in July when the average depth of rooting was 3.5 feet. Two days after irrigation he takes a soil auger to the field and by boring holes into the soil determines that the average depth of penetration in each acre of the 5 acres is as follows:

Average Depth of Penetration Feet
2.7
3.2
3.6
4.0
3.5

(a) What is the water-distribution efficiency?

(b) What is the water-storage efficiency?

7. Determine the water-application efficiency, the water-storage efficiency, and the water-distribution efficiency for the following conditions:

Stream of 3 cfs delivered to the field for 2 hours.

Runoff averaged 1.5 cfs for 1 hour.

Depth of root zone was 5 feet.

Depth of penetration varied linearly from 5 feet at one end to 2 feet at the other end of the field.

8. A stream of 5 cfs was diverted from the river and 3 cfs were delivered to the field. An area of 4 acres was irrigated in 8 hours. The root-zone depth was 6 feet. The runoff averaged 1.5 cfs for 3 hours. The depth of water penetration varied linearly from 6 feet at the head of the field to 4 feet at the end of the field. Determine the water-conveyance efficiency, the water-application efficiency, the water-storage efficiency, and the water-distribution efficiency.

CHAPTER 14

1. Why does the method of irrigation used depend upon the value of the crop being produced, the cost of water, and the general economy of the area?

2. Which of the methods of irrigating is the oldest and the least efficient?

3. What crops are generally irrigated by flooding?

4. What are the essential points of difference between border-strip flooding and check flooding?

5. What are the essential points of difference between corrugations and furrows used for irrigation?

6. Level borders have been used for irrigating relatively flat lands in areas not strictly arid or humid. Explain why heavy rains would create a problem.

7. What natural conditions favor sub-irrigation?

8. List the reasons why the hydraulics of surface irrigation are much more complicated than the hydraulics of open-channel flow.

9. What ten criteria must be considered when designing an irrigation system?

CHAPTER 15

1. (a) What are the major functions of diversion structures?

(b) What forces are permanent diversion structures required to resist?

(c) Are the dimensions of farm diversion structures, i.e., lengths, depths, and widths, influenced by the soils in which they are built? Explain.

2. In the selection of a permanent farm conveyance structure, give the conditions which would influence your choice between a flume, ditch, surface pipe, and underground pipe.

3. (a) Why is it necessary to have a larger bottom width and depth in order to convey a given quantity of water through an earth ditch than through a concrete flume of the same slope?

(b) If you were going to build an 8-inch pipe on a slope of 2 feet per mile and you wanted to get the largest possible quantity of water through it, neglecting differences in cost, what kind of pipe would you select? Why?

4. Using the roughness coefficients given in Chapter 5, illustrate the importance of keeping ditches and canals clean.

5. (a) What major objectives should influence the irrigator's selection of irrigation water-distribution structures?

(b) Does the cost of water influence the selection of a distribution structure?

(c) Do the soil properties influence the selection of distribution structures? Explain briefly.

6. In measuring water to a farmer from a variable canal stream will a submerged orifice or a weir be most helpful toward delivery of a flow as nearly uniform as practical? Why?

7. To what extent are shop facilities needed to maintain and repair irrigation implements and structures?

8. What implements are required for irrigated farming that are not needed for non-irrigated farming?

9. Explain why a good plowing job is essential if the land is to be irrigated.

10. (a) What crops might be damaged by deep furrowing as a result of tearing up shallow roots?

(b) Would deep cultivation also damage the same crops?

11. Explain the statement that good water control and therefore adequate water-control structures are essential before irrigation water can be used efficiently.

12. Discuss the relative practicality and economics of using wood, concrete, brick and mortar, metal, and prefabricated components for water-control structures in your locality.

13. What are the advantages of using buried pipe for water conveyance?

14. How would you start water flowing in a 2-inch siphon used to convey water from a ditch to a furrow?

CHAPTER 16

1. How do excessive intake rates interfere with good surface irrigation?

2. Explain how an irrigation stream too small to irrigate efficiently using surface methods can often be applied very efficiently by sprinkler irrigation.

3. Explain why less skilled labor can be used for sprinkler irrigation than for surface irrigation.
4. Why is it easier to apply small depths of water more efficiently by sprinkler than by surface irrigation methods?
5. For what reasons would frequent light irrigations be desirable?
6. What precautions should be taken when using sprinkler irrigation for frost protection?
7. List the possible advantages in applying fertilizers and soil amendments to the soil through sprinkler systems.
8. Make sketches of the ways in which fertilizers can be injected into a sprinkler system.
9. Why is it advisable and often necessary to operate the sprinklers for a few minutes before and after fertilizer is applied? Consider the system, the crop, and the soil.
10. Which of the three general types of sprinklers is used most extensively? Why?
11. Tabulate the average pressures and rates of application for the three sprinkler types.
12. What are the advantages of medium-pressure sprinklers?
13. What crops would most nearly justify use of semi-permanent sprinkler sytems?
14. What are the advantages of portable sprinkler systems?
15. What natural conditions would favor use of a gravity sprinkler system?
16. Why are silt and debris in irrigation water objectionable when using sprinkler systems?
17. Why is debris removal from irrigation water difficult?
18. Discuss the relative importance of initial and annual costs of irrigation systems.
19. Larger sprinkler nozzles and higher pressures require more power and less labor for irrigation. How would this fact influence sprinkler design in an underdeveloped country?
20. What information regarding soils is needed for a good sprinkler irrigation design?
21. (a) When is the soil most susceptible to puddling and crusting from droplet impact from sprinklers?
    (b) What can the irrigator do to minimize damage from a given sprinkler system?
22. Explain how the cropping pattern which will develop when the sprinkler system is in full use should influence the design of the sytem.

CHAPTER 17

1. Discuss the statement that irrigation and drainage are complementary practices in arid regions.
2. What is the relative importance of surface and of subsurface drainage in arid and in humid climates?

3. Explain why the need for surface drainage increases as the annual rainfall increases.

4. List the benefits of drainage.

5. How can knowledge of the source of irrigation water be useful in solving a drainage problem?

6. Under what conditions would you use piezometers instead of auger holes to measure the hydraulic characteristics of the ground water?

7. What practical means can frequently be taken to reduce the severity of drainage problems by controlling the source of the drainage waters?

8. Discuss the statement that shallower water-table depths are economically feasible under excellent management.

9. A field investigation of drainage deals with what four subjects? Why is each one important?

10. The soil excavated from an open drain is all placed upon one side. Make assumptions regarding side slopes and widths and prepare a graph showing the value of land removed from production by a 1000-foot drain excavated to 5-, 10-, and 15-foot depths.

11. List in sequence the operations performed by a large trenching machine, including the use of related equipment such as a tractor or bulldozer.

12. What can be done to maintain a tile drainage system in good operating condition?

13. Why is re-use of drainage water important?

14. Prepare a list of advantages and disadvantages of tile drains vs. open drains.

15. In Equation 17.2 list the quantities R, k, H, L, and h in the order of ease of measurement, placing first those that are most easily measured.

16. Refer to Fig. 17.7 and Equation 17.2 and explain why the flow of ground water to the drain is proportional (approximately) to the square of the effective depth of drain.

17. A new open drain is not drawing enough water from the soil to lower the water table sufficiently. To increase the drain discharge would you increase the bed slope, make the drain wider, deeper, longer, or use a combination of these remedies? Give reasons for your answer.

18. Determine the flow from the soil into a 10-foot-depth open drain 400 feet long when the drains are spaced 150 feet apart. The depth of pervious stratum is 15 feet, and the depth of the water table midway between drains is 3 feet below the ground surface. The average permeability of the pervious stratum is 4.5×10^{-4} fps. The depth of water in the drain is negligible.

19. For a soil of great depth and uniform permeability, with all other conditions as in Problem 6, determine the flow into 6-inch diameter tile drain flowing two-thirds full, by means of Equation 17.7.

20. In Problem 6, what will be the change in flow q, toward the drain, if all other factors remain unchanged, but: (a) the spacing of the drain lines S is doubled? (b) the permeability of soil k is doubled? (c) the depth of the drain is increased from 10 to 12 feet?

21. Sandy loam soil to be drained is 50 feet deep and has a permeability of 1×10^{-4} fps. The water table is 4 feet below the ground surface. Seven-foot-depth tile drains 500 feet long have a slope of 1/2000. Determine the discharge of each drain and the spacing for 6-inch-diameter tile flowing half full. Assume the flow toward the drain is through semicylindrical surfaces.

22. In a field drainage experiment by pumping from an artesian aquifer the following data were obtained:
    (a) Flow of water to well or pump discharge $Q = 4.2$ cfs.
    (b) Radius at maximum pressure head $R = 1480$ feet.
    (c) Radius at minimum pressure head $r = 18$ feet.
    (d) Pressure head at maximum radius $H = 26$ feet.
    (e) Depth of water-bearing aquifer $D = 16$ feet.
    Find the permeability in feet per second.

CHAPTER 18

1. Why are most successful irrigation and drainage enterprises owned and operated by the water users?

2. Why has the commercial irrigation corporation been less influential than the mutual irrigation corporation?

3. Differentiate between private, quasi-public, and public irrigation and drainage enterprises.

4. What are the distinctive features of a mutual irrigation company?

5. How does a mutual irrigation company differ from an irrigation district?

6. Can a mutual irrigation corporation sell the land owned by its delinquent stockholders in order to collect payments of irrigation assessments?

7. Can an irrigation district sell the land owned by its delinquent members in order to collect payments of irrigation assessments?

8. What are the two conflicting water right doctrines? How do they differ?

9. What are the salient features of the (a) doctrine of riparian rights, (b) doctrine of appropriation, (c) common law doctrine, (d) reasonable use concept?

10. What creates priority under (a) riparian right doctrine, (b) appropriation right doctrine?

11. Why is a drainage district normally necessary before land can be adequately drained?

12. What powers are vested in a drainage district?

13. What advantages would result from combining an irrigation district with the drainage district organized for the same land?

14. Explain how settlers ultimately gain ownership of a Bureau of Reclamation project.

INDEX

Abraham, 1
Absorption blocks, gravimetric, 179
Absorption of water by plants, rate of, 234
Acre-foot, 99
Acre-inch, 99
Aerial maps, 30
Afghanistan, 10
Africa, 3, 249, 282
Albertson, M. L., 80
Alfalfa, 267, 268, 269, 270, 271, 275, 297
Alkali, black, 214
  concentration of, 84
  white, 213
Alkaline, defined, 210
Alkali soil, defined, 210
All American canal as an example of canal lining, 85
Amraphel, 1
Andes mountains, 3
Apparent specific gravity, 150
Appropriation, doctrine of, 379
Arabian peninsula, 3
Areas irrigated in world, 9, 10
Argentina, 10
Arid regions, defined, 3
Aromatic solvents, 92
Asia Minor, 3
Assyria, Queen of, 2
Atmometers, 254
Atmosphere, 164
Auger, post-hole, 173
  spiral bit, 173
Australia, 3, 10

Babylon, 1
Backwashing, 42

Baffle boards, 118
Bailer, 37
Barton, James R., 80
Bengo, Lucala, and Cuanza rivers, Africa, 282
Bernoulli energy equation, 76, 186, 191
Bible, irrigation mentioned, 1, 2
Blaney, Harry F., 238
Blaney-Criddle formula, 252, 256
  limitations of, 253
Boise, Idaho, 248, 249
Borders, 302
  implements for making, 315
  strips, 300, 301
Boron content of irrigation waters, 224, 225, 226
Boundary form, effect on coefficient of discharge, 112
Boundary geometry, 183
Bouyoucos, G. J., 179
Brazil, 3
Bulk density, 150
Bulkheads, 119, 121
Burma, 10

Cable tool method, 36
  basic tools for, 36, 37
Cache Valley, Utah, sub-irrigation in, 307
Calcium, 154
Calibrated gates, 104
California Experiment Station, 173
California State Durham Colony, 300
Canals, Assyria, 2
  best hydraulic cross section of, 82
  cleaning of, 91, 92
  Damanum, 1